Quantitative Chromatographic Analysis

CHROMATOGRAPHIC SCIENCE SERIES

A Series of Textbooks and Reference Books

Editor: JACK CAZES

1. Dynamics of Chromatography, *J. Calvin Giddings*
2. Gas Chromatographic Analysis of Drugs and Pesticides, *Benjamin J. Gudzinowicz*
3. Principles of Adsorption Chromatography: The Separation of Nonionic Organic Compounds, *Lloyd R. Snyder*
4. Multicomponent Chromatography: Theory of Interference, *Friedrich Helfferich and Gerhard Klein*
5. Quantitative Analysis by Gas Chromatography, *Josef Novák*
6. High-Speed Liquid Chromatography, *Peter M. Rajcsanyi and Elisabeth Rajcsanyi*
7. Fundamentals of Integrated GC-MS (in three parts), *Benjamin J. Gudzinowicz, Michael J. Gudzinowicz, and Horace F. Martin*
8. Liquid Chromatography of Polymers and Related Materials, *Jack Cazes*
9. GLC and HPLC Determination of Therapeutic Agents (in three parts), *Part 1 edited by Kiyoshi Tsuji and Walter Morozowich, Parts 2 and 3 edited by Kiyoshi Tsuji*
10. Biological/Biomedical Applications of Liquid Chromatography, *edited by Gerald L. Hawk*
11. Chromatography in Petroleum Analysis, *edited by Klaus H. Altgelt and T. H. Gouw*
12. Biological/Biomedical Applications of Liquid Chromatography II, *edited by Gerald L. Hawk*
13. Liquid Chromatography of Polymers and Related Materials II, *edited by Jack Cazes and Xavier Delamare*
14. Introduction to Analytical Gas Chromatography: History, Principles, and Practice, *John A. Perry*
15. Applications of Glass Capillary Gas Chromatography, *edited by Walter G. Jennings*
16. Steroid Analysis by HPLC: Recent Applications, *edited by Marie P. Kautsky*
17. Thin-Layer Chromatography: Techniques and Applications, *Bernard Fried and Joseph Sherma*
18. Biological/Biomedical Applications of Liquid Chromatography III, *edited by Gerald L. Hawk*
19. Liquid Chromatography of Polymers and Related Materials III, *edited by Jack Cazes*
20. Biological/Biomedical Applications of Liquid Chromatography, *edited by Gerald L. Hawk*
21. Chromatographic Separation and Extraction with Foamed Plastics and Rubbers, *G. J. Moody and J. D. R. Thomas*
22. Analytical Pyrolysis: A Comprehensive Guide, *William J. Irwin*
23. Liquid Chromatography Detectors, *edited by Thomas M. Vickrey*
24. High-Performance Liquid Chromatography in Forensic Chemistry, *edited by Ira S. Lurie and John D. Wittwer, Jr.*
25. Steric Exclusion Liquid Chromatography of Polymers, *edited by Josef Janca*
26. HPLC Analysis of Biological Compounds: A Laboratory Guide, *William S. Hancock and James T. Sparrow*

27. Affinity Chromatography: Template Chromatography of Nucleic Acids and Proteins, *Herbert Schott*
28. HPLC in Nucleic Acid Research: Methods and Applications, *edited by Phyllis R. Brown*
29. Pyrolysis and GC in Polymer Analysis, *edited by S. A. Liebman and E. J. Levy*
30. Modern Chromatographic Analysis of the Vitamins, *edited by André P. De Leenheer, Willy E. Lambert, and Marcel G. M. De Ruyter*
31. Ion-Pair Chromatography, *edited by Milton T. W. Hearn*
32. Therapeutic Drug Monitoring and Toxicology by Liquid Chromatography, *edited by Steven H. Y. Wong*
33. Affinity Chromatography: Practical and Theoretical Aspects, *Peter Mohr and Klaus Pommerening*
34. Reaction Detection in Liquid Chromatography, *edited by Ira S. Krull*
35. Thin-Layer Chromatography: Techniques and Applications. Second Edition, Revised and Expanded, *Bernard Fried and Joseph Sherma*
36. Quantitative Thin-Layer Chromatography and Its Industrial Applications, *edited by Laszlo R. Treiber*
37. Ion Chromatography, *edited by James G. Tarter*
38. Chromatographic Theory and Basic Principles, *edited by Jan Åke Jönsson*
39. Field-Flow Fractionation: Analysis of Macromolecules and Particles, *Josef Janca*
40. Chromatographic Chiral Separations, *edited by Morris Zief and Laura J. Crane*
41. Quantitative Analysis by Gas Chromatography: Second Edition, Revised and Expanded, *Josef Novák*
42. Flow Perturbation Gas Chromatography, *N. A. Katsanos*
43. Ion-Exchange Chromatography of Proteins, *Shuichi Yamamoto, Kazuhiro Nakanishi, and Ryuichi Matsuno*
44. Countercurrent Chromatography: Theory and Practice, *edited by N. Bhushan Mandava and Yoichiro Ito*
45. Microbore Column Chromatography: A Unified Approach to Chromatography, *edited by Frank J. Yang*
46. Preparative-Scale Chromatography, *edited by Eli Grushka*
47. Packings and Stationary Phases in Chromatographic Techniques, *edited by Klaus K. Unger*
48. Detection-Oriented Derivatization Techniques in Liquid Chromatography, *edited by Henk Lingeman and Willy J. M. Underberg*
49. Chromatographic Analysis of Pharmaceuticals, *edited by John A. Adamovics*
50. Multidimensional Chromatography: Techniques and Applications, *edited by Hernan Cortes*
51. HPLC of Biological Macromolecules: Methods and Applications, *edited by Karen M. Gooding and Fred E. Regnier*
52. Modern Thin-Layer Chromatography, *edited by Nelu Grinberg*
53. Chromatographic Analysis of Alkaloids, *Milan Popl, Jan Fähnrich, and Vlastimil Tatar*
54. HPLC in Clinical Chemistry, *I. N. Papadoyannis*
55. Handbook of Thin-Layer Chromatography, *edited by Joseph Sherma and Bernard Fried*
56. Gas–Liquid–Solid Chromatography, *V. G. Berezkin*
57. Complexation Chromatography, *edited by D. Cagniant*
58. Liquid Chromatography–Mass Spectrometry, *W. M. A. Niessen and Jan van der Greef*
59. Trace Analysis with Microcolumn Liquid Chromatography, *Milos Krejcl*

60. Modern Chromatographic Analysis of Vitamins: Second Edition, *edited by André P. De Leenheer, Willy E. Lambert, and Hans J. Nelis*
61. Preparative and Production Scale Chromatography, *edited by G. Ganetsos and P. E. Barker*
62. Diode Array Detection in HPLC, *edited by Ludwig Huber and Stephan A. George*
63. Handbook of Affinity Chromatography, *edited by Toni Kline*
64. Capillary Electrophoresis Technology, *edited by Norberto A. Guzman*
65. Lipid Chromatographic Analysis, *edited by Takayuki Shibamoto*
66. Thin-Layer Chromatography: Techniques and Applications, Third Edition, Revised and Expanded, *Bernard Fried and Joseph Sherma*
67. Liquid Chromatography for the Analyst, *Raymond P. W. Scott*
68. Centrifugal Partition Chromatography, *edited by Alain P. Foucault*
69. Handbook of Size Exclusion Chromatography, *edited by Chi-san Wu*
70. Techniques and Practice of Chromatography, *Raymond P. W. Scott*
71. Handbook of Thin-Layer Chromatography: Second Edition, Revised and Expanded, *edited by Joseph Sherma and Bernard Fried*
72. Liquid Chromatography of Oligomers, *Constantin V. Uglea*
73. Chromatographic Detectors: Design, Function, and Operation, *Raymond P. W. Scott*
74. Chromatographic Analysis of Pharmaceuticals: Second Edition, Revised and Expanded, *edited by John A. Adamovics*
75. Supercritical Fluid Chromatography with Packed Columns: Techniques and Applications, *edited by Klaus Anton and Claire Berger*
76. Introduction to Analytical Gas Chromatography: Second Edition, Revised and Expanded, *Raymond P. W. Scott*
77. Chromatographic Analysis of Environmental and Food Toxicants, *edited by Takayuki Shibamoto*
78. Handbook of HPLC, *edited by Elena Katz, Roy Eksteen, Peter Schoenmakers, and Neil Miller*
79. Liquid Chromatography–Mass Spectrometry: Second Edition, Revised and Expanded, *W. M. A. Niessen*
80. Capillary Electrophoresis of Proteins, *Tim Wehr, Roberto Rodríguez-Díaz, and Mingde Zhu*
81. Thin-Layer Chromatography: Fourth Edition, Revised and Expanded, Bernard Fried and Joseph Sherma
82. Countercurrent Chromatography, *edited by Jean-Michel Menet and Didier Thiébaut*
83. Micellar Liquid Chromatography, *Alain Berthod and Celia García-Alvarez-Coque*
84. Modern Chromatographic Analysis of Vitamins, Third Edition, Revised and Expanded, *edited by André P. De Leenheer, Willy E. Lambert, and Jan F. Van Bocxlaer*
85. Quantitative Chromatographic Analysis, *Thomas E. Beesley, Benjamin Buglio, and Raymond P. W. Scott*

ADDITIONAL VOLUMES IN PREPARATION

Current Practice of Gas Chromatography–Mass Spectrometry, *edited by W. M. A. Niessen*

Quantitative Chromatographic Analysis

Thomas E. Beesley
Advanced Separation Technologies, Inc.
Whippany, New Jersey

Benjamin Buglio
Hoffmann-La Roche, Inc.
Nutley, New Jersey

Raymond P. W. Scott
Georgetown University
Washington, D.C, and
Birkbeck College, University of London
London, England

MARCEL DEKKER, INC. NEW YORK • BASEL

Library of Congress Cataloging-in-Publication Data

Beesley, Thomas E.
Quantitative chromatographic analysis / Thomas E. Beesley, Benjamin Buglio, Raymond P.W. Scott.
p. cm. -- (Chromatographic science series ; 85)
Includes bibliographical references and index.
ISBN 0-8247-0503-3 (acid-free paper)
1. Chromatographic analysis. 2. Chemistry, Analytic--Quantitative. I. Buglio, Benjamin. II. Scott, Raymond P. W. (Raymond Peter William). III. Title. IV. Chromatographic science ; v. 85.
QD117.C5 B35 2001
545'.89--dc21 00-047575

This book is printed on acid-free paper.

Headquarters
Marcel Dekker, Inc.
270 Madison Avenue, New York, NY 10016
tel: 212-696-9000; fax: 212-685-4540

Eastern Hemisphere Distribution
Marcel Dekker AG
Hutgasse 4, Postfach 812, CH-4001 Basel, Switzerland
tel: 41-61-261-8482; fax: 41-61-261-8896

World Wide Web
http://www.dekker.com

The publisher offers discounts on this book when ordered in bulk quantities. For more information, write to Special Sales/Professional Marketing at the headquarters address above.

Current printing (last digit):
10 9 8 7 6 5 4 3 2 1

PRINTED IN THE UNITED STATES OF AMERICA

Preface

The most popular techniques employed in the modern analytical laboratory are, without doubt, chromatography based. Among its numerous forms, chromatography offers the highest sensitivity, the greatest resolution and the fastest analyses; at the same time, it can provide highly accurate and precise quantitative results. In fact, the very tight specifications mandated, for example, by the Environmental Protection Agency, are possible only because chromatography provides analytical techniques that can detect and quantitatively estimate contaminants at the levels stipulated. It follows that the quality of everyday living in some part is a direct result of the high analytical performance obtainable from chromatographic techniques.

This book has been written for the benefit of both analysts new to chromatography and those who are highly experienced in the technique. It has been written to provide a fundamental understanding of the chromatographic techniques that are commonly used in general analysis and to give practical guidance in the many analytical procedures that are employed. It is hoped that this book will act as a directory to help the analyst achieve the necessary resolution in the time required, and also obtain quantitative results with the essential high accuracy and precision.

Many practical examples are included, and these have been selected not only to illustrate the scientific principles behind the analytical procedures, but also to demonstrate the wide diversity of sample types that can be analyzed by chromatographic techniques.

Acknowledgments

We would like to thank the many manufacturers of chromatography instruments, supplies and chemicals for their help in providing data for inclusion in this book. In particular, special thanks are due to Supelco, Inc., for providing many of the application examples that are given and to Perkin Elmer, Inc., Hewlett-Packard, Inc., ASTEC, Inc. and Valco, Inc. for details about their more recent instruments and techniques.

Finally, we would like to thank the Royal Society of Chemistry for permission to reproduce figures from the "Analyst", the American Chemical Society for permission to reproduce figures from "Analytical Chemistry" and the Elsevier Scientific Publishing company for permission to reproduce diagrams from their "Journal of Chromatography".

Thomas E. Beesley
Benjamin Buglio
Raymond P. W. Scott

Contents

Part 1

Introduction to Quantitative Chromatographic Analysis

Chapter 1

The Critical Factors that Govern a Successful Quantitative Chromatographic Analysis

Historical Introduction

The early work in liquid chromatography from Tswett onward involved almost no quantitative assays, but was largely used as a preparative technique, quantitative evaluations being carried out off-line using separate analytical methods. Actual quantitative assays made directly by monitoring the column eluent commenced with gas chromatography (GC), first used in this way by the inventors of the technique, James and Martin, in 1952 [1] for the analysis of fatty acid mixtures. In fact, the need to determine the composition of mixed fatty acids extracted from plant tissue, to help elucidate their synthetic pathways, was the actual incentive that provoked the development of the technique in the first place. In the first instrument, the column eluent was bubbled through a suitable aqueous liquid to absorb the eluted acids. The solution contained an indicator, the color of which changed as each solute was eluted, and the solution was then manually titrated. Later the titration process was automated by the inventors (probably the first automatic titration apparatus to be made and certainly the only one available at that time), and an integral chromatogram was formed by plotting the volume of base solution added against time. The resulting *integram* displayed each substance as a step, the height of which was proportional to the amount of fatty acid eluted. Subsequently, Martin developed the density balance detector [2], the first inline detector, which had a very useful linear response. The gas density balance was an extremely complicated and ingenious device consisting of a Wheatstone network of capillary tubes that were drilled out of a high conductivity copper block. Consequently, the sensing device was fairly compact. In the block, there were two columns of gas, one containing the column eluent, the other a reference gas. When solute vapor was present in the column eluent, the pressure difference across the two columns, due to the differing densities of the gases, was arranged to cause a flow of gas over two heated thermocouples, cooling one and heating the other. The output from the thermocouples was fed to an appropriate recording milliammeter. The detector was linear over about three orders of magnitude of concentration and had a sensitivity (minimum detectable concentration) of about 5×10^{-7} g/ml (*n*-heptane). The detector output provided a *differential* output that displayed solute peaks in the conventional manner and could be assessed quantitatively using peak areas or peak heights (methods will be discussed later).

In the early days of gas chromatography little attention was paid to sample preparation, precision sampling, standard selection, etc. Chemists were so elated to be able, quite unexpectedly, to obtain apparently miraculous separations of hitherto completely unresolvable mixtures in a relatively short period of time that the accuracy and precision of the quantitative measurements initially became of somewhat secondary importance. Sensitivity, however, was of great interest and a number of new detectors were rapidly developed. The next GC detector to be developed was the *katharometer* detector [3] in 1954 (now also known as the *thermal conductivity detector* and the *hot wire detector*). This was another, relatively low sensitivity detector (about the same as the density balance) and was quickly followed by another detector of similar sensitivity, the *flame thermocouple* detector, developed by Scott in 1956 [4]. The last of the early detectors of limited sensitivity was the first of the ionization detectors, the *cross-section detector*, described by Boer also in 1956

[5]. Subsequent to 1956, the age of the high sensitivity GC detectors began and the first to make its appearance was the ubiquitous *flame ionization detector* (FID) described by McWilliams [6] in 1958. This was to become the workhorse of all GC analyses, having an extremely high sensitivity and a linear dynamic range exceeding five orders of magnitude. The FID was followed by the relatively specific *flame luminosity detector* by Grant [7] and finally the exciting family of ionization detectors summarized by Lovelock [8] in 1960. These detectors comprised the *macro-argon detector*, *micro-argon detector* and probably the most sensitive detector available, beitmay a specific detector, the *electron capture detector*. Correctly designed and operated the argon detectors can have sensitivities at least one order of magnitude greater than the FID and the electron capture detector nearly two orders of magnitude greater than the FID.

Subsequent to the burst of innovation that provided most of the detectors in common use today for quantitative gas chromatography, attention was at last turned to the other parts of the gas chromatograph that had an impact on quantitative accuracy. Although some other detectors were developed (*e.g.*, the *nitrogen phosphorus detector*, a modification of the FID), consideration was now given to the design of accurate and reproducible sampling systems, columns which were appropriately inert to the samples that they were to separate and finally to the methods required for processing the data provided by the detector

The development of quantitative liquid chromatography (LC) has been much slower and initially lagged behind GC by almost a decade. Quantitative LC had to await the development of first, sensitive detectors, second, columns with adequate efficiency, third, high pressure sampling systems that could be used with such columns and fourth, mobile phase supply systems that could provide accurate solvent flow rates at the necessary high pressures. These were all serious engineering challenges, the development of suitable detectors being the least of the problems. High pressure sampling was solved by the introduction of the internal and external loop valves. These were rotary valves consisting of two discs with finely machined contact faces held together by very strong springs. This permitted the valve to be rotated to allow the different ports to be matched without leaking. Columns with the necessary high efficiencies were eventually produced by packing tubes with very small particles (initially 10 µm in diameter) and operating them at very high pressures. The construction of such columns also demanded the development of slurry packing methods to achieve stable beds. High pressure pumps were also required that could be manufactured at a reasonable cost. As a result, the single stroke reciprocating piston pump was developed, which was made from a stainless steel cylinder and a sapphire piston fitted with non-return valves consisting of sapphire balls and seats. Eventually, dual pumps were produced with specially devised driving cams to reduce pump pulsation.

The LC detectors that were developed were many orders of magnitude less sensitive than their GC counterparts and had significantly smaller linear dynamic ranges. The early detectors were, nevertheless, sufficiently sensitive and linear to allow accurate quantitative analysis to be carried out, and also to aid in the development of better LC columns. It is interesting to note that the first LC detector to be developed was by the inventor of the first GC detectors, A. J. P. Martin but in this case with his coworker S. S. Randall [9]. Martin's device was an electrical conductivity detector which was used with the old type Tswett columns.

The first practical detector that could be used for quantitative work was the refractive index detector developed by Tiselius and Claesson [10] that monitored the change in the refractive index of the column eluent when a solute was present. The design of this detector was extended by Zaukelies and Frost [11] and Vandenheuval and Sipas [12]. Modifications based on the work of Christiansen [13] and the interference detector first described by Bakken and Stenberg [14] were also developed into refractive index-based detectors. Today, despite the many forms of the refractive index detector, including the thermal lens detector of Gorden *et al.* [15], the most common design used for quantitative analysis is very similar to that originally devised by Vandenheuval and Sipas. The refractive index detector has limited sensitivity and a restricted linear dynamic range but still survives midst the modern technology of LC detectors, largely on account of its catholic response. It is still used, on occasion, to detect the many substances to which other detectors do not respond.

The detector that is most often used in quantitative LC is the UV detector similar to the basic design originally described by Horvath and Lipsky [16]. UV absorption detectors respond to those substances that absorb light in the range 180 to 350 nm. A great number of substances absorb light in this wavelength range, including those substances having one or more double bonds (π electrons) and substances having unshared (unbonded) electrons, *e.g.* all olefins, all aromatics and compounds, for example, containing $>C=O$, $>C=S$, $-N=N-$ groups. The UV light passes through a cell carrying the column eluent and falls on a photo–electric cell (or array) the output of which is conveyed to an appropriate signal modifying amplifier and thence to a recorder or data acquisition system. There are a number of different types of UV detector the details of which will be discussed later. This detector, like the FID in gas chromatography, is the workhorse of quantitative LC analysis. It has reasonable sensitivity (many orders less than the FID) and a linear dynamic range of about three orders of magnitude.

The next detector to be developed was the fluorescent detector, which has a relatively high sensitivity (compared with the UV detector but not the FID). When light is absorbed by a molecule, a transition to a higher electronic state takes place and this absorption is highly specific for the molecules concerned If the excess energy is not dissipated rapidly by collision with other molecules or by other means, the electron will return to the ground state with the emission of light at a lower frequency and the substance is said to fluoresce. As some energy is always lost before emission occurs, the wavelength of the fluorescent light is always greater than the incident light [17]. The popularity of this detector resides largely in its specificity and consequently high sensitivity, which, to some extent, arises from the very low background signal (noise level) that is inherent with this type of detector. Unfortunately its linear dynamic range is little greater than two orders of magnitude.

The most recently developed LC detectors are the light scattering detectors which, although sensitive to virtually all involatile solutes, have relatively poor sensitivity (about that of the refractive index detector) and a non-linear response, which is sometimes difficult to compensate by electronic means. The evaporative light scattering detector does, however, provide complete flexibility in the choice of solvent providing the solvents are volatile, which gives it a versatility that no other commonly used detectors possess. In addition, it can tolerate gradient elution while maintaining a completely stable baseline.

Thin layer chromatography has survived in the face of other chromatography techniques because of its low cost and capacity for multiple separations. In addition, due to its unique development characteristics, it can be made to exhibit quite unique selectivity often difficult or impossible to emulate with other techniques. However, it is still used largely for qualitative or semiquantitative analytical work. One of the major difficulties is the quantitative estimation of the thin layer spot. This problem has been eased by the introduction of high performance plates and advanced methods of coating, but accurate spot measurement is still difficult without expensive scanning devices. By using fluorescing derivatives, the spot can be uniquely scanned against a relatively noise-free background but, even with such sophisticated scanning systems, the precision and accuracy of quantitative TLC do not generally compare with those obtained from other chromatography techniques.

In all chromatography techniques, both detectors and the overall chromatographic apparatus has now been developed to a level where accurate quantitative analyses can be readily carried out [18]. However, the function of the equipment, the sample preparation and data processing must be well understood if the potential high accuracy and precision are to be achieved.

The Importance of Chromatography as an Analytical Technique

Chromatography is probably the most versatile and widespread technique employed in analytical chemistry today. There are a number of reasons for this. Firstly, as a result of the very sensitive methods of detection that are available to all types of chromatography, very small quantities of material can be separated, identified and analyzed quantitatively. It follows that only minute samples are required and merely a few micrograms of sample (at the extreme, even less than a nanogram) will be all that is necessary to ensure adequate accuracy. Secondly, chromatographic separations are usually relatively fast and an average analysis can regularly be completed in a few minutes and, under some circumstances, even in a few seconds. The high speed analyses achievable in chromatography have resulted directly from the considerable amount of research and development that has been carried out over the last 25 years in column technology and the production of synthetic phases. Another advantage to chromatography, that has made it so popular, is its relative simplicity and ease of operation compared with other instrumental analytical techniques. In addition, even the more sophisticated equipment can still be relatively inexpensive. Finally, if the established procedure is well controlled and the apparatus well maintained, good accuracy and precision can be maintained. However, if the established analytical protocol is not very carefully adhered to, there is evidence that analytical reproducibility between different laboratories can vary and sometimes leave a lot to be desired.

Quantitative chromatographic analysis is used extensively in almost all areas that involve chemical testing; specific examples are too numerous to mention individually. The high sensitivity makes the techniques invaluable for the analysis of environmental samples such as soil and water contamination and atmospheric pollution [19]. The high sensitivity of the technique is also exploited for forensic purposes such as testing for drug residues in blood and urine, flammable materials in arson samples and potential poisons or toxic materials.

The technique is also extensively employed by the pharmaceutical industry for both research purposes and in quality control. In a similar manner the biochemical and biotechnology industries widely use chromatographic methodology, and in many cases, such techniques are indeed essential, as alternative methods do not exist [20]. The most important, and more recent, applications are those involving chiral separations and the resolution of biopolymers [21], where the individual solutes can be very similar and demand the high resolving power of the modern column, the selectivity of the new chiral stationary phases and the ability to quantitatively detect trace components in a mixture. The technique is also an essential analytical tool in the agrochemical industry, not only for raw material analysis and quality control, but also for soil analysis and monitoring potential water pollution.

Critical Factors Involved in a Successful Chromatographic Analysis

There are a number of critical components to a chromatographic analysis, beitmay GC, LC, TLC or CEC, and each component will be discussed in considerable detail in various sections of this book. These essential elements to a successful analysis will be briefly outlined here, partly to introduce them to the reader, and partly to illustrate the way the subject will be presented and organized in Part 1 of this book.

A chromatographic analysis will involve the following essential procedures:

1. Collecting a representative sample
2. Transporting the sample to the laboratory
3. Storing the sample until analysis is possible
4. Preparing the sample for analysis
5. Performing the chromatographic analysis
6. Processing the analytical data
7. Reporting the results

Each of the above procedures will differ in many ways depending on a number of variables in the analysis. For example, within each of the above groups, the following are some of the factors that will determine the nature and choice of procedure.

1. The substance that is to be sampled may be a gas, liquid or solid and each physical form will require different sampling procedures.

2. The physical form and nature of the material to be analyzed will determine its manner of transport and method of storage.

3. The nature of the material will also influence the way the sample is prepared and the selection of the appropriate chromatographic technique for its analysis.

4. The level at which the materials of interest are present in the sample (trace components or major components) will also influence the choice of technique and the choice of apparatus.

Procuring a Representative Sample

The sample that is examined must be truly representative of the bulk to be analyzed; otherwise the results will be virtually meaningless.

The results of a chromatographic analysis can only be as reliable as the integrity of the sample that is taken.

It follows that as much care should be taken over the sampling procedure as is taken over the analysis. In many cases the analyst does not take the sample and has no control over the sampling procedure. Nevertheless, even in these cases, the analyst does have the responsibility for trying to ascertain the sampling method and other pertinent sampling details in order to access the validity of the consequent analytical results. In the case of a container of a homogeneous liquid, it is a fairly simple to obtain a sample that is representative of the total contents. However, if it is a sample of water from a flowing river, a sample from a silo of corn or even a sample from a drum of aspirin tablets, the way to obtain a representative sample is not immediately obvious and will require special sampling techniques.

There are, however, certain instances where a representative sample is impossible to obtain and, indeed, may not even be necessary. For example if the sample consists of a deposit scraped from the surface of a incinerated container, then the significance and even the meaning of the term *representative* becomes lost. In this case the analysis may only have a *qualitative significance* for forensic purposes. Any quantitative pertinence will lie in the relative proportions of the constituents within the sample not in their absolute values.

Sample Transportation and Storage

The method of transportation and storage will depend on the physical form of the sample, its sensitivity to the environment and to changes in ambient conditions. Merely sealing the sample in an air-tight container may not be sufficient. If the substance is thermally labile or of a biological origin, and thus susceptible to denaturing and bacterial breakdown, the sample may need to be refrigerated until ready for preparation. If the sample is susceptible to atmospheric oxidation, and may be stored for some time before analysis, then it may need to be stored under nitrogen or helium. If the substance contains only traces of the material of interest (*e.g.*, pesticides in a water sample), then the container must be made of a material that will not adsorb significant quantities of the pesticide. In general, sample containers must be chemically and physically inert to both the materials of interest and the sample matrix. This may demand the use of glass, polythene, Teflon (polytetrafluoroethylene) or stainless steel, etc., as the material of construction. Thus, each container must be made of a specific material appropriate for the sample. The individual sample containers should then be packed into a suitable box, well insulated against physical shock that might rupture the containers, and if necessary, thermally insulated from excess heat or refrigerated. On arriving at the laboratory the samples should be stored in an appropriate (preferably locked) enclosure that is readily available to the analyst when required for processing.

Sample Preparation

Sample preparation can involve a wide variety of techniques, each applicable to particular types of sample. Gas samples will require special handling and special sampling devices. Liquid samples may need to be extracted (for example, pesticides from water), solvent extracts evaporated (or concentrated by some other means). In some cases a preliminary separation may be necessary which may include filtration or distillation Samples that are inherently involatile but still need to be separated by GC will require to be derivatized. In trace analysis, certain materials may need to be sensed by selective detection, by, for example, the use of the fluorescence detector. Under such circumstances fluorescent derivatives will need to be prepared. Irrespective of the special skills involved in the successful operation of a chromatograph, sample preparation, an essential part of all chromatographic analyses, will also demand many other skills not directly associated with chromatography. Such skills include many of the laboratory operations normally associated with general chemical analysis and, in particular, the techniques of microanalysis.

Analytical Procedures

The choice of the analytical procedure can be quite difficult, as many samples can be analyzed, quite satisfactorily by more than one chromatographic technique. The choice of technique will be discussed in detail later but some indication of the basis of choice is appropriate here. Gas solid chromatography (GSC) is the most useful technique for the separation of permanent gases, *e.g.*, nitrogen, oxygen, hydrogen, the inert gases, and the low molecular weight hydrocarbon gases. Some of the corrosive gases, *e.g.*, the halogens, hydrogen chloride, sulfur dioxide, are also often separated by GSC but the complete conduit system of the chromatograph must be constructed of appropriately inert materials. This will include all parts of the conduit system, including the detector sensor that may come in contact with the corrosive sample.

Volatile materials, *e.g.*, essential oils, petroleum fractions, and solvents, are usually separated by gas liquid chromatography (GC) in preference to LC, as higher efficiencies are available and separations can be much faster. Some samples can be separated by both GC and LC and the decision must hinge on the nature of the sample. If the components of the sample are relatively involatile and thermally labile, LC might appear the technique of choice. However, if trace materials are important or there is very limited sample, then the higher sensitivities of the GC detectors may be necessary. Thus the sample will need to be derivatized to increase the volatility of the constituents to be analyzed by GC. Conversely, if the components are thermally labile, the sample may need to be separated by LC, using a high sensitivity detector such as the fluorescence detector, and the sample constituents will need to be carefully derivatized with a fluorescing reagent. It should be noted where both GC and LC are practical possibilities, GC will usually provide higher sensitivities, higher efficiencies and faster analyses, but the substances of interest must be either volatile or readily reacted to form volatile derivatives. In contrast, LC will handle completely involatile samples and thermally labile samples and, because of the freedom to modify relative retention by both changing the stationary phase and the mobile phase, it offers far greater selectivity.

Liquid chromatography is probably the most versatile of all the chromatography techniques and about 65% of all chromatographic analyses are carried out employing LC techniques. One of its widest fields of application is that of biotechnology where it can be used to separate many biopolymers, including the polypeptides, proteins, carbohydrates, lipids, etc. LC has the advantage over GC in that it can separate materials on the basis of ionic interactions as well as dispersive and polar interactions, which gives it an extra degree of selectivity. In addition, LC can separate molecules on the basis of size by exclusion techniques. In general, LC is the more useful technique, as, with the exception of the permanent gasses, LC can separate those substances that are normally separated by GC, albeit perhaps less efficiently. However, the converse does not apply.

TLC is the inexpensive, cost effective cousin of LC. Many separations that are achievable by LC can also be successfully resolved on a thin layer plate. The spot on a thin layer plate, however, suffers significant dispersion which, even with high efficiency plates, is difficult to control. Consequently, the efficiencies obtained from a plate are much less than those obtainable from an LC column. Nevertheless, due to the unique method of development [22], which results from a type of frontal analysis of the solvent system as it moves up the plate, TLC can be highly selective providing the sample is not too complex. This, in fact, tends to compensate for the relatively poor plate efficiency. However, as will be seen later, the big disadvantage of TLC is its poor quantitative accuracy and precision compared with that of LC. Thus, a less expensive, simpler procedure is realized at the expense of quantitative accuracy and precision. Notwithstanding, there are many samples for which the quantitative accuracy provided by TLC is still sufficient and the technique should always be selected where appropriate.

Capillary electro-chromatography (CEC) in principle is very similar to LC but the mobile phase is driven through the column by electro-osmotic flow and not by hydraulic pressure. Thus, for the technique to be effective, the sample should be soluble in an electrolyte, usually aqueous, which is necessary for electro-osmotic flow. The capillary has an exceedingly small diameter, and as the diameter usually constitutes the path length of the sensor, the detector is relatively insensitive. Consequently, relatively high concentrations of solute need to be measured and the dynamic range of the analysis is small. CEC is a relatively new technique, and although heralded by enthusiasts, has yet to be established as a versatile form of chromatography with a wide field of application. The technique requires a very small sample but, with present sampling techniques, it requires a significant amount of material to place the small sample on the column. The areas of application of this technique are still not clearly defined and there are, to date, a very limited number of publications that give a clear indication of the accuracy and precision that can be expected from quantitative analyses made by CEC.

Data Processing

Analytical data was originally obtained by direct measurements taken from the chromatogram. Manual data processing involves the physical measurement of peak heights, retention distances, peak widths and, if appropriate equipment is available, peak areas. Despite the introduction of the computer to analytical instrumentation in the mid-1960s, manual measurement is still carried out today in those laboratories with limited funds or in university laboratories employing laboratory-made equipment. However, the majority of contemporary chromatographs are fitted with computer data acquisition and data handling systems which automatically measure the necessary chromatographic parameters and, after correcting for individual response factors, calculate the results and print out the analysis. The data from each analysis is stored on disk, thus eliminating much tedious bookkeeping, and the computer assures a high level of calculating accuracy. Unfortunately, the accuracy of the calculation is the least likely and least important source of error in most chromatographic analysis; it is the assumptions that are made in the calculations, direct or implied, that can lead to serious error. Nevertheless, certain procedures for obtaining the data need to be followed, if the results are to be meaningful. In addition, special precautions need to be taken if the peaks of interest are incompletely resolved from their neighbors. It follows that the protocol used for data processing needs to be known and its limitations understood.

Analytical Reports

The report of a chromatographic analysis can range in complexity from the entry of a date, a sample number and a percentage figure onto a standard printed form, to a report of fifteen pages or more.

The former would be typical of a process control analysis, the latter more typical of a forensic or pollution report. In the former the method of sampling, the sample preparation, the chromatographic conditions, the method of calculation, the confidence levels would all have been firmly established for the sample and would not be reported on the form. However, in the latter case the complete analytical details, from the method, time and place of sampling to the accuracy and precision of the method, would need to be recorded and presented with the analytical results. In forensic reports, particularly for cases where litigation is anticipated or in progress, every pertinent detail needs to be included and the report may need to be prepared with the aid of an attorney. Most analytical reports will be more akin to that of the process control but, between the two extremes, the analytical details that should be included will vary significantly from one type of report to another. However, it must be emphasized that an analytical report is designed to *inform* and *not confuse*, and thus pertinent information, whatever the amount, should be presented clearly and precisely. In addition, the report should be couched in terms so that those unfamiliar with chromatography will still understand the significance of the results.

Synopsis

Initially, chromatography was merely used as a separation technique. On the discovery of gas chromatography and sensitive in-line detectors, however, chromatography began to be used for quantitative analysis, and by the time high performance liquid chromatography was developed, the majority of chromatographic separations were carried out for quantitative purposes. Chromatography is one of the most important and popular analytical techniques available as it offers high resolution, fast separations and good quantitative accuracy. Nevertheless, there are a number of important factors affecting the accuracy of a chromatographic analysis that must be understood. The sample must be representative of the bulk being analyzed and must be transported and stored in an appropriate manner. The sample must be prepared correctly for analysis, and the chromatographic separation carried out in a manner that ensures that the solutes of interest are resolved and the integrity of the sample is maintained. The physical form of the sample, gas, liquid or solid, and its chemical nature will influence all sampling and chromatographic procedures so there will a unique set of analytical conditions for each sample type. The nature of the sample, its complexity, the amount of material in the sample, the cost of the analysis and the required accuracy will all determine the type of chromatography that is most appropriate (*i.e.*, GC, LC, TLC or CEC). Data is still manually processed but the majority of analytical results are now acquired and processed by a computer. The analytical report may be simple or extensive depending on the source of the sample and the reason for the analysis. The report should be concise and accurate, and should be presented in a style that will allow those unfamiliar with the technique to understand the significance of the results.

References

1. A. T. James and A. J. P. Martin, *Biochem. J.*, **50**(1952)679.
2. A. T. James, The Times Science Review, **Summer** (1955)8.
3. N. H. Ray, *J. Appl. Chem.*, **4**(1954)21.
4. R. P. W. Scott, "*Vapor Phase Chromatography*" (Ed. D. H. Desty and C.L. A. Harbourn), Butterworths Scientific Publications, (1957)131.
5. H. Boer, "*Vapor Phase Chromatography*" (Ed. D. H. Desty and C.L. A. Harbourn), Butterworths Scientific Publications (1957)169.
6. I. G. McWilliams and R. A. Dewer, *"Gas Chromatography 1958"* (Ed. D. H. Desty), Butterworths Scientific Publications, (1957)142.
7. D. W. Grant, *"Gas Chromatography 1958"* (Ed. D. H. Desty), Butterworths Scientific Publications, (1957)153.
8. J. E. Lovelock, "Gas Chromatography 1960", (Ed. R. P. W. Scott) Butterworths Scientific Publications Ltd., London, (1960)9.
9. A. J. P. Martin and S. S. Randall, *Biochem. J.* **49**(1951)293.
10. A. Tiselius and D. Claesson, *Ark. Kemi. Mineral. Geol.* **15B(No 18)**(1942).
11. 2. D. Zaukelies and A. A. Frost, *Anal. Chem.* **21**(1949)743.
12. F. A. Vandenheuval and E. Sipas, *Anal. Chem.*, **33**(1961)286.
13. C. Christiansen, *Ann. Phys. Chem.*, **3**(1884)298.
14. M. Bakken and V. J. Stenberg, *J. Chromatogr. Sci.*, **9**(1971)603.

15. J. P. Gorden, R. C. C. Leite, R. S. Moore, S. P. S. Posto, J. R. Whinnery, *Bull. Am. Phys. Soc.*, **(2) 9**(1964)501.

16. C. G. Horvath and S. R. Lipsky, *Nature*, **211**(1966)748.

17. A. T. Rhys-Williams, *Fluorescence Detection in Liquid Chromatography,* Perkin Elmer Ltd., Beaconsfield, England (1980).

18. R. P. W. Scott, *Chromatographic Detectors*, Marcel Dekker Inc., New York-Basle (1996).

19. R. P. W. Scott, *Introduction to Analytical Gas Chromatography,* Marcel Dekker Inc., New York-Basle (1996).

20. E. D. Katz, *High Performance Liquid Chromatography: Principles and Methods in Biotechnology*, John Wiley and Sons, Chichester-New York (1996).

21. T. E. Beesley and R. P. W. Scott, *Chiral Chromatography*, John Wiley and Sons, Chichester-New York (1998).

22. R. P. W. Scott, *Techniques and Practice of Chromatography*, Marcel Dekker Inc., New York, (1995)19.

Chapter 2

Sample Collection, Transport and Storage

Sampling methods can be broadly separated into three groups based on the physical form of the specimen, *viz.* gases, liquids and solids. Each type obviously demands quite different sampling techniques. In addition, within each sample type, the sampling method will also vary, depending on the level of the components of interest. It is impossible in a book such as this to cover all the sample classes that might arise and the intent is, therefore, to cover the major types that the analyst might meet and, where possible, give references to other applications that may be appropriate.

Gas Samples

Gas samples usually fall into two categories, one where the sample contains a limited number of major components, for example flue gases, which will contain mainly nitrogen. oxygen and carbon dioxide together, perhaps, with some carbon monoxide. Alternatively, the second type of sample may contain a large number of trace vapors such as the essential oils in the atmosphere around a flower, or hydrocarbon vapors in the vicinity of a gasoline pump. These two, quite different examples must be sampled by very diverse procedures. They also represent two quite different motives for analysis. The first is obviously important for the efficient control of combustion in a furnace and the hydrocarbons in the vicinity of a fuel pump would be an example of pollution analysis. The procedures used for the two extreme types of sample will be considered.

Sampling for the Major Components of a Mixture of Permanent Gases

Gases are compressible and consequently expand and contract with temperature and pressure. It follows that it will be important to take the sample under known and measured conditions, particularly if the sample is contained in a flue duct which is not likely to be at ambient temperature or pressure. Ideally, the sample vessel should be thermostatted and the sample taken at atmospheric pressure. However, if the detector has a known or calibrated response to each component, sample expansion or contraction will occur identically with all the sample components, assuming they are all permanent gases. Consequently, the relative proportions of each component will not change and so under some circumstances, there may not be a need to control or measure the temperature and pressure.

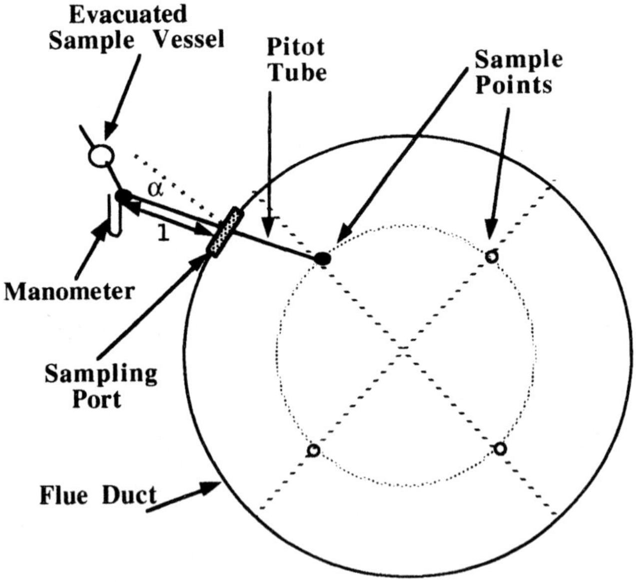

Figure 1 Sampling a Flue Duct

However, in large flue ducts, neither the flow velocity, the temperature or the sample composition is likely to be constant over the *cross section* of the duct, irrespective of its shape, nor is flow profile likely to be Newtonian. This means that samples must be taken across the duct and, at the same time, the flow velocity, temperature and pressure must also be taken at the same sample points. Consider a flue duct, the cross section of which is depicted in figure 1. By using a Pitot tube and a manometer in the standard manner, the flow velocity and the absolute pressure in the duct can be measured at any given point. Knowing the total length (L) of the Pitot tube, the length (l) exposed outside the flue duct and the angle (α), the position of the sampling point can be easily calculated. The Pitot tube senses the pressure at apertures normal to the flow of gas and tangential to the flow of gas. The pressure difference is a function of the kinetic energy of the gas striking the normal aperture and thus is a function of the gas velocity. Consequently, the pressure difference gives a means of measuring the gas velocity. In addition, with a simple valving system, and by using an evacuated sample bulb as a container, a sample can be taken from the Pitot tube at the same time and position that the flow rate and pressure measurements are made. The tube and sampling arrangement are shown in figure 2.

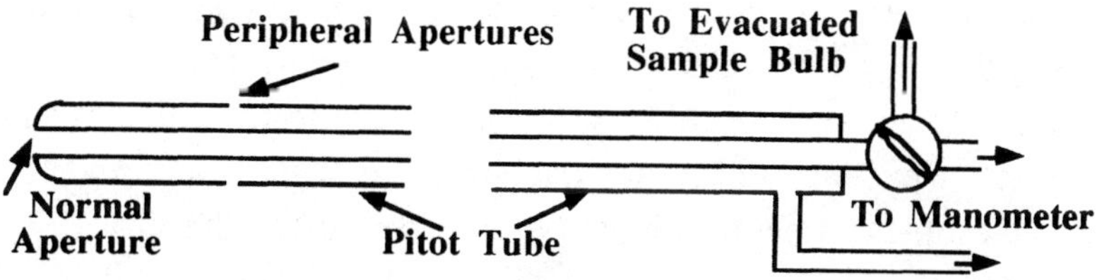

Figure 2 The Pitot Tube

The gas velocity (V) is given by the expression, $V = C\sqrt{2g\Delta H}$

where (C) is a constant,
(g) is the acceleration due to gravity,
and (ΔH) is the pressure difference between the impact pressure and the peripheral pressure.

After the gas flow has been measured from the pressure difference between the central aperture of the Pitot tube and the peripheral apertures, the valve is rotated. The central tube is then connected directly to the evacuated gas sample container and a sample withdrawn down the center tube for analysis.

Figure 1 shows that four symmetrically located sample points are employed, but if the flow varies widely across the duct, or the duct is not circular in cross section, more than four sampling points will probably be necessary. It should be noted that only one sample opening in the duct is required for the Pitot tube, as, providing it is long enough, it can be placed at any point across the duct. As the flow velocity and pressure will be known, and the gas composition obtained from the analysis, the total flux of each component along the tube can be easily calculated. Such data is often necessary to carry out a heat balance on a combustion unit.

Gas samples are typically collected in glass containers with taps at either end, but stainless steel and Teflon® containers can be equally effective as long as the gases do not react with these materials. The sample tube is usually evacuated before being used for sampling. However, if this procedure is not adopted, it is important that the sample tube is well washed with the sample gas before the actual sample is taken. The elution profile of the gas (air), contained in the sample tube, by the sample gas itself will be logarithmic and thus at least 10 volumes of sample must be passed through the tube before it can be considered free of air. If the sample tube is evacuated, then the tap can be opened in the presence of the gas to be sampled, the pressure in the sample tube allowed to come into equilibrium with the sample atmosphere and the taps then closed.

Solid particles, such as dust in the sample, can easily contaminate the chromatograph but are unlikely to affect the accuracy or precision of the analysis where *major components* are concerned. Therefore, a filter should be inserted between the sample tube and the sample inlet to remove any relatively large solid particles that may be present. Another precaution needs to be taken if one or more of the sample components are significantly soluble in water and the gas is wet. Under such circumstances, a drying tube should be inserted in line with the filter. However, any drying agent that is used must be chosen so that it does not disproportionately adsorb any one of the sample components as this can also introduce a significant error in the analytical results. To ensure accuracy and avoid the effect of sample variation, preferably three separate samples should be taken and each sample analyzed in duplicate.

When the sample arrives in the laboratory, the sample can be sucked directly from the sample container into the gas sample loop of the sample valve by means of a pump. An alternative, but somewhat archaic procedure is to displace the gas sample, from the sample vessel, into the sample valve with mercury, and this technique is, at times, still employed. It is somewhat inconvenient, but allows the sample loop to be well purged with sample before analysis. It also allows the residual sample to be kept at atmospheric pressure for extended periods of time without risk of contamination by external air. Despite the advertised toxicity of mercury, this process, if carried out carefully, with the necessary precautions being taken, need not be hazardous and has been used for decades with little concern.

Sampling for Minor Components of a Gas Mixture

The determination of trace quantities of volatile substances in air or other permanent gases requires quite different sampling techniques. Some examples of this type of gas sampling applications will be given here, but others will be included in the section on GC applications. One common example is the measurement of a toxic vapor (*e.g.*, benzene vapor) in air as an environmental test, another is the analysis of the 'spectrum' of essential oils generated by a flower to identify the optimum time for harvesting and finally, the analysis of the head space over foods for quality control or to identify product deterioration.

Sampling Benzene Vapor in Air

This type of sampling procedure may be required in the neighborhood of an oil refinery, gasoline storage tanks, possible even in the vicinity of gasoline pumps. The concentration range for benzene in air may extend from toxic levels of 1 ppm or more to relatively safe levels of 1 ppb or less. Consequently, at the lower, safe concentrations, the amount of benzene present in one cubic centimeter of air will be extremely small. Thus to obtain sufficient material for analysis, it is necessary to trap and concentrate the benzene by some suitable procedure from a significant volume of gas and regenerate the benzene for analysis. There are a number of ways of doing this; some techniques require temperature desorption to recover the benzene, others regenerate the benzene by solvent displacement. A diagram of the sampling apparatus that might be used for the determination of benzene in air is shown in figure 3.

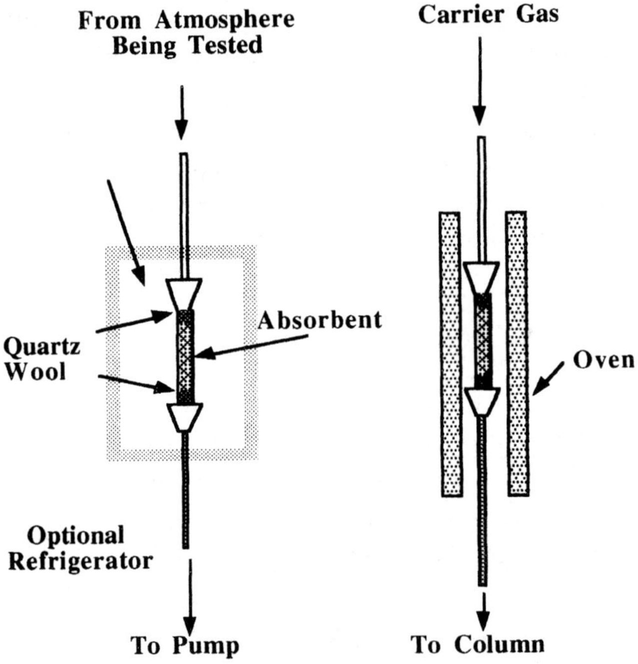

Figure 3 Apparatus for Sampling Trace Amounts of Benzene in Air for Subsequent GC Analysis

Traps used with temperature desorption are usually packed with graphitized carbon, and are 2–4 mm I.D. and about 10 cm long

(Supelco Carbotrap 300 is an example). The air sample is usually aspirated through the trap by reduced pressure applied to the outlet by a pump or syringe. The volume of air sample taken may range from 500 ml to several liters and will depend on the contamination level of the benzene. The tube is often refrigerated to improve the trapping efficiency but this is normally only necessary for large sample volumes. Cooling avoids the possibility of some of the benzene being eluted from the trap. However, if cooling is employed, the trap may well also contain water that may cause sample preparation difficulties or interference in the subsequent GC analysis. This problem may be reduced, although not eliminated, if the trap temperature is maintained at 2°C or above whenever possible.

When the sampling procedure is complete, the trap is sealed by capping and placed in a thermally insulated box. The box should also be lined with some appropriate packing to protect the sample from mechanical shock during transport. If the transport has refrigeration facilities the box should be kept refrigerated at about 2°C. On arriving at the laboratory the sealed trap should be kept in a refrigerator until ready for analysis.

When ready for analysis, the tube of adsorbent is placed in an oven and connected directly to the inlet of the chromatographic column. The column is either cooled to about 10°C or to room temperature and the trap temperature programmed up to about 300°C. A flow of carrier gas (of a few ml/min.) sweeps the desorbed benzene onto the column and is focused on the cool front of the column.

After the benzene has been completely desorbed, the column is programmed up to about 100°C (the precise temperature will depend on the type of column that is used) and the benzene peak monitored in the normal way.

If the benzene is to be regenerated with solvent, a different type of trap is used which can be packed with coconut charcoal or similar adsorbent and is usually about 7 cm long and 6 mm I.D. (Supelco Charcoal trap 32 *small* is an example). The sample is taken in a similar manner to that described previously and stored and transported in the same way. However, when ready for analysis, the benzene is displaced from the trap with about 2 ml of carbon disulfide in a vial which is sealed with a septum cap. The vial with a septum cap is commonly used for storing GC samples after being prepared or derivatized and a diagram of a vial is shown in figure 4.

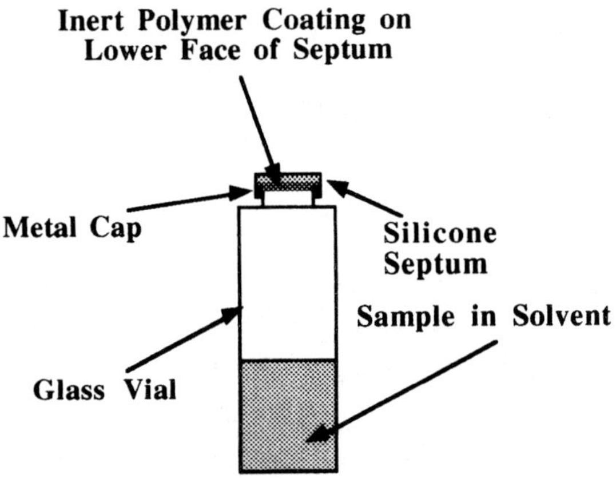

Figure 4 Sample Storage Vial

The vial should be fitted with a metal cap that can be crimp-sealed to the glass tube and have a hole in the top that exposes the silicone septum. The lower face of the silicone septum may need to be protected from the sample by a film of inert plastic (Teflon or some other suitable material). Such septums are readily available. The vials are usually about 1, 2 or 5 ml in capacity. The sample syringe is made to pierce the septum and then to dip into the sample solution. The appropriate amount is then withdrawn into the syringe. This procedure can be manual or, if the vial is situated in the carousel of an automatic sampler, can be actuated automatically.

There is little to choose between the accuracies that can be obtained by the two methods, but temperature desorption of the sample is probably the simpler procedure. However, in using the temperature desorption technique, the thermal stability of the components of the mixture at the desorption temperature must always be considered. Thermal desorption may not be suitable for essential oils which will often contain a number of terpene based substances that can be thermally labile.

The Analysis of Flower Fragrances

Not surprisingly, there is a unique period in the growth of a flower, when both the composition and yield of the essential oil is at an optimum. It is less obvious that, for some flowers, there is also a particular time of day when the yield and composition is best and ideal for harvesting. It follows that the precise composition of the essential oils evolved from flowers over their total blooming period and during the day would be particularly valuable to the flavor and perfume industries. GC is clearly the ideal analytical technique for this specific type of assay.

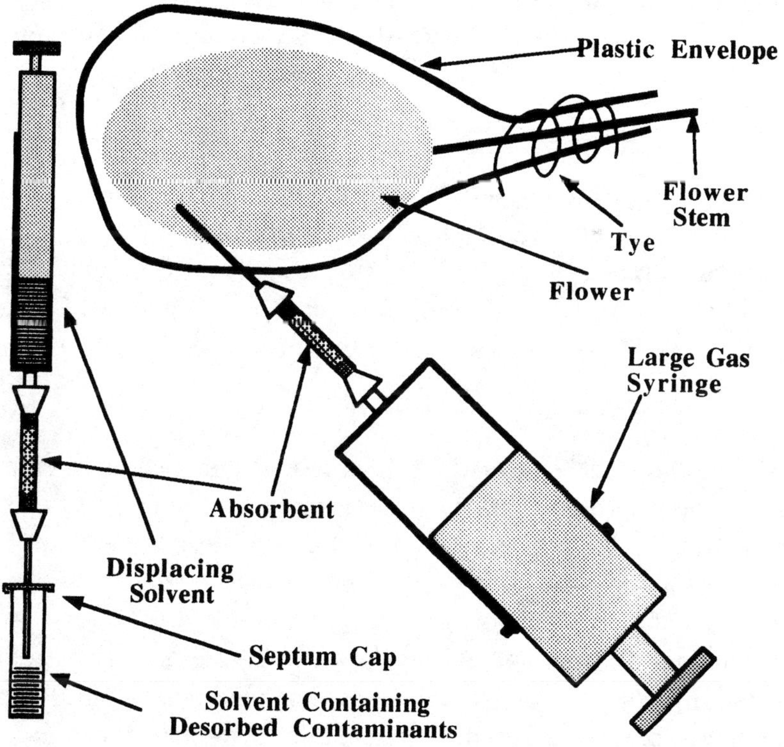

Figure 5 Apparatus for Sampling Flower Fragrance for Subsequent GC Analysis

The procedure is relatively simple and the apparatus used is shown in figure 5. The flower head is covered with a plastic envelope that must be made of an inert material through which the essential oils cannot diffuse or, at least, diffuse very slowly. Thin Teflon sheets or other inert plastic materials can be used. When a sample is taken, the neck of the envelope is tied round the stem of the flower to temporarily isolate the flower from the surrounding atmosphere.

The envelope is left round the flower for about 10 minutes to ensure equilibrium is established between the flower and surrounding atmosphere, then the sample is taken. This is carried out by inserting a syringe needle attached to the adsorption tube through the envelope and extracting a known volume of equilibrated air (*ca* 100-200 cc) through the adsorption tube as shown in the diagram. The tube is then capped and preferably stored in a refrigerator. The sample tubes are then placed in a padded, thermally insulated box and preferably transported to the laboratory while refrigerated at about 2°C.

There are a number of adsorbents that can be used in the trap and the choice will depend on the nature of the essential oil. Typical adsorbents are carbon, silica gel, macroporous polymer adsorbents or a reversed phase. A typical adsorption trap (*e.g.*, Supelpack 20P) would be 6 mm O.D., 7 cm long carrying about 100 mg of macroporous polymer. The sample is usually regenerated by solvent displacement.

After sampling, one end of the trap (the end the sample entered) is connected to a syringe containing the displacing solvent, and the other end (the trap exit) is connected to a hypodermic needle that is made to pierce the septum of a sample vial. The solvent is slowly discharged through the trap by the syringe into the sample vial, desorbing the essentials oils into the solvent as shown in figure 5. It is extremely important that the displacing solvent is absolutely clean and it should be passed through a reversed phase cleaning column prior to use. An appropriate volume of the extract is then injected onto the GC column and the components of the essential oil are separated. Obviously this technique can be used for a number of different materials containing traces of volatile compounds such as fruits, herbs, tree bark and other natural products. However, these type of samples are usually brought back to the laboratory for sampling.

Head Space Analysis

Another type of gas sampling technique which has been used successfully for similar samples is called *head space analysis.* This sampling technique is also used to monitor trace organic components in air or some other permanent gas which has been allowed to come into equilibrium with a parent sample matrix. This technique has been used to examine the condition of food stuffs, specifically to detect food deterioration. The deterioration of certain foods during storage is often accompanied by the characteristic generation of certain volatile products such as low molecular weight organic acids, alcohols and ketones. In addition, the same sampling technique can be used to identify the essential oils from fruits and vegetables, and the pattern of peaks can often help identify their source.

Due to the type of apparatus needed for head space analysis, when assaying materials such as foodstuff, herbs or plant tissue, the sample must be collected on site and then brought back to the laboratory for subsequent extraction and analysis. It is therefore exceedingly important to ensure that the material does not deteriorate, or change, between the time of sampling and the actual time of analysis. This situation can be particularly crucial if the sample consists of foodstuff that is suspected of spoilage as, unless careful precautions are taken, the spoilage will continue during the waiting period between tissue collection and analysis and consequently give false results. Accordingly, after the sample of foodstuff or plant tissue has been selected it should be placed in a sealed plastic bag made from opaque material (this is necessary to prevent photochemical decomposition) and refrigerated at about 2°C. At this temperature bacterial action will be strongly retarded and the condition of the sample will remain reasonably stable for several hours. If the head space sampling and analysis cannot be accomplished within a few hours, however, then the sample should be frozen immediately after selection. Under circumstances where the composition of the sample does change with time, and it is impossible to carry out the analysis immediately, the sample can be analyzed regularly after known time periods. The results are then plotted against time and the analysis extrapolated back to zero time when the sample was taken.

The basic sampling procedure is similar to that employed previously, but the essential oils or volatile organic contaminants are usually regenerated by thermal desorption as opposed to solvent displacement.

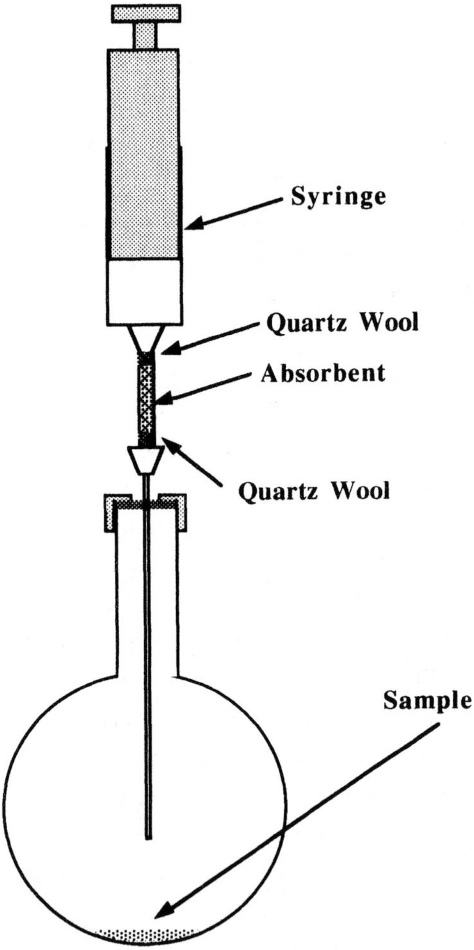

Figure 6 Head Space Sampling

A basic form of the apparatus is shown in figure 6, which demonstrates the simplicity of the technique and in fact, the apparatus can be fabricated in the laboratory if so desired. Despite its basic simplicity, this technique has been automated by a number of manufacturers and as a result the instrumentation has become quite complex. The material under examination is usually transferred to a flask, but the head space of the container in which the material is stored, if suitable, can also be used as the sample source. To increase the distribution of the volatile substances in the gas phase the flask may be warmed to about 40°C before sampling is commenced. A defined volume of the air above the material is withdrawn through an adsorption tube by means of a gas syringe. Graphitized carbon is often used as the adsorbing material although other substances such as porous polymers can also be employed. Carbon adsorbents that have relatively large surface areas (*ca* 100 m^2/g) can be used for adsorbing small molecular weight materials, whereas adsorbents of lower surface areas (*ca* 5 m^2/g) are more suitable for adsorbing larger molecules. Choosing the adsorbent surface area that is appropriate for the substances to be sampled is important, as it allows the larger molecules, that have lower volatility and higher boiling points, to be thermally desorbed more easily. In addition, desorption of high molecular weight materials from the lower surface area adsorbents reduces the risk of thermal degradation.

To desorb the sample, the trap is placed in an oven and connected to the chromatograph. The column is maintained at a low temperature (50°C or less) to allow the desorbed solutes to be concentrated on the front of the column. The trap is heated rapidly to about 300°C and a stream of carrier gas then sweeps the desorbed solutes onto the column. When desorption is complete, the temperature of the column is programmed up to an appropriate temperature and the components of the head space sample are separated and quantitatively assayed. Employing head space analysis, off–flavors that develop in processed foods during very early spoilage can often be detected and identified.

Solid Phase Micro-Extraction

The solid phase micro-extraction technique operates on exactly the same principle as the head space sampler but is much smaller and thus can examine very small samples. The apparatus used is shown in figure 7. The basic extraction device consists of a length of fused silica fiber, coated with a suitable polymeric adsorbent, which is attached to the steel plunger contained in a protective holder. The steps that are taken to sample a vapor are included in figure 7. The sample is placed in a small head space vial which may be heated if necessary and allowed to come to equilibrium with the air (1). The needle of the syringe containing the fiber is made to pierce the cap, and the plunger pressed to expose the fiber to the head space vapor. The fiber is left in contact with air above the sample for periods that can range from 3 to 60 minutes, depending on the nature of the sample (2).

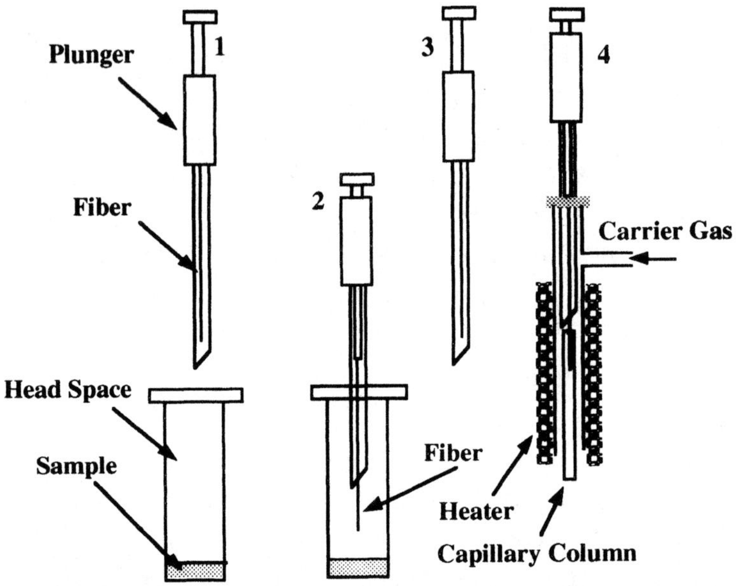

Figure 7 The Solid Phase Micro-Extraction Apparatus

The fiber is then withdrawn into its holder and removed from the vials (3) and passed through the septum of the injection system of the gas chromatograph into the region surrounded by a heater (4). The plunger is again depressed, and the fiber, now protruding into the heater, is rapidly heated to desorb the sample onto the GC column. In most cases the column is kept cool so the components concentrate on the front of the column. When desorption is complete (a few seconds), the column can then be temperature programmed to separate the components of the sample. A chromatogram of the head space sample taken over tobacco is shown in figure 8. The sample consisted of 1 g of tobacco (12% moisture) contained in a 20 ml head space vial to which had been added 3.0 ml of 3 M potassium chloride solution. The fiber was coated with a 100 μm film of polydimethyl siloxane (a highly dispersive adsorbent). The vial was heated to 95°C and the fiber was left in contact with the head space for 30 min. The sample was then desorbed from the fiber for one minute at 259°C. The separation was carried out on a column 30 m long, 250 μm I.D. carrying a 0.25 μm thick film of 5% phenylmethylsiloxane.

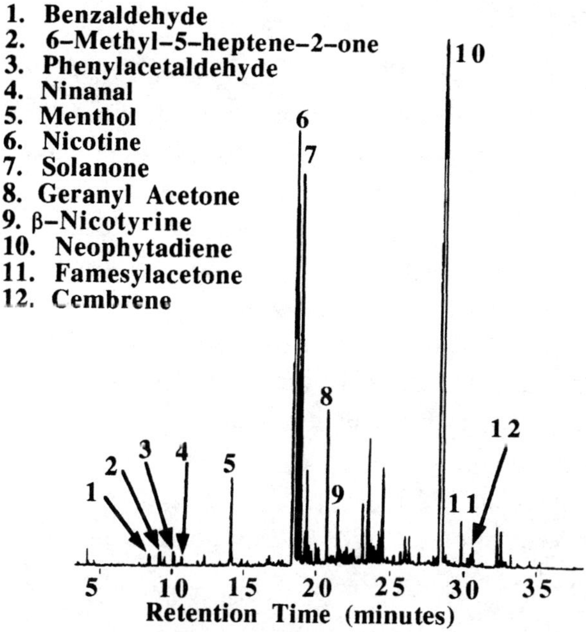

Courtesy of Supelco Inc.

Figure 8 A Chromatogram of Tobacco Head Space

The column was held isothermally at 40°C for one minute and then programmed to 250°C at 6°C/min. and then held at 250°C for 2 min. It is seen that a clean separation of the components of the tobacco head space is obtained and the resolution is quite adequate to compare tobaccos from different sources. This sampling technique has been widely applied to many vapor-in-air problems, ranging from the analysis of the volatile components from blood and urine for diagnostic purposes to monitoring soil pollution, and combustion research. The technique is reliable, extremely sensitive, and can provide accurate *proportional* analysis. However, as the concentration of the vapor of each solute in the air will depend on its distribution coefficient between the sample matrix and air, the value for each will differ and thus *absolute* concentrations in the sample matrix can *not* be assayed.

Liquid Sampling

Sampling a liquid is a relatively straightforward procedure. However, problems may arise if the sample is not homogeneous. Solid particles must be removed by filtration or centrifugation. Further, if trace components are important, any solids that are removed may contain some of the trace materials adsorbed on their surface. Consequently, the separated solids must be extracted with appropriate solvents and the solution added to the extracted trace materials prior to analysis. If the sample consists of an emulsion of two liquids, then the mixture must be centrifuged and the two liquids separated. In most cases the two liquids will need to be analyzed separately. If one liquid (usually the lower) also contains solids, then these must also be removed by filtration. Samples should be sealed in dark glass vials or bottles for storage to prevent photochemical decomposition and refrigerated to prevent bacterial decomposition.

Sampling bottles or drums of liquids is relatively simple as in most cases the sample will be homogeneous. If the sample is heterogeneous, however, care must be taken to mix the bulk material well before a sample is taken so that a truly representative sample is obtained. Problems can arise when sampling large masses of moving material such as sea water from the sea shore and water from the banks of a swiftly flowing river. These type of samples are very common in pollution monitoring and pollution studies. The composition of the sample will almost certainly differ with the distance from the shore or bank, with the water velocity and probably with the depth at which the sample is taken. It follows that a number of samples must be taken at different depths and at different distances from the bank or beach. In addition some effort should be made to measure the apparent water velocity at the sampling points in order to relate the analytical results to water movement. Even when sampling 'still' water such as ponds and lakes, multiple samples need to be taken at different positions in the lake or pond and at different depths, if a valid analysis is to be achieved.

Water samples should be stored in dark glass bottles or stainless steel containers to reduce photochemical reactions and refrigerated at about 2°C to retard any bacteria or algae growth. The level of oxygen is often important in such samples, and may change the amount, or nature, of the contaminants, so the bottles or containers should be completely filled so that there is the minimum of air between the water surface and the stopper. In all samples where biological activity can occur, the analysis should be carried out as soon as possible after the sample has been taken.

Major liquid deliveries comprising petroleum products or milk often arrive in a number of large containers such as drums or tanks. If each container is not to be assayed separately, then to obtain a single representative sample of the total delivery, a volume should be taken from each container that is proportional to the volume of the respective container. All the samples are then bulked together. Samples should be stored in bottles or cans made from appropriately inert material: glass, stainless steel, plastic etc. If the product is chemically and thermally stable it can be transported under ambient conditions, but if susceptible to biological or biochemical change (*e.g.* milk, whey, corn syrup), then the sample should be refrigerated and kept at about 2°C until assayed. Liquids of biological origin should not be stored for more than 24 hr. before analysis even at 2°C, otherwise the sample may change significantly both physically and chemically. Heterogeneous liquids, such as paint or emulsions, must be well mixed before a sample is taken. If it can be ascertained that all the cans of a particular delivery were filled from the same batch after manufacture or blending, then the contents of one or two cans can be expected to be representative of the total batch.

Corrosive liquids should be handled according to the protocols published for the respective material. Samples can be stored in glass, in stainless steel or, for extremely corrosive materials, in Teflon containers. Recommended materials for storage containers are published for most highly corrosive substances and the pertinent references should be obtained. The containers must be well protected against mechanical shock in transport and appropriately labeled in case of accidents. Due to the hazardous nature of such materials the volume or mass of material taken for analysis should be kept to the minimum required.

Solid Sampling

The sampling of solids is far more difficult than sampling liquids or gases as much larger sampling errors may arise due to the lack of uniformity of the material. In table 1, a guide is given for the minimum amount of sample that should be taken related to total size of the bulk material.

Table 1 Minimum Sample Size Relative to the Bulk Material

Material Bulk Weight tons (long or metric)	Sample Weight (kg)
up to 0.5	2.5
0.5–1.0	4
1.0–2.0	6
2.0–5.0	10
5.0–10.0	15
1--25	25
25–50	40
50–100	60

If the uniformity of the bulk is found to be particularly large (a delivery of iron ore), then the sample quantities will have to be increased.

Powder, Granules or Crystals

Free flowing powders or granules can be sampled using a sampling 'spear', a diagram of which is shown in figure 9.

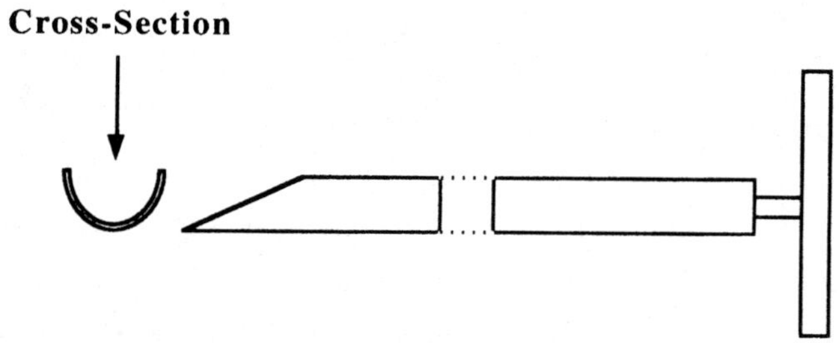

Figure 9 A Sampling Spear

The spear is thrust at an angle into the material with the opening underneath, rotated two or three times and then carefully withdrawn with the opening uppermost. The contents are then discharged into the sample container.

Tablets and Capsules

A number of tablets or capsules should be randomly selected (a minimum of 20 is recommended by the United States Pharmacopoeia) and the total weight of tablets measured. In the case of caplets, the contents of each capsule should be emptied and mixed and the total mass weighed. In this way the average weight of product per tablet or capsule can be calculated. A portion of the powdered tablets, or a portion of the mixed capsule contents, is taken as representative of the product. After analysis the individual components are either presented as a percentage of the contents of the average tablet or capsule, or as the actual mass present in an average tablet or capsule.

Bulk Stocks

Bulk stocks should be flattened as much as possible and samples taken at numerous places with the spear. Very large stocks cannot be sampled *in situ* but should be sampled either as they are accumulated or as the material is withdrawn.

Bags and Sacks

The spear should be inserted in a place where the aperture can easily be closed after sampling. Care must be taken to ensure that no fragments of the container contaminate the sample.

Casks and Kegs

If the top of the container cannot be removed, holes should be bored to allow the spear to enter. Samples should be taken from top to bottom and from one side of the container to the other and the two samples combined. Again any fragments of the container must be removed before placing the sample in its container. The final sample for analysis must be obtained by riffling or quartering which is described below.

Oily Solids

Stiff pastes should be sampled with a sampling auger and complete cores from different parts of each container should be placed in a suitable bowl. The contents of the bowl can be thoroughly mixed with a spatula made from a suitably inert material, taking care not to lose any liquid. A solid sector is then cut from the paste and stored in the sample container. Pastes that cannot be satisfactorily liquefied by heating should be mixed with a paddle and a portion taken with ladle. The mixing and sampling should be repeated at least twice, the sample bulked and placed in the container.

Coarse or Lumpy Solids

This type of material is the most likely to exhibit the greatest variation in property within its container. Great care must be taken, therefore, to ensure that a representative sample is taken. The final sample should be prepared by reducing the size of the lumps, followed by riffling or quartering.

Small Containers

The whole of the container contents should be spread out on a clean surface and a number of lumps and fines removed in proportion to their distribution in the bulk sample. It must be remembered that in the heap formed when emptying the containers the finer particles will remain at the center and the large pieces roll to the edge.

Road or Rail Vehicles

Sufficient sample must be removed from all parts of the load so that it represents the size distribution of all the material in the vehicle. It should be noted that the movement of the vehicle will cause the larger pieces to segregate and migrate to the top of the vehicle.

Bulk or Heaps

The heap should be flattened as far as possible and two trenches dug at right angles to one another across the heap. The bulk sample is that removed from the trenches. Very large bulk stocks cannot be sampled satisfactorily *in situ* and should be sampled either as they accumulate or as the material is withdrawn.

Ships Holds

Sampling should take place during loading or unloading. Material representative of the size distribution should be taken from the conveying plant or trucks at regular intervals throughout the transfer operation and combined to provide the final bulk sample

Reduction of the Sample to Analytical Size

There are two common methods of reducing bulk samples to a practical size, and they are riffling and quartering. A diagram of a riffle is shown in figure 10. The sample is fed onto the top of the riffle and as it falls through the device the sample is divided equally into two bins. When the operation is complete the contents of one bin is discarded and the other passed through the riffle again. In this way the sample is progressively and randomly halved until its bulk is reduced to that required for laboratory work-up. The riffle, due to its design, is extremely difficult to clean and therefore should be permanently kept for use with a specific material.

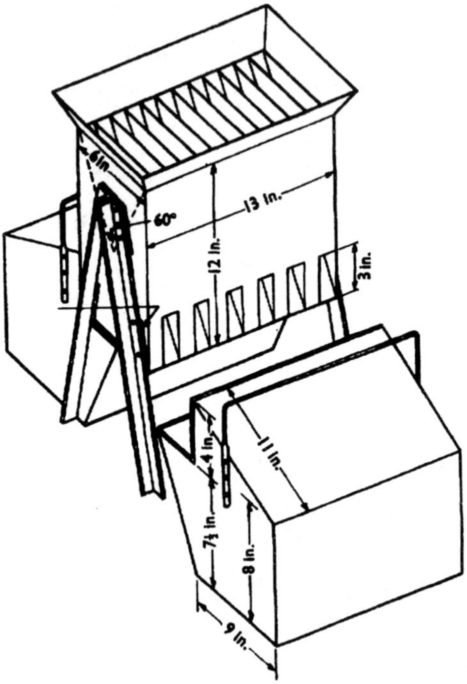

Figure 10 A Sampling Riffle

The quartering procedure is diagramatically represented in figure 11.

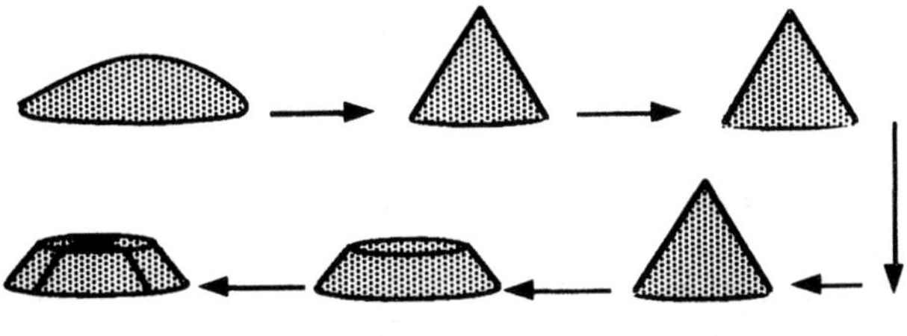

Figure 11 Sample Quartering

The sample is placed on a clean impervious surface and formed into a cone by shoveling. Each shovelful is placed on the top of the last so that the material runs down the side of the pile and is thereby distributed around the sides of the pile as evenly as possible. If larger pieces of material roll away from the base of the cone, they must be pushed back to the edge. From the first cone two successive cones are made in the same manner. The third cone is then flattened to a uniform thickness and then quartered along two diameters. One diagonal pair of quarters is shoveled out and rejected. This four stage procedure is repeated until a sample of the required size is obtained. If the initial bulk sample is very coarse then the average particle diameter may need to be reduced by grinding between each quartering procedure.

Many materials have standard sampling and storage specifications that have been developed by producers and customers and should be used wherever pertinent. Such regulation apply, for example, to building materials, petroleum products, and pharmaceutical products medical products and, more recently, regulations for sampling have also been published for certain biotechnology products.

Sample Size

Finally, the size of the sample that is required for the actual chromatographic analysis need to be mentioned. Some samples, *e.g.*, forensic samples such as material from an incinerated container, are obviously limited in size, but it is generally important to obtain an adequate quantity of sample. If bulk material is available, then the sample size is no problem. If the amount of material available is limited either by cost or availability, however, then a quantity sufficient for at least three replicate analyses should be taken. If possible, enough extra sample should be taken for reference purposes and to allow at least three additional samples to be analyzed should the initial results become challenged. These sample sizes are based on the assumption that a separation procedure has been developed and the appropriate column and operating conditions are known. If the separating conditions need to be determined from the sample, then at least ten times the amount suggested may be required.

Synopsis

Sampling methods can be separated into three groups: those used for gases, liquids and solids. Within these groups different techniques are used for different types of sample. Gas analyses, where the major constituents are permanent gases, are sampled using syringes or evacuated containers. Tests for minor constituents in pollution studies or for forensic purposes involve trapping the sample in short tubes of adsorbent through which the gas sample is drawn. The materials of interest are then either desorbed by heat and passed directly onto the column, or extracted with solvent and the solution is injected onto the column. Vapors in gases can also be sampled by head space analysis. The sample is heated in a container and the head space exposed to an adsorbent. The adsorbent is then transferred to a small heater which desorbs the materials directly onto the chromatography column. Liquids are sampled directly and if delivered in different containers a representative sample is taken from each container. Bulk liquid samples such as rivers and lakes for pollution tests require specific sampling procedures that are usually defined by the agency. All samples should be stored in containers made from appropriately inert material. If the substances that are sampled are thermally labile, subject to bacteriological activity, or chemically unstable, they should be stored and transported at low temperatures and if necessary under an inert gas such as nitrogen. Solids are the most difficult from which to obtain a representative sample. Sampling techniques vary, depending on whether the material is a powder or in bulk form; whether the material is in bags, sacks, casks or kegs; whether the solids are dry or oily, coarse or lumpy or are in small containers, in road or rail vehicles or in a ships hold. The amount of sample taken depends on the size of the bulk material and may vary from 2 to 60 kilos. When the sample is large it must be reduced by prescribed riffling or quartering techniques to ensure the ultimate sample is representative of the bulk. In most chromatographic analyses the sample size is not limited but if there are restrictions, sufficient should be taken for at least three analyses and, preferably, sufficient for three more should be stored for confirmation if required.

Recommended Reading

1. R. P. W. Scott, *Introduction to Analytical Gas Chromatography*, Marcel Dekker Inc, New York (1998)65.

2. R. P. W. Scott, *Techniques and Practice of Chromatography,* Marcel Dekker

Inc, New York (1995)27.

3. Chromatographic Analysis of Pharmaceuticals (Ed. J. A. Adamovics), Marcel Dekker Inc, New York (1997).
4. Sample Pretreatment and Separation (Ed. N. B. Chapman) John Wiley and Sons, Chichester-New York (1991)

Chapter 3

Sample Preparation

Introduction to Sample Preparation

Specific experimental details of the various methods of sample preparation will be given when particular applications are described, but it must be emphasized that sample processing may demand as much skill and experience as the chromatographic analysis itself. Sample processing can range widely in complexity from simple dissolution in an appropriate solvent to a complicated extraction procedure, followed by derivatization and further extraction. In addition, most samples will require the addition, or preparation, of an internal or external standard. Likewise all these operations must be accomplished quantitatively, with no material loss, so the integrity of the final analysis is not prejudiced. Many of the techniques that are used are borrowed from micro-chemistry and, in particular, from micro-analysis, and some of the more useful procedures employed in sample preparation will be described here.

Extraction Techniques

The determination of trace materials in bulk solids and liquids, *e.g.* pesticides in soil and drug metabolites in blood serum requires the use of appropriate extraction and concentration procedures. These techniques are many and varied and often provoked by the specific nature of the particular sample. Broadly there are three basic extraction methods, solvent extraction, solid phase extraction and supercritical liquid extraction. However, the effective use of these techniques often entails the preliminary use of precipitation procedures, crude separation processes together with subsequent concentration methods. Typical pre-extraction procedures might include centrifugation, precipitation and dialysis and are usually provoked by the presence of bulk materials in the sample that interfere with the separation and measurement of the substances of interest.

Pre-Extraction Practices

There is a large number of pre-extraction techniques that have been reported in the literature, many of which are exclusively specific to particular samples and it would not be appropriate to describe them all. Consequently, only the more common methods and procedures will be detailed here.

Ultra Centrifugation

Ultra centrifugation is a technique for concentrating large, high molecular weight materials (*e.g.* proteins) by centrifugal forces developed at extremely high rotational speeds. In most cases proteins are the *interfering materials* in the sample and it is the supernatant liquid that contains the materials of interest for chromatographic analysis. As a secondary use in this context, the ultracentrifuge can also be employed to determine the molecular weight of the proteins from their sedimentation rate under centrifugal forces. The sedimentation rate (S), or $\left(\frac{dr}{dt}\right)$ ((r) is the radius and (t) the time) is given by,

$$S = \frac{dr}{dt} = \frac{M(1 - \nu\rho)}{N_A f}$$

where (N_A) is Avogadro's Number,
(ρ) is the solution density,
(ν) is the reciprocal of the protein density,
(f) is the frictional coefficient of the protein molecule (which will depend on molecular geometry)

Ultra centrifugation is usually only employed as a last resort, and when other techniques have proved ineffective. After centrifugation it can be assumed that the concentration of the lower molecular weight solutes of interest are evenly distributed throughout the now, heterogeneous, sample and thus the supernatant liquid can be assumed to provide a representative sample of the materials to be assayed. A sample of the supernatant liquid is then taken for subsequent extraction.

Precipitation

Precipitation is another technique commonly used to remove proteins from biological samples (*e.g.* blood serum, urine samples) but may be used in any cases where the interfering materials can be selectively precipitated by appropriate regents. Care must be taken, however, to ensure that the materials of interest are not co-precipitated, adsorbed or occluded by the precipitate and thus make the analysis unreliable. Proteins can be particularly damaging to LC columns as they may be precipitated by the mobile phase and block the interstitial voids in the column which can increase the flow impedance of the column to the point where the flow is actually arrested. Precipitated protein can also change both the retentive capacity and the selectivity of the column. The best method for protein removal is to reduce the pH by adding trifluoroacetic acid or perchloric acid, which causes denaturation and almost instant protein precipitation. The protein is then separated by filtration, or better, by centrifugation and the upper layer used for subsequent extraction or analysis. Proteins can also be precipitated from aqueous media with polar solvents such as methanol or acetonitrile. The efficiency of such solvents for this purpose varies considerably with the sample/solvent ratio and, because of this, is best avoided if acid precipitation is possible. When removing bulk materials from samples by precipitation, precautions must be taken to avoid the possibility of co-precipitation and/or adsorption removing some of the solutes of interest. Such possibilities can always be examined by spiking sample with the materials of interest and by determining the amount recovered. It is important to appreciate that 100% recovery is not essential as long as the actual recovery is constant and known. However, for certain extraction procedures, for example trace quantities of heavy metals from aqueous media, co-precipitation can be employed usefully as an isolation technique. The heavy metals are then removed in a concentrated form from the precipitate.

Dialysis

Dialysis is one of the extraction techniques that depends on the selectivity of semi-permeable membranes. The membranes are fabricated to have specific pore sizes. During dialysis the material is transported across the membrane by means of a concentration gradient. The system can either concentrate a sample or selectively extract and dilute it, depending on whether the material of interest is smaller than the pores and thus passes through the membrane or is larger than the pores and is selectively held behind the membrane and concentrated. A simple example would be a very dilute aqueous solution of fairly large molecules on one side of the membrane and a strong sugar solution on the other. Water would pass through the membrane to dilute the sugar solution (because of osmotic pressure) and, as a result, the material would be concentrated. If the solution contained salt and pure water was on the other side of the membrane, then salt would pass though the membrane and the sample would be desalted. Dialysis is often used as a desalting procedure, particularly after protein precipitation with salt, to lower the ionic strength of the sample ready for a ion exchange separation. Membranes having a wide range of pores are readily available. Dialysis can also be carried out in the continuous mode by passing the sample though fibers having semi-permeable walls and situated in a countercurrent extraction fluid.

Ultra-Filtration

Ultra-filtration is also a preliminary separation step that utilizes a semi-permeable membrane to, in effect, filter the sample. However, ultra-filtration differs from dialysis in that the material is transferred across the membrane by hydraulic pressure, as opposed to a concentration gradient. Ultra-filtration usually results in sample concentration and can also be carried out on a continuous basis. Due to the relatively high pressures involved, and the relatively fragile nature of the membranes, they are usually required to be supported on an appropriate web structure to provide adequate mechanical strength. Both dialysis and ultra-filtration are relatively gentle processes and are useful for concentrating thermally labile materials that can not be heated. Of the two methods dialysis is more easily miniaturized and easier to automate and is probably in more common use than ultra-filtration.

Separation by Exclusion

Exclusion chromatography is also a simple and effective means of removing large molecular weight materials from a complex sample matrix. Employing a stationary phase of appropriate pores size, those substances, larger than the pores, will be eluted at the interstitial volume of the column and those very small molecules that can penetrate all the pores will be eluted at the dead volume of the column. Thus proteins will be eluted first, at the solvent front, and the smaller molecules eluted at a retention volume equivalent to the interstitial volume plus the pore volume. The procedure can only work satisfactorily if the solvent is chosen so that the only mechanism of retention is exclusion, and no interaction between the solutes and the stationary phase takes place. The method of sample clean-up is frequently used in the analysis of materials of biological origin which often contain a wide variety of biopolymers. The selective removal of large molecules by exclusion techniques using either silica based stationary phases or macro-porous polymers is also a very gentle process that can be easily automated.

Solvent Extraction

Solvent extraction can be used for both solids and liquids but has been largely superseded for liquids by solid state extraction. Continuous solvent extraction usually involves elevated temperatures which can cause thermally labile substances to decompose or undergo molecular rearrangement. Nevertheless, the techniques are still used for the removal of relatively insoluble components. It is clear that for the extraction of liquids, the solvent must be immiscible with the sample and consequently there must be two types of continuous extractor. One type would be appropriate for a solvent that was denser than the sample liquid and one would be suitable for a solvent that was lighter than the sample liquid. These devices have been miniaturized and an example of the two types of continuous extractors are shown in figures 1 and 2.

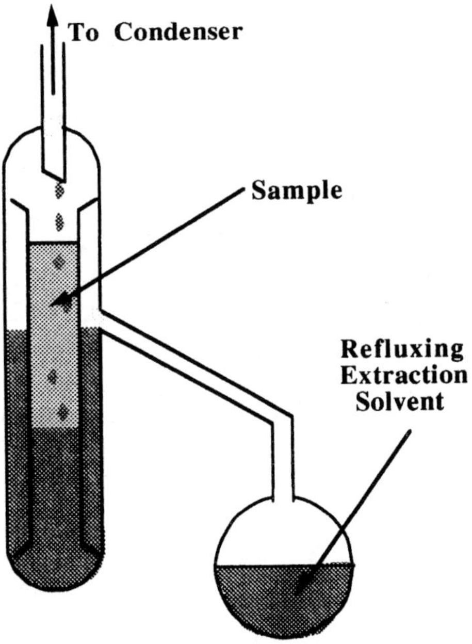

Figure 1 Continuous Liquid Extractor Where the Liquid Sample Is Less Dense than the Extraction Solvent

In figure 1, the sample liquid is less dense than the extraction solvent and the refluxed liquid falls through the sample liquid, extracting the materials of interest, while falling to the solvent pool at the bottom of the extraction tube. The extract then passes up the outside of the tube and siphons through the side tube back to the boiling flask. The sample solvent is rarely completely insoluble in the extraction solvent and so some of the sample solvent will continually accumulate in the flask along with the material of interest. It follows that the extraction time should be limited to that which allows the material to be extracted, but is accompanied by the minimum amount of sample liquid.

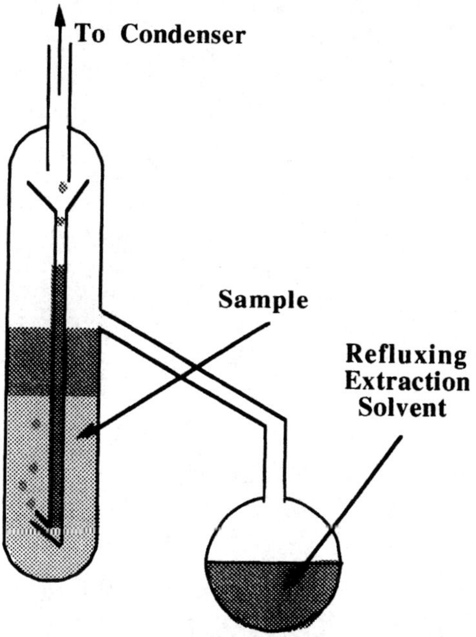

Figure 2 Continuous Liquid Extractor Where the Liquid Sample Is Denser than the Extraction Solvent

In figure 2, the converse situation is depicted where the sample liquid is more dense than the sample solvent, which is more often the case. In this system, the sample liquid resides in the bottom of the extraction tube and it is arranged that a head of the lighter liquid is built up to force a flow through the sample liquid which then passes out of the side arm and back to the extract flask. The same problem arises from the inevitable slight solubility of the sample liquid in the extraction solvent and so the extraction process must be carefully limited to ensure that all the material of interest is extracted, but the amount of sample liquid in the extract is minimized.

Solvent Extraction by Segmented Flow

Continuous extraction can also be carried out using segmented flow. The sample is mixed with, or dissolved in, a carrier solvent which is then mixed with an immiscible extraction solvent. Hence a heterogeneous mixture is produced and, as a consequence, segmented flow is formed. The segmented flow is then passed though an extraction coil, where, due to the 'bolus' flow that occurs within each segment of solvent, the material is extracted very efficiently. (Bolus flow results from the liquid at the walls of the tube being held stationary while the center of the liquid segment moves forward. As a result a circular flow is formed within the segment from its center to the outside.) A diagram of a segmented flow system is shown in figure 3.

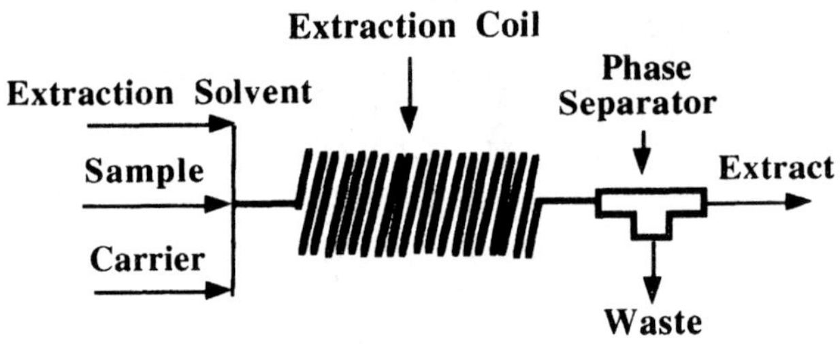

Figure 3 Extraction by Segmented Flow

This method of extraction, although effective and novel, is difficult to control in practice, and its use is limited. As the extration is not sequential, it will only function well if the distribution coefficients of the solutes of interest are strongly in favor of the extracting solvent. For example, if the distribution coefficient of a solute between the carrier and the extractant is 50, the maximum amount of solute that could be recovered would be about 95% v/v.

Discrete solvent extraction, beitmay solid or liquid, using a separation funnel can often be very effective depending on the distribution coefficient of the material between the sample matrix and the extracting solvent. Even if the distribution coefficient is only unity, then four extractions using the same volume of solvent as that of the sample will remove over 93% of the material. If the distribution coefficient is 2, then four extractions using the same volume of solvent as that of the sample will remove over 99% of the material. Conversely, if the distribution coefficient is less than unity, then discrete solvent extraction can be very inefficient and time consuming and, in addition, a very large bulk of solvent will need to be concentrated after the extraction is complete. Nevertheless, providing the distribution coefficient is greater than unity and the sample volume one milliliter or less, then the ultimate volume of solution that will need concentrating will probably be less than 5 ml.

Solid Phase Extraction

Extraction of liquids by solvents, although possible, is a somewhat clumsy procedure and solid phase extraction is now the chosen alternative. In fact, one of its most common uses is in environmental chemistry and pollution studies for concentrating trace materials dissolved in natural waters to assess pesticide contamination. However, it also has significant relevance to the extraction of drugs and drug metabolites from biological fluids such as blood serum and urine. Solid phase extraction is also used in forensic chemistry and by biochemists for studying the metabolic pathways of drugs and other physiologically active materials. The procedure is relatively simple and is carried out in a short tube packed with an appropriate adsorbent such as silica, silica bonded with a reversed phase, or macro-porous polymer beads, etc. The choice of the adsorbent will be determined by the interactive nature of the liquid matrix, relative to the interactive character of the materials of interest. For example, if the solutes are dispersive in character and the matrix is water (*i.e.*, very polar), then a dispersive adsorbent would be appropriate (*e.g.*, a reversed phase). Conversely, if the solutes are polar and contained in a dispersive matrix such as gasoline, then a polar adsorbent such as silica would be more suitable. The solutes of interest are extracted as a result of competitive molecular interactions between the forces that hold the solutes in the matrix and the stronger forces that hold solute on the surface of the adsorbent. After extraction, the solutes of interest can be displaced from the adsorbent by selecting a solvent that interacts more strongly with the adsorbent than the solutes themselves. For example if the solutes of interest are extracted from water onto a reversed phase, then they would be displaced by a more dispersive solvent such as n-hexane, methylene dichloride or THF. If, however, the materials of interest were selectively adsorbed on silica from a dispersive liquid such as petroleum, then they would be displaced by more polar solvents such as ethanol, methanol or even water. The diagram of a simple solid phase extraction tube is shown in figure 4.

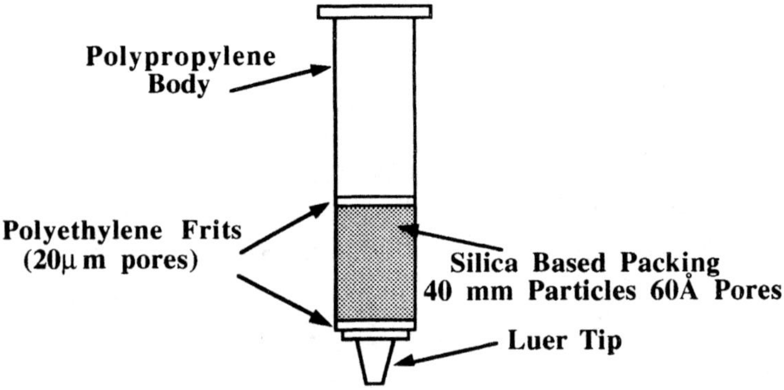

Courtesy of Supelco Inc.

Figure 4 A Solid Phase Extraction Tube

Solid phase extraction tubes are usually made of an inert plastic such as polypropylene and commonly have 1, 2, or 5 ml capacities. The tube is about one-fifth filled with adsorbent which is contained in the tube by a plastic frit situated at either end. The upper, empty part of the tube acts as a funnel or container for the liquid to be extracted, which is allowed to percolate through the adsorbent bed and, after extraction, will contain the washing and displacement solvents. In some cases the lower end of the tube is connected to a vacuum or the top to a gas supply to increase the flow through the adsorbent bed. The extract contained in the displacing solvent can be diluted to a known volume (250 μl, 500 μl, or 1 ml) and an aliquot used for analysis. Alternatively, the extract can be concentrated by evaporation and the total concentrate employed for analysis to increase the sample mass.

The optimum size of the tube depends on a number of factors, such as the sample volume, the level of the materials to be extracted and the nature of the sample matrix. In general, if the sample is 1 ml or less, then a 1 ml tube would be appropriate and a 3 ml or 5 ml tube, if the sample volume is between 2 and 250 ml. If the level of the solutes to be extracted is relatively high, then a larger tube may be necessary to provide sufficient adsorbent capacity for the extract. The packing should be conditioned by an appropriate solvent, or solvents, before use and the solvent will depend on the tube packing and the nature of the solutes. Common conditioning solvents are methanol, methyltributylether (MTBE), methylethylketone (MET), deionized water and potassium chloride solutions. It is important to prevent the packing drying during extraction, washing, or when desorbing the extract. The sample is either forced through the bed under pressure or sucked through the bed with a vacuum. The adsorbent activation can deteriorate if too large a sample is used and for a 3 ml tube the sample volume should be kept below 250 ml. The sample should not be passed through the bed too rapidly and a maximum flow rate of 5 ml/min. is recommended.

After extraction, the adsorbent bed should be washed with an appropriate solvent. For example, if the sample matrix was water, then the bed might be washed with deionized water before displacing the extracts with a solvent. One tube volume of washing solvent is all that is usually necessary. The extracted material is then displaced with another solvent, the nature of which will depend on the type of solutes extracted and the packing that is used. Suitable displacing solvents are often recommended by the manufacturers of the tubes. A volume of between 200 µl and 2 ml is usually sufficient to elute all the adsorbed solutes.

However, trace analysis has a number of problems unique to the isolation and measurement of very small quantities of material. One problem that is frequently encountered is the retention of certain solutes on the surfaces of the material of which parts of the solid phase extraction apparatus are constructed. This adsorption effect can be further complicated by the fact that some substances of biological importance are liable to decompose, denature or thermally rearrange on such surfaces. To avoid chemical changes in labile materials, a totally inert extraction apparatus can be constructed from Teflon. A diagram of such an apparatus, produced by Alltech, is shown in figure 5, which even includes a Teflon hypodermic needle.

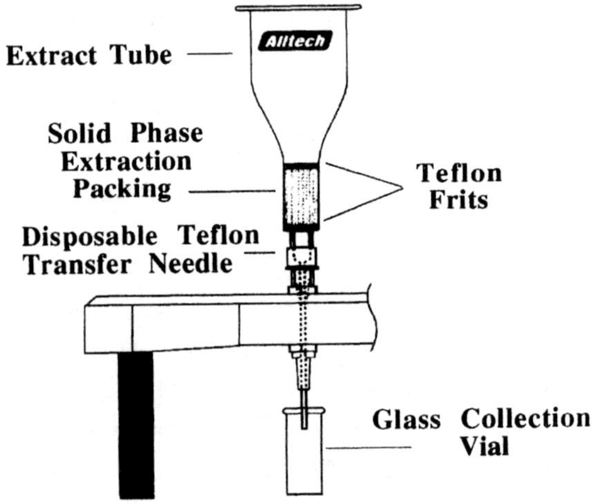

Figure 5 An All–Teflon Solid Phase Extraction Apparatus

This type of extraction system is used extensively in the biotechnology and essential oil industries where many compounds occur that undergo molecular rearrangement and chemical change when in contact with active surfaces (*e.g.*, certain terpenes and other similar cyclic compounds). An example of the use of solid phase extraction to determine trace amounts (5 ppb) of some chlorinated pesticides in drinking water is shown in figure 6. The solid phase extraction tube was the Novo-Clean C18, 47 mm tube with a membrane manifold supplied by Alltech Inc. The materials were removed from the water sample by the strong dispersive forces between the solutes and the C18 reversed phase in the manner already discussed. The extraction tube was conditioned before use with 10 ml of methanol, 10 ml of MTBE, 15 ml of methanol and finally 125 ml of deionized water. The water sample was then pumped through the extraction tube at a rate of 100 ml/min. The solutes removed were displaced from the extraction tube with 10 ml of methanol followed by 10 ml of MTBE and dried over anhydrous sodium sulfate. It is seen that all the chlorinated pesticides were extracted and concentrations down to 1 ppb could be easily identified.

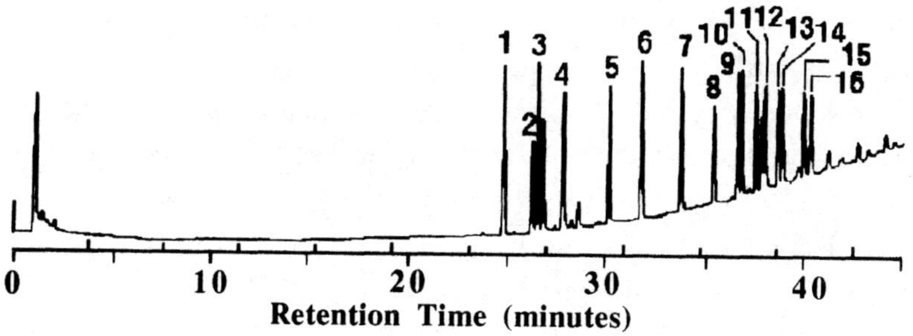

1. Lindane–α
2. Lindane–β
3. Lindane–γ
4. Lindane-Δ
5. Heptachlor
6. Aldrin
7. Heptachlo Epoxide
8. Dieldrin
9. *p,p'*-2-chlorophenyl,2-*p*-chlorphenyl 1,1 dichloroethylene
10. Endrin
11. *p,p'*-2,2-*bis p*-chloropheny1 chlor-ethylene
12. Endrin aldehyde
13. Ensodulfan Sulfate
14 *p,p'*-1'1'1-trichlor, 2,2-*bis p*-chloro-phenyl ethane
15. Endosulfan II

Courtesy of Alltech Inc.

Figure 6 Separation of Some Chlorinated Pesticides Removed from Drinking Water by Solid State Extraction

Instruments for automatic, multiple extractions are also readily available. The solid phase extraction tubes are mounted in racks of 12, 24 or even 96 and vacuum or pressure manifolds are also obtainable to facilitate the extraction process. The conditioning, extraction and regeneration can be automated and the racks used in conjunction with an automatic sampler to process batches of assays at one time.

The Solvent Extraction of Solids

In general, hard or brittle solid samples must be well ground and preferably reduced to a mean particle size of less than 20 μm using the simple pestle and mortar or the so-called 'diamond' mill. This will usually provide sufficient surface area for efficient extraction. If, however, the material of interest is present in very small quantities, and the matrix is hard and not porous, then the particle size may need to be further reduced. If the particle size is not small enough, then a significant amount of the material may be trapped in the particle and out of contact with the extracting solvent. Softer materials can be ground in a laboratory ball mill and very soft materials of biological origin, (*e.g.*, animal and vegetable tissue, processed foods, etc.) can be minced, crushed or shredded in a suitable blender. Substances such as plastics and rubber can be ground after freezing in liquid nitrogen or sometimes ground with solid carbon dioxide.

As already discussed, if the material of interest is preferentially distributed in the extracting solvent, the sample can be shaken or stirred with sequential amounts of solvent. Extraction can be aided by direct heating or by means of a microwave generator. In addition, solution of the materials of interest can often be accelerated by sonication with a sonic probe or in a supersonic bath. Sonication will also break open cells by disrupting their cell walls, and release the contents for extraction.

The continuous solvent extraction of solids is frequently carried out using the traditional Soxlet extractor and filter thimble. However, the established technique has been improved by immersing the thimble in the boiling solvent resulting in a much faster extraction and a smaller consumption of solvent. Traditional continuous Soxlet extraction is frequently used for very insoluble materials where continual exposure to pure solvent is necessary to isolate all the material. If micro apparatus is employed, the volume of initial solvent is set (that contained by the reservoir flask) and so solvent consumption is minimal and constant and subsequent extract concentration restricted to a practical level. Another interesting modified form of solvent extraction is fluid extraction under pressure.

Fluid Extraction under Pressure

Fluid extraction under pressure is similar in nature to super-critical fluid extraction except normal solvents are employed and the extraction rate is increased by operating at elevated temperatures and pressures. A diagram of the apparatus used for high pressure fluid extraction is shown in figure 7.

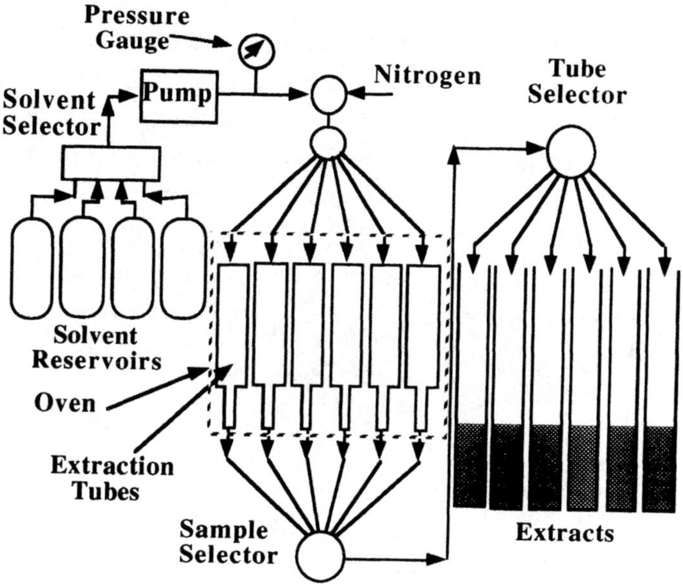

Figure 7 High Pressure Fluid Extraction Apparatus

The extraction is arranged to take place in a heated extraction tube and the solvents, selected from a group of solvent reservoirs, is pumped into the extraction tube under pressure. When the extraction is complete, the extract is drained off into sample tubes and concentrated by evaporation. The same type of solvents are used as in Soxlet extractions and about 10 to 30 g of material is extracted with about 20 to 40 ml of solvent. Extraction temperatures range from 100°C to 180°C at pressures that span from 1500 psi to 2000 psi. This method has been extensively used to extract herbicides and insecticides such as organo-phosphorus and organo-chlorine compounds from soil samples. The method is most effective when the solid samples have been carefully dried before extraction. It is seen that the instrumentation used for high pressure fluid extraction is very similar to that used for super-critical fluid extraction.

Super-Critical Fluid Extraction

One of the more recent extraction methods to have been developed is super-critical fluid extraction which employs a gas above its critical temperature to extract the material. This technique was developed with a view to increasing the extraction efficiency, to eliminate the use of solvents and to simplify subsequent concentration procedures. Unfortunately, the technique has not been shown to be as effective as was hoped. A diagram of the apparatus used for super-critical fluid extraction is shown in figure 8.

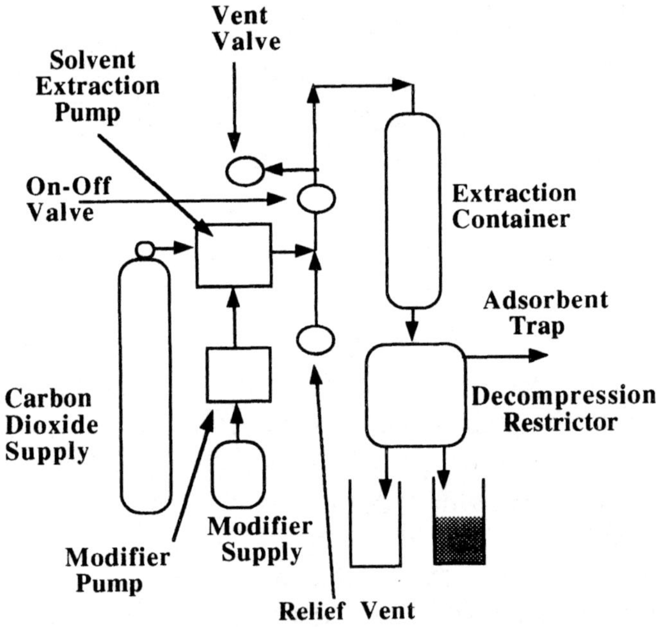

Figure 8 Super-Critical Fluid Extraction Apparatus

Virtually all modern super-critical fluid extraction instruments use carbon dioxide as the primary super-critical fluid, largely due to the fact that its critical temperature and pressure are, conveniently,

31.4°C and about 1000 psi respectively and the material is readily available. Above the critical temperature carbon dioxide behaves as a gas at all pressures and thus transfer rates are fast and extraction proceeds very rapidly. At low pressures liquid carbon dioxide interacts dispersively with a solute much like *n*–hexane, but at high pressures the dispersive interactions increase in strength and are more akin to those of methylene dichloride or chloroform. Unfortunately, carbon dioxide has little or no polar character and thus to dissolve polar materials from the solid matrix of a sample, a solvent modifier must be used. The need for solvent modifiers is a distinct disadvantage as it not only involves the use of solvents and their disposal, it also introduces another concentration and recovery step in the extraction process. There are a number of common solvents that can be used as solvent modifiers, among those are *n*-hexane, benzene, toluene, methylene dichloride, nitromethane, ethyl acetate, tetrahydrofuran, acetonitrile, propanol, ethanol and methanol. The solvents are listed from strongly dispersive to strongly polar. The liquid carbon dioxide is mixed with the modifying solvent in the appropriate proportions (the solvent feed is controlled by a pump) and passed to the extraction vessel. After extraction, the solution passes through a restrictor, where the pressure is reduced and the carbon dioxide allowed to bleed away. The residual solvent containing the extract is then allowed to drain into tubes and is subsequently concentrated. Although originally hailed as the ideal extraction technique, like super-critical fluid chromatography, it has not proved to be as useful as first thought. Besides needing modifying solvents to extract polar or semie-polar solutes, the extraction procedure is complicated to develop, the apparatus is expensive (almost as costly as the chromatograph) and the work-up procedure is no simpler than by normal solvent extraction. There are certain types of samples for which it works well, but unless prescribed for particular types of analysis, the technique should only be used as a last resort.

Derivatization

Derivatization procedures are employed in both GC and LC and to some extent in TLC, but usually for quite different reasons. In GC, derivatization is generally carried out to render the solutes volatile so that they can be separated by development in a gaseous mobile phase. However, in LC, derivatization is largely carried out to add UV chromaphores, fluorophores, etc., to solute molecules to aid in detection. Before LC had been developed to its present highly efficient level, sample derivatization was often carried out merely to make gas chromatography amenable to the analysis of involatile and strongly polar materials. Before 1965, GC was the more accurate and reliable technique, so GC was to be preferred, even though LC might be the more appropriate technique. However, as LC became as accurate and dependable as GC, the reverse occurred. Volatile substances that were quite suitable for separation by GC were derivatized to make them detectable by LC because LC was the simpler and more popular technique. It is almost unbelievable, but the trend now appears to be changing once again. LC requires the use of large amounts of solvent, the disposal of which is often difficult and expensive. In contrast GC has no waste disposal problems and a clear economic advantage. Thus, derivatization is again being used to make solutes volatile and amenable to analysis by GC.

To render highly polar materials sufficiently volatile, to allow their elution at high temperatures without thermal decomposition or molecular change, requires derivatization. Examples of highly polar, involatile materials that need to be derivatized prior to separation are organic acids, amides, polyhydroxy compounds and amino acids, etc. Such materials are either esterified, silanated or acetylated, but there are a number of different methods used for synthesizing each of the three classes of derivative.

Esterification

A simple method for the esterification of acids is to treat them with a suitable alcohol in the presence of an inorganic acid to catalyze the reaction. Hydrochloric acid was originally the favored catalyst because it was sufficiently strong and could be removed relatively easily. Sulfuric acid is less favored as it is difficult to remove and can cause charring. Other esterification catalysts that have been successfully employed are trifluoroacetic acid, dichloroacetic acid, benzene sulphonic acid, *p*-toluene sulphonic acids and sulphuryl and thionyl chlorides. A volatile acid such as hydrochloric acid or thionyl chloride is recommended. However, the derivative must not be too volatile, or it will be lost while removing the excess alcohol and the hydrochloric acid and, in the case of thionyl chloride, the removal of the catalyst itself. A fairly standard method for esterification would be as follows. One or two milligrams of the sample contained in a small vial are heated with 125 µl of either methanol or ethanol that contains 3M hydrochloric acid at 65°C for about 35 minutes. The alcohol is removed with a stream of nitrogen leaving the residual ester. It is important to ascertain that the derivatized ester is sufficiently high boiling, or has an adequately low vapor pressure, to be sure that none is lost in the alcohol removal process. Preliminary experiments must be carried out to ensure that the esterification is complete and, if not, the procedure must be modified by extending the esterifying time or increasing the temperature.

Amino acids, which are more difficult to derivatize than most simple acids, can also be esterified in a very similar manner. A few milligrams of the amino acid mixture is mixed with 2 ml of 4M alcoholic methanol and heated at 70°C for 2 hours. The excess methanol is removed by evaporation in a stream of nitrogen. Entrained water can be removed by adding methylene dichloride (*ca* 150 µl) and evaporating the solvent. The derivative is the hydrochloride of the amino acid, and thus the free base must be released without saponifying the ester to facilitate separation on the GC column.

Another useful reagent for forming esters from acids is the Lewis acid boron trifluoride or the analogous reagent boron trichloride. Boron trifluoride is commercially available as a 14% solution in methanol for forming ester derivatives. Boron trifluoride catalyzed reactions carried out on a water bath are very rapid and can be complete in a few minutes. 1 to 15 mg of the acid are placed in a vial fitted with a ground glass stopper and 1 ml of 14% boron trifluoride in methanol is added. The mixture is heated in a water bath for 2 minutes and then cooled. The esters can be extracted into n–heptane, but vigorous shaking is essential to ensure that all the ester is extracted.

Thionyl chloride is another useful catalyst that can be used for the esterification of acids. The procedure is simple but the thionyl chloride reagent should be purified by distilling from linseed oil before use. To 10 to 50 mg of acid, contained in a vial fitted with a ground glass stopper, is added 200 μl of dry methanol and cooled in a solid carbon dioxide-acetone cooling bath. 20 μl of thionyl chloride is then added with shaking and the vial warmed to 40°C and maintained at that temperature for two hours. The solution is then evaporated to dryness in a stream of nitrogen.

The esterification of an alcohol is the complementary form of ester derivatization. The most common reagent used for this purpose is an acid anhydride which, in fact, supports the reaction by removing the water as the esterification proceeds. Although there is some competition between the alcohol and the water for the anhydride, under optimum conditions the anhydride reacts preferentially with the water. 10 to 100 mg of acid contained in a stoppered vial is reacted with 1 molar equivalent of the alcohol in the presence of 1.2 to 1.4 equivalents of trifluoroacetic anhydride. The mixture is warmed and the reaction proceeds rapidly and is complete in about 10 minutes.

One of the more popular reagents for the esterification of acids is diazomethane. *However, it must be emphasized that diazomethane is carcinogenic and can be extremely unstable.* All reactions should be carried out in a fume hood and any stored solutions of diazomethane in diethyl ether should *be restricted to a maximum* of 100 ml and kept in a refrigerator. The materials must *not be overheated* as there is a risk of explosion. Nevertheless, despite the hazards, the reagent is extremely effective and, if its use is restricted to microscale reactions, it is generally safe to use. It is in common use in most analytical laboratories employing gas chromatographic techniques where the usual safety precautions are taken. Diazomethane is employed as a ethereal solution but, in fact, is a yellow gas. Its reaction with an organic acid is as follows:

$$R\text{—}COOH + CH_2N_2 \longrightarrow R\text{—}COO\text{—}CH_3 + N_2$$

The yellow color persists when the reaction is complete and thus the reagent acts as its own indicator. An apparatus developed by Schlenk and Gellerman (1) for esterification with diazomethane is shown in figure 9.

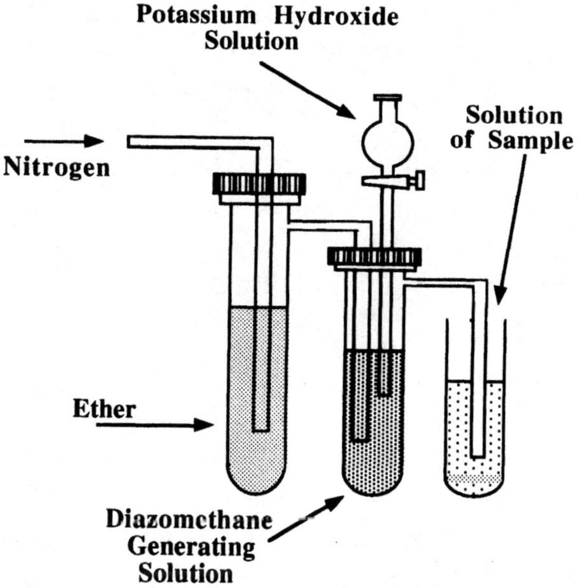

Figure 9 Apparatus for Generating Diazomethane for Esterification

Nitrogen is first bubbled through ether and then through a vessel which contains a solution of *N*-methyl-*N*-nitroso-*p*-toluene sulfonamide which, when treated with potassium hydroxide, added through a dropping funnel, generates the diazomethane. The diazomethane is passed into the solution of the sample being derivatized until a yellow color persists. 0.5 to 30 mg of acid can be used, which is dissolved in 2 ml of a 10% solution of methanol in diethyl ether. The nitrogen flow is about 4 ml/min and subsequent to the reaction being completed, the ether methanol mixture is removed by evaporation, leaving the ester derivative ready for analysis. The derivatization procedure does not work well with phenolic acids as the hydroxyl group is also methylated. Diazoethane can also be used employing a similar technique.

Both alcohols and acids react with appropriate silyl reagents to form trimethylsilyl ethers and trimethylsilyl esters respectively. These derivatives are volatile and allow easy separation of the derivatives. The two reagents, N,N–bis(trimethylsilyl)trifluoroacetamide (BSTFA) and bis(trimethylsilyl)–acetamide (BSA), react rapidly with organic acids and give high yields; the latter reagent is used if an electron capture detector is being employed. A few milligrams of the acid is placed in a vial and about 50 µl of BSA or BSTFA added. Reaction is claimed to be complete on solution of the acid, but some recommend the mixture be heated for 5 to 10 min. at 60°C to ensure that reaction is really complete. The mixture can then be injected directly onto the gas chromatograph.

Tert-butyldimethylsilyl esters (TBDMS) are also used, particularly for GC/MS. TBDMS esters are prepared by dissolving about 5 mg of the acid in 100 µl of dimethylformamide containing 20 µmol of imidazole and 10 µmol of TBDMS. The mixture is heated to 60°C for about 15 minutes, an equal volume of 5% NaCl is added and the esters are extracted with 1 ml of ether.

The halogenated silyl esters are likewise useful for producing silyl esters, for example, the chloromethyldimethyl silyl (CMDMS) and bromo-methyldimethylsill (BMDS) esters. A few milligrams of the acid are dissolved in 600 µl of pyridine (a scavenger for HCl that is released) and 200 µl of di(chloro-methyl)tetra methyldisilazane and 100 µl of chloromethyldimethylsilyl chloride is added. Reaction is allowed to proceed for 30 minutes at room temperature.

Acylation Reactions

Acylation is also widely used to provide volatile derivatives of highly polar and involatile organic materials. The technique, however, also has a number of other advantages. In addition to improving volatility, acylation reduces the polarity of the substance and thus improves the quality of the separation and, in particular, reduces peak tailing. As a result amide esters are usually well separated with symmetrical peaks. By inserting protecting groups into the molecule, acylation also improves the stability of those compounds that are inherently thermally labile. Acylation can also render extremely polar materials such as sugars that are amenable to separation by GC and are a practical alternative to the silyl derivatives. In particular acylation is used to provide derivatives of amines, amides, alcohols, thiols, phenols, enols, and glycols. A typical example of acylation is the reaction of acetyl chloride with an alcohol.

$$R\text{–}CO\text{–}Cl + R'\text{–}O\text{–}H = R\text{–}CO\text{–}O\text{–}R' + HCl$$

A common method of acylation is to heat the acid (*ca* 5 mg) dissolved in 5 ml of chloroform together with 0.5 ml of acetic anhydride and 1 ml of acetic acid for 2 to 16 hours at 50°C. The excess reagents are removed under vacuum and the residue is dissolved in chloroform and injected directly onto the column. Sodium acetate can be used as an alternative to acetic acid. The reaction mixture consists of 0.3 ml of acetic anhydride and 12 mg of sodium acetate. Derivatization is carried out at 100°C for about an hour, excess reagent is removed by evaporation and the residue taken up in a suitable solvent for analysis. Another reagent that might appear useful is acetyl chloride but, in fact, the added problem of the removal of the hydrochloric acid, formed during acylation, has rendered this reagent unpopular. Anhydrides are usually employed for acylation unless special circumstances demand alternative reagents.

An excellent handbook describing a wide range of procedures used to produce derivatives for chromatographic analysis has been compiled by Blau and Halket (2).

In LC, derivatization is almost solely used to improve the detection limits of specific substances, *e.g.*, peptides, proteins, etc. There are various classes of reagents that link specific chemical groups to the solute molecule, *e.g.*, UV chromophores or fluorophores, and thus would be termed UV chromaphore reagents, fluorescence reagents, etc. Each reagent class can be divided into groups similar to the GC reagents based on the chemical nature of the solutes with which they react. For example, there are fluorescent reagents that are used to specifically derivatize different solutes, *e.g.*, acids, alcohols, amino acids, etc.

Two of the most common UV absorbing reagents used to derivatize amino groups are benzoyl chloride and *m*–toluoyl chloride. Both reagents add a benzene ring to the molecule that contributes a strong UV absorbing chromaphore. The *p*–nitrobenzoyl chloride reagent imparts an even stronger UV chromaphore than the simple aromatic ring and is sometimes preferred for this reason. The reaction is fairly rapid and takes place quite satisfactorily at room temperature. Pyridine is usually employed either as, or together with, a solvent to scavenge the hydrochloric acid that is released. Similar reagents are the toluene-sulphonyl chloride and the benzene sulphonyl chloride, which both produce sulfonamides with the amines. As an example, gentamicin, a polyfunctional amino compound, can be analyzed by reverse phase chromatography after labeling with benzene sulphonyl chloride. The derivatization is completed in about 10 minutes at 75°C. 1-Fluoro-2,4-dinitrobenzene is also commonly used for derivatizing amino compounds, for example, the amino glycosides such as neomycin, fortimicin, amikacin, tobramicin, gentamicin and sissomicin.

Carboxylic acids are a large group of naturally occurring compounds such as fatty acids, prostaglandins, bile acids and other organic acids all of which have relatively weak UV and visible absorption. Common reagents used to introduce a UV chromaphore into such compounds are phenacyl bromide and naphthacyl bromide. Benzoyl chloride, *m*-toluoyl chloride and *p*–nitrobenzoyl chloride are also used to derivatize compounds with hydroxy groups and, to introduce a really strong UV chromaphore, 3,5-dinitrobenzoyl chloride can be used. The two most common reagents for derivatizing substances that contain a carbonyl group are 2,4-dinitrophenylhydrazine and *p*–nitrobenzyl-hydoxylamine. For example, these reagents are commonly used to derivatize aldehydes, ketones, ketosteroids and sugars. *p*–Nitrobenzyl-hydoxylamine has been used successfully to insert UV chromaphores into prostaglandins extracted from biological samples.

Fluorescent reagents. One of the most popular fluorescent reagents is 5-dimethyl aminonaphthalene-1-sulphonyl chloride (dansyl chloride, DNS-chloride or DNS-Cl). Dansyl chloride reacts with phenols and primary and secondary amines under slightly basic conditions to form a fluorescent sulfonate ester or sulfonamide. The quantum efficiency of dansyl derivatives is high; whereas the reagent itself does not fluoresce. Unfortunately, the hydrolysis product, dansylic acid, is very fluorescent, causing interference with water-soluble derivatives. The interfering derivatives, however, are often removed by the subsequent chromatographic process. The detection limits of the dansyl derivatives are often in the low nanogram range (*ca* 1×10^{-9} g/ml) and the excitation and emission maxima can vary between 350 and 370 nm for excitation and 490 and 540 nm for emission. This reagent has been used successfully in the analysis of amino acids, alkaloids, barbiturates and pesticides.

4-Chloro-7-nitrobenz-2,1,3-oxadiazole (NBD chloride) reacts with aliphatic primary and secondary amines to form highly fluorescent derivatives. Aromatic amines, phenols and thiols yield weakly or non-fluorescent derivatives; consequently, the reagent is specific for aliphatic amines. The reaction is carried out under basic conditions and the products are extractable from aqueous mixtures by solvents such as benzene or ethyl acetate. The fluorescence can be significantly reduced by the presence of water and so the solution should be dry and the reagent can obviously not be used to form derivatives for reversed phase chromatography. Detection limits are in the fraction of a nanogram range ($2–5 \times 10^{-10}$ g/ml). The advantage of this reagent over dansyl chloride is that both the reagent and its hydrolysis products are *non-fluorescent*. The excitation and emission wavelengths are also higher (480 nm excitation and 530 nm emission). NBD chloride derivatives have been used for the analysis of amino acids, amphetamines, alkaloids and nitrosamines.

Fluorescamine (4-phenylspiro(furan-2-(3H),1'-phthalan)3,3'-dione) is also a commonly used fluorescence reagent. It reacts almost instantly and selectively with primary amines, while the excess of the reagent is hydrolyzed to a non-fluorescent product. The reagent itself is non-fluorescent. The reaction is carried out in aqueous acetone at a pH of about 8–9 and the derivatives can be chromatographed directly. The excitation and emission wavelengths are 390 nm and 475 nm respectively. Two disadvantages of the reagent are its cost and the fact the products are less stable, cannot be stored and should be injected onto the column immediately after formation. Fluorescamine has been employed in the analysis of polyamines, catecholamines and amino acids.

A less costly alternative to fluorescamine is *o*-phthaldehyde (OPT), the derivatives of which are more stable and consequently can be stored overnight if necessary. It is used in a similar manner to fluorescamine, the detection limits being about 0.1 ng (*ca.* 4×10^{-10} g/ml). OPT has been used in the analysis of dopamine, catecholamines and histamines. Other fluorescence reagents that are sometimes used include 4-bromoethyl-7-methoxycoumarin, diphenylindene, sulphonyl chloride, dansyl-hydrazine and a number of fluorescent isocyanates. For further information on GC derivatizing reagents the reader is strongly recommended to refer to the *Handbook of Derivatives for Chromatography* edited by Blau and Halket [2].

Post-Column Derivatization

Post-column derivatization does not merely require the selection of the most appropriate reagent to react with the solute to render it detectable, but also involves the modification of the chromatographic system to allow the reaction to take place prior to entering the detector. This necessitates the insertion of a post-column reactor between the column and the detector. Such a reactor can easily interfere with the resolution obtained from the column and consequently the reactor system must be designed with some care to minimize extra-column dispersion. The post-column reactor is required to fulfill the following functions:

1/ Provide a source of reagent and a means of mixing it efficiently with the column eluent.

2/ Ensure the reaction is complete before the derivatized product enters the detector.

3/ Minimize the dispersion that takes place in the reactor so that the integrity of the separation achieved by the column is maintained.

A diagram of a post-column reactor system is shown in figure 10.

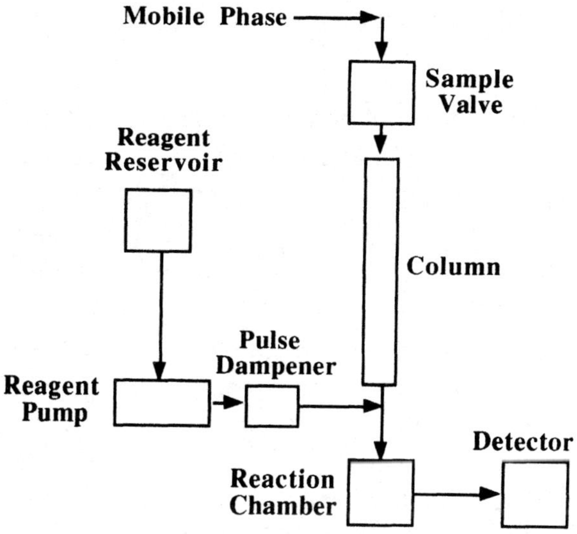

Figure 10 Chromatograph with Post-Column Reactor

The post-column reactor consists essentially of four parts, a reagent reservoir, a pump, a mixing "T" and a reaction chamber. The pump should provide a pulse-free flow of reagent, as there is little resistance downstream to the detector and most detectors are flow sensitive to a greater or lesser extent. Preferably, either a syringe pump should be employed or, if a reciprocating pump is used, it should be fitted with an efficient pulse dampener. The former is preferable, but syringe pumps can be clumsy to refill.

The design of the mixing T can be critical and different kinds of mixing systems have been studied by Scholten *et al.* [4]. The mixing T must ensure that the column eluent is intimately mixed so that there are no localized volumes of mobile phase that do not contain sufficient reagent to allow complete derivatization. Furthermore, the mixing T must not cause any peak dispersion and impair the separation achieved by the column. The reaction chamber provides a hold-up volume to give time for the slower derivatizing reactions to proceed to completion. This volume must also be carefully designed to reduce any peak dispersion to a minimum. Low dispersion tubing such as serpentine tubing can be used [5], or if a relatively large reaction time is necessary, then a short length of column packed with glass beads might be suitable.

The reagents used in post-column derivatization will be specific for the solute of interest and will often be the same as or similar to those used in pre-column derivatization. Reagents that produce derivatives that absorb in the visible light range are popular, as relatively simple detectors can be employed. In general, derivatization, particularly post-column derivatization, is only used as a last resort to provide sensitivity or selectivity, which is difficult or impossible to obtain by any other means. Post-column reactors complicate an already complex instrument, render it more difficult to operate and make it more expensive. In addition, however well the post-column reactor is designed, it will inevitably impair, to some extent, the separation achieved by the column.

In quantitative chromatographic analysis the efficiency of the derivatization and its reproducibility are extremely important. The derivatization procedure must be carefully developed so that the reaction can be carried out in a highly reproducible manner. The reaction need not proceed to completion, but must consistently react with the same proportion of sample. In addition, it may be necessary to derivatize the standard and thus this reaction must also be highly reproducible. This aspect of derivatization will be discussed again when the preparation of standards (internal and external) is considered.

Concentration Techniques

Extraction, derivatization and many other sample preparation procedures often conclude with the sample being significantly diluted in a solvent and consequently the solution will probably need to be concentrated. However, the concentration procedure can be fraught with likely sources of error, particularly if the sample is small and has even a slight, but significant vapor pressure at the evaporation temperature.

Consider a substance having a molecular weight of 200 and a vapor pressure was 760 mm of mercury at 0°C (*i.e.*, it boils at 0°C); then the mass of material in 1 cu cm of vapor would be $\frac{200}{22400} = 8.93\,\text{mg}$ (the molecular weight of a substance occupying 22,4 liters as a vapor at STP).

Now consider the material was very involatile and had a vapor pressure of only 0.001 mm of mercury at 0°C and was heated to 50°C during a solvent recovery procedure; then, applying the simple gas laws, the mass of material per cu cm of gas would be

$$\frac{200 \times 0.001 \times 323.4}{22400 \times 760 \times 273.4} = 1.39 \times 10^{-8}\,\text{g or } 13.9\text{ ng.}$$

This means that if a pollution analysis was carried out, and 1 μg of the material was extracted into a solvent which was finally evaporated in a stream of nitrogen at 50°C in a drying tube, after the solute was dry, the total material would be lost as vapor when only 72 ml of nitrogen was passed through the tube. In practice, however, this loss would not just start when the solvent was removed. Losses would increase progressively as the solution became more concentrated and as the partial pressure of the solute vapor in the gas and solvent vapor increased, reaching a maximum rate of loss when the material was dry. Bearing in mind that 1 micron of mercury pressure (0.001 mm) is a very low vapor pressure, it is clear that many common substances have vapor pressures that exceed this value. Consequently, any material that is to be satisfactorily concentrated from a solvent solution without loss must have a very low vapor pressure indeed at the concentrating temperature. Most ionic materials, strongly polar substances and biopolymers will meet this requirement but the lower weight compounds and semie-polar and dispersive (hydrophobic) materials, which may well include many pesticides, will not. Thus, concentration by evaporation must be carried out with extreme caution and blank runs carried out, using exactly the same volume of scavenging gas, and at the same temperature, to ensure that no loss occurs. Assuming that concentration by evaporation is possible without loss, then there are a number of devices available that will concentrate solutions by evaporation. However, many of them, the rotary evaporators, turbo evaporators, etc., are largely used for bulk evaporation in preparative work, and are not generally useful for analytical or micro-analytical procedures. The Kurderna-Danish concentrators, however (in the micro form), are often used for evaporation, when relatively large solvent volumes have been found necessary for efficient extraction. However, the most common form of evaporative concentrator used in chromatographic analysis is shown in figure 11. It consists of a simple tube contained in a heated block with provision for passing a stream of filtered nitrogen or argon over the surface. The block can be thermostatted at any desired temperature and can have multiple apertures for drying tubes and a nitrogen manifold to evaporate 25 or 96 samples at a time.

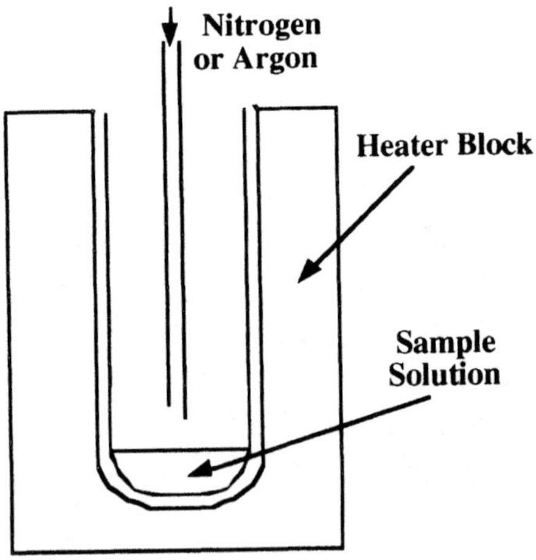

Figure 11 A Micro-Evaporator

One of the more gentle and safe methods of solvent removal, particularly for aqueous solutions, is lyophilization. The solution is frozen and the solid diluent removed under vacuum. There will be a finite vapor pressure over the solid (as already discussed) and providing this is large enough, all the solid diluent can be removed as vapor under vacuum. Due to the relatively low vapor pressure of a solid, the process often takes a number of hours to complete. As this technique is commonly used in biotechnology equipment, it is readily available fitted with micro-containers that can be used for sample preparation in chromatographic analysis.

Another practical sample-concentration technique that is suitable for labile materials is exclusion chromatography. If the solute molecules are large, and the solution is passed through a dry exclusion column, then the solvent molecules will preferentially enter the pores, and the large solute molecules will be carried along with the solvent front. This procedure is, in effect, a drying system, but can be carried out at room temperature without incurring sample loss or decomposition. It can easily concentrate the solution by an order of magnitude or more. Extraction techniques that do not produce a diluted sample are to be preferred and it is this aspect of super-critical fluid extraction that makes it an attractive alternative for those samples with which it can be used effectively.

Synopsis

Sample preparation can be as critical as the actual chromatographic analysis itself and can embrace a number of other analytical techniques, many of which have been borrowed from micro-chemistry and micro-analysis. One of the more important procedures is that of liquid and solid extraction. Before extraction, however, liquid samples may need ultra-centrifugation or precipitation to remove large molecular weight interfering materials such as proteins or other biopolymers. Proteins are usually removed by denaturing with acids, or, by dialysis or ultra-filtration, though an appropriate membrane. As opposed to ultra-filtration, dialysis is the simpler procedure and more easily automated. Exclusion chromatography can also be used to remove large molecular weight materials by passing the sample though a column packed with particles of the appropriate porosity.

The large molecular weight materials are eluted first at the sample front. Solvent extraction can be used for both solids and liquids, but for liquids, the technique has been largely replaced by solid state extraction. Continuous solvent extraction of solids and liquids are usually carried out using commercially available devices based on the Soxlet extractor. However, apparatus employing segmented flow extraction followed by a phase separating device is also available. Solid phase extraction, employing selective adsorbents, is the most effective way of efficiently extracting trace materials from liquids. The adsorbent can be silica gel, a bonded silica or macro-porous polymer beads. The adsorbent is usually activated by pre-washing the packing and the sample is then passed through the bed. After extraction the bed is washed and the adsorbed solute displaced with an appropriate solvent. Tubes packed with a wide range of adsorbents and of different sizes are commercially available, some designed specifically for particular types of analysis. Solid state extraction equipment is also fabricated from inert materials such as Teflon to eliminate solutes present at the trace level being adsorbed on the surfaces of the apparatus. The solvent extraction of solid samples requires preliminary grinding to an appropriate particle size in a diamond mill or ball mill, depending on the hardness of the sample. Soft solids such as biological tissue can be crushed, shredded or blended and plastic materials can be ground after freezing in liquid nitrogen or with solid carbon dioxide. Extraction can be aided and accelerated by, direct heating or microwave heating and by sonication. Probably, Soxlet extraction is still the most common method for solid extraction but it is claimed that the extraction rate is improved by immersing the thimble in the boiling solvent. The apparatus has been miniaturized and thus solvent volume can be kept small. Solvent extraction can also be carried out at elevated temperatures and pressures to increase the efficiency of extraction and the extraction rate. This technique is more effective if the sample is well dried before extraction. Supercritical fluid extraction, which employs apparatus similar to that used for high pressure extraction, uses carbon dioxide above its critical temperature for extraction. It is claimed transfer rates are improved and subsequent concentration by other molecules largely by dispersion forces and so a polar solvent needs to be added to extract polar solutes. The need for additional solvents tends to defeat the original objectives of the technique and, despite popular acclaim, may well have limited areas of application. Another exceedingly important sample preparation procedure is sample derivatization. In GC, derivatization is carried out to render highly polar substances such as alcohols, acids and amino acids sufficiently volatile to be amenable to distribution between a gas and a liquid phase without thermal decomposition. Common derivatizing procedures for this purpose are esterification, silanization and acetylation. In LC derivatization is usually carried out for a different reason, that is, to aid in detection. Aliphatic alcohols and acids, for example, are often derivatized to introduce a benzene or dinitrobenzene group into the molecule to provide a UV chromaphore. A number of reagents are available that will add a fluorescing group to the solute molecules to make them detectable by fluorescence. To maintain quantitative accuracy, the synthetic procedure must either achieve complete derivatization or a constant and reproducible degree of derivatization. Consequently, the procedure must be very carefully developed and the conditions rigorously controlled, if accurate quantitative results are to be obtained. Derivatization in LC can also be carried out post-column, that is between the column and the detector. The reagent is added to the column eluent through a "T" tube and the mixture passed down a reaction tube to the detector. The reaction tube provides sufficient delay for the reaction to be completed, but must be carefully designed to provide the minimum peak dispersion and thus conserve the resolution. Extraction, derivatization and other sample preparation procedures will usually result in the sample becoming diluted and thus the solution will need to be concentrated. Straightforward concentration by evaporation must be undertaken with great caution. If the materials that have been reacted or extracted are of the order of 1 μg (common in trace analysis and pollution studied) and the material has a partial vapor pressure of merely 1 micron of mercury, the total sample can be lost by passing just a few cubic centimeters of a gas over the surface, even at a very moderate temperature (50°C). Many biopolymers and strongly polar materials will often have sufficiently low vapor pressures to allow them to be recovered from solution by evaporation. Concentration by this means can be carried out simultaneously for many samples using thermostatted metal blocks. The blocks have multiple apertures for 25, 96 or even more drying tubes and a nitrogen manifold is provided to supply a current of gas over the solvent surface. However, evaporation losses must be carefully calculated or measured before this technique can be used with confidence. Lyophilization is a somewhat safer technique as it is carried out at much lower temperatures and thus the solutes will have a significantly lower vapor pressure. The solution is frozen and the solid diluent removed under vacuum. This technique is commonly used in biotechnology and thus lyophilizing equipment is readily available. Exclusion chromatography is also a very effective sample concentrating technique. It can be operated at room temperature, used confidently with labile compounds, has virtually no possibility of sample loss and yet can give a very high concentration ratio; it is also relatively easy to automate.

References

1. H. Schlenk and J. L. Gellerman, *Anal. Chem.*, **32**(1960)1412.
2. K. Blau and J. Halket, *Handbook of Derivatives for Chromatography*, John Wiley and Sons (1993).
3. D. Knapp, ed. *Handbook of Derivatization Reactions* , John Wiley and Sons, New York (1979).
4. A. M. M. T. Scholten, U. A. Brinkman and R. W. Frei, *J. Chromatogr.*, **218**(1981)3.
5. E. D. Katz and R. P. W. Scott, *J. Chromatogr.*, **268**(1983)169.

Chapter 4

The Chromatography Detector

Introduction

There are two essential elements to a quantitative chromatographic analysis. Firstly, the solutes of interest must be adequately resolved from one another, so that measurements can be taken that will disclose the relative amounts of each component present. Secondly, the eluted solutes must be monitored by a device that will give an output that is some known function of the mass or concentration of the solute eluted per unit time. The former required a suitable column and phase system to be chosen to ensure adequate resolution. The latter requires a detector with a known and predictable response and which has a sensitivity commensurate with the mass of sample being analyzed. The majority of chromatographic detectors that are commercially available are suitable for quantitative analysis, but they all differ, to some extent, in their actual performance specifications.

A list of detector specifications are given in table 1 and those essential to performing accurate and precise quantitative analysis are highlighted. The importance of detector specifications generally are discussed elsewhere [1] and only those important for quantitative analysis will be considered here. The five specifications that must be considered are those that determine the linearity and linear dynamic range of the detector and those that determine the sensitivity or minimum detectable mass or concentration.

Table 1 Detector Specifications

Specifications	Units
Dynamic Range	(D_R) g/ml (*e.g.* 3 x 10^{-9} to 6 x 10^{-5})
Response Index	(r) dimensionless
Linear Dynamic Range	(D_{LR}) g/ml (*e.g.* 1 x 10^{-8} to 2 x 10^{-5})
Detector Response	(R_C) Volts/g or (specific units of measurement/g)
Detector Noise Level	(N_D) usually in millivolts but may be in specific units (*e.g.* Refractive Index Units)
Sensitivity or Minimum Detectable Concentration	(X_D) g/ml (*e.g.* 3 x 10^{-8}) but may be in specific units (*e.g.* Absorption Units)
Total Detection System Dispersion	(σ_D^2) ml^2 often µl^2
Cell Dimensions	(length (l), and radius (r)), cm
Cell Volume	(V_D), µl
Overall Time Constant (sensor and electronics)	(T_D), seconds (sometimes milliseconds)
Pressure Sensitivity	(D_P) usually in the USA, p.s.i. in Europe MPa
Flow Rate Sensitivity	(D_Q) usually in ml/min
Temperature Range	^oC

The Dynamic Range of the Detector

The dynamic range of a detector extends from the minimum detectable concentration (*i.e.* the sensitivity) to that where the detector no longer responds to any concentration increase. The dynamic range is not usually pertinent to quantitative chromatographic analysis, but may be of some concern in preparative chromatography where much higher solute concentrations are employed. In preparative chromatography, high sensitivities are often a *disadvantage* and the most popular detectors for this purpose are the least sensitive, which is in complete contrast to those used for analytical use.

The dynamic range of a detector (D_R) is usually given in the form

$$D_R = 6 \times 10^5$$

The manufacturer should give the dynamic range in dimensionless units as it is a ratio of concentrations and independent of the units used. Employing the minimum detectable concentration, in conjunction with the dynamic range, the analyst can then calculate the *maximum* concentration detectable.

Detector Linearity

Detector linearity is probably the most important specification for any detector that is to be used for quantitative analysis. It is defined as the concentration range over which the detector response is linearly related to the concentration of solute passing through it. That is, over the *linear dynamic range*,

$$V = A C_m$$

where (V) is the detector output,
(C_m) is the concentration of solute sensed by the detector,
and (A) is a constant.

Because of the imperfections in mechanical and electrical devices, true linearity is actually an hypothetical concept and practical detectors can only approach this ideal response. Consequently, it is important for the analyst to have some measure of linearity that can be specified in numerical terms, so that comparisons can be made between detectors, and the proximity of the detector to true linearity understood. Fowlis and Scott [2] proposed a method of measuring detector linearity which could also allow corrections to be made for any slight non-linearity that might exist in the detectors response. Fowlis and Scott assumed, that for a closely linear detector, the response could be described by the following power function:

$$V = A C_m^r \qquad (1)$$

where (r) is defined as the *Response Index* and the other symbols have the meanings previously ascribed to them.

It follows from equation (1) that for a truly linear detector, r = 1, and the proximity of (r) to unity will indicate the extent to which the response of the detector approaches true linearity.

The response of some detectors having different values for (r) are shown as curves relating the detector output (V) to solute concentration (C_m^r) in figure 1.

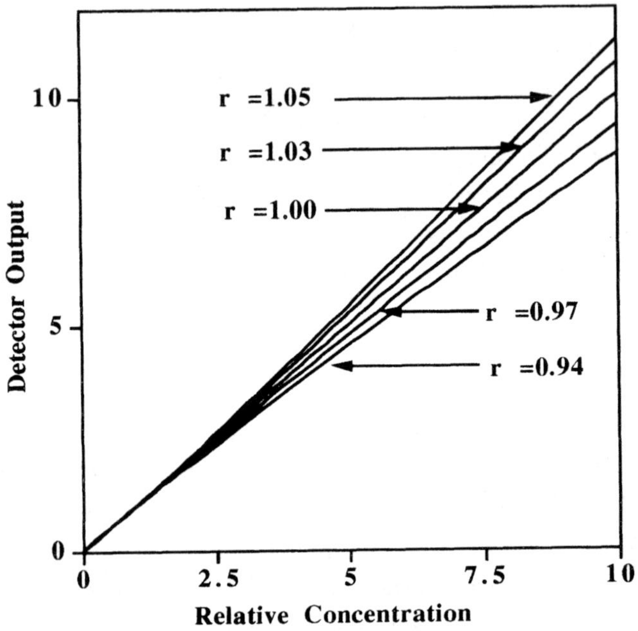

Figure 1 Graph of Detector Output against Solute Concentration for Detectors with Different Response Indices

It is seen that the individual curves appear as straight lines but there can be significant errors that arise if true linearity is assumed. Consider the analysis of a binary mixture containing 10%w/w of one component and 90%w/w of the second component. By assuming all five detectors with responses indices defined in figure 1 are in fact linear, the errors involved can be calculated and the results are shown in table 2.

Table 2 The Analysis of a Binary Mixture Employing Detectors with Different Response Indices (r)

Solute	r = 0.94	r = 0.97	r = 1.00	r = 1.03	r = 1.05
1	11.25%	10.60%	10.00%	9.42%	9.05%
2	88.75%	89.40%	90.00%	90.58%	90.95%

Examination of table 2 shows that errors in the level of the smaller component can be as much as 12.5% (1.25% absolute) for r = 0.94

and 9.5% (0.95% absolute) for r = 1.05. In general analytical work, if reasonable linearity is to be assumed and no corrections made, then,

$$0.98 < r < 1.03$$

There are two basic advantages to defining linearity in this way. Firstly, a claim of linearity can be substantiated, secondly, if the detector is not perfectly linear, and the value for (r) is known, then a correction can be applied to accommodate the non-linearity and still provide accurate results. There are alternative methods for defining linearity which, in the authors' opinion, are less precise and less useful. The recommendations of the ASTM, E19 committee [3] on linearity measurement are typically given for a photometric detector as follows:

'the linear range of a PD (photometric detector) is that concentration range of the test substance over which the response of the detector is constant to within 5% as determined from a linearity plot, – the linear range should be expressed as the ratio of the highest concentration on the linearity curve to the minimum detectable concentration'

This method for defining detector linearity ensures a minimum linear performance from the detector and, for most analyses, an acceptable quantitative accuracy. However, the specification is significantly 'looser' than that given above and there is no means of correcting for any non–linearity that may exist as there is no correction factor given that is equivalent to the response index. It is strongly advised that the response index of all detectors (GC and LC) that are to be used for quantitative analysis are experimentally determined even if a value is provided by the manufacturer. In most cases, (r) need only be determined once, unless the detector experiences some catastrophic event, in which case (r) may need to be checked again.

The Determination of the Response Index of a Detector

There are two methods that can be used to measure the response index of a detector, the *incremental method of measurement* and the *logarithmic dilution method* of measurement. The former requires no special apparatus but the latter requires a log-dilution vessel which fortunately, however, is relatively easy to fabricate.

The Incremental Method of Linearity Measurement

The apparatus necessary is the detector itself, with its associated electronics and recorder or computer system, a mobile phase supply, an appropriate sampling device and preferably the column that is to be used for the subsequent analyses, although this is not essential. Ideally, the chromatograph itself which is to be used for the subsequent analyses can be employed for measuring the response index. The solute is chosen as typical of the type of substances that will be analyzed and the mobile phase or operating temperature (depending on whether it is a gas chromatograph or a liquid chromatograph) is chosen that will elute the solute in a reasonable time. Initial sample concentrations are chosen that will be appropriate for the sensitivity of the detector under examination.

Duplicate samples are placed on the column and the eluted peaks either recorded or the data stored by the computer. The sample solution is then diluted by a factor of three and again duplicate samples are placed on the column and the eluted peaks either recorded or the data stored by the computer. This procedure is repeated increasing the detector sensitivity setting where necessary to accommodate the increased sample size. The dilution procedure is continued until the height of the eluted peak is commensurate with that of the noise level.

If the detector has no data acquisition and processing facilities, and peak areas or peak heights cannot be recalled from a computer, then the peaks from the chart recorder must be used to provide the calibration data. The width of the peak at 0.607 of the peak height is measured and from the chart speed and the mobile phase flow rate the peak volume can be calculated. The concentration of solute at the peak maximum will be twice the average peak concentration and can be calculated from the following equation:

$$C_p = \frac{ms}{wQ}$$

where (C_p) is the concentration of solute in the mobile phase at the peak height in (g/ml),
(m) is the mass of solute injected,
(w) is the peak width at 0.6067 of the peak height,
(s) is the chart speed of the recorder or printer,
and (Q) is the flow rate in ml/min.

The log of the peak height (y) (where y is the peak height in millivolts or whatever units in which the output is defined) is then plotted against the log of the solute concentration at the peak maximum (C_p). From equation (1),

$$\log (y) = \log A + (r) \log (C_p)$$

Thus the slope of the log/log curve will give the value of the response index (r). If the detector is truly linear, r = 1 (*i.e.* the slope of the curve will be sin π/4 = 1). Alternatively, if suitable software is available, the experimental data can be curve-fit directly to a power function and the value of (r) extracted from the results. The same data can be employed to determine the linear range as defined by the ASTM E19 committee. In this case, however, a linear plot of detector output against solute concentration at the peak maximum should be used and the point where the line deviates from 45° by 5% determines the limit of the linear dynamic range. The Incremental Method for determining the response index is probably the simplest and has the advantage of including any characteristics of the chromatograph that may affect the results. However, the logarithmic dilution method is easier to use over a very wide linear dynamic range and may be more suitable for GC detectors.

The Logarithmic Dilution Method of Linearity Measurement

The logarithmic dilution method for detector calibration was introduced by Lovelock [4] for the calibration of GC detectors and was modified by Fowlis and Scott [2] for the calibration of LC detectors. The system provides a continuous flow of gas or liquid through the detector that contains a solute, the concentration of which decreases exponentially with time. The theoretical diagram of a log dilution apparatus is shown in figure 2.

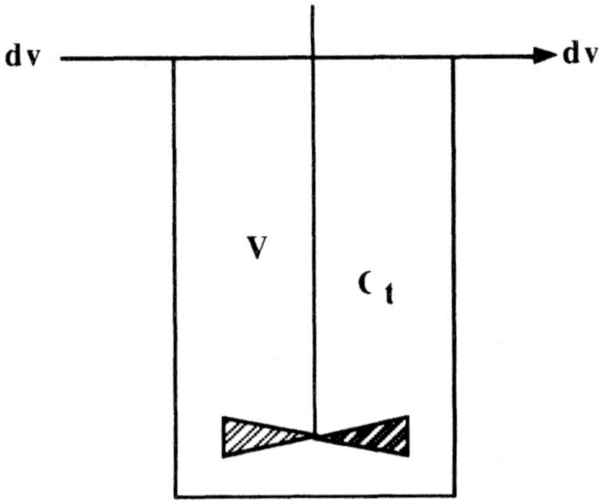

Figure 2 Diagram of a Logarithmic Dilution Vessel

A known mass of solute is introduced into a well–stirred vessel through which a flow of gas or solvent continuously passes. The mixture is thus continuously diluted and the concentration of the solute in the exit flow progressively monitored by means of the detector.

Let the vessel have a volume (V), and let the concentration of the solute in the vessel after time (t) be (C_t). Let a volume (dv) of pure solvent enter the vessel and displace a similar volume (dv) from the vessel.

The mass of solute removed (dm) is given by

$$dm = C_t dv$$

Now the change in mass (dm) in the vessel will result in a change in concentration (dC_t).

Thus, $$V\,dC_t + C_t\,dv = 0$$

and $$\frac{dC_t}{C_t} = \frac{-dv}{V}$$

Integrating $$\log C_t = \frac{-v}{V} + k$$

$$= \frac{-Qt}{V} + k$$

where (Q) is the flow rate,
(v) is the volume flow of mobile phase through the system after time (t)
and (k) is the integration constant.

Now when $t = 0$, $C_t = C_o$, where C_o is the initial concentration of solute.

Thus, $k = \log C_o$ and $$\log C_t = \frac{-v}{V} + \log C_o$$

or $$C_t = C_o e^{\frac{-Qt}{V}}$$

Thus if the logarithm of the detector output is plotted against time, then for a truly linear detector, a straight line will be produced having a slope (–Q/V). If the detector has a response index of (r) and the slope of the line is (φ), then

$$\varphi = \frac{-Qr}{V} \quad \text{or} \quad r = \frac{-\varphi V}{Q}$$

Thus, the response index (r) can be easily determined. However the accuracy of the measurement will depend on the flow rate remaining constant throughout the calibration, and consequently, for a GC detector a precision flow controller must be employed and for an LC detector, a good quality solvent pump. Manufacturers do not usually provide the response indices for their detectors and so it is left to the analysts to measure it for themselves. It should be pointed out that the logarithmic dilution method should *not* be used if the linearity is to be measured by the method recommended by the E19 committee of the ASTM.

The Linear Dynamic Range of a Detector

The linearity of most detectors deteriorates at high concentrations and thus the *linear dynamic range* of a detector will always be less than the *dynamic range.* The symbol for the linear dynamic range is usually taken as (D_{LR}) and might be specified in the following form for the FID as an example:

$$D_{LR} = 2 \times 10^{-3} \quad \text{for } 0.98 < r < 1.02$$

The lowest concentration of the linear dynamic range is usually equal to the *minimum detectable concentration* or the *sensitivity* of the detector. The largest concentration in the linear dynamic range would be that where the response factor (r) falls outside the range specified. Unfortunately, many manufacturers do not differentiate between (D_R) and (D_{LR}) and even if a value for (r) is specified, a range over which the response index (r) remains constant is not given.

Detector Response

Detector response can be defined in two ways. It can be taken as the voltage output for unit change in solute concentration in which case, and in a similar way to detector sensitivity and dynamic range, the solute used for measurement must be specified. Alternatively, it can be taken as the voltage output that would result from unit change in the physical property that the detector measures, *e.g.* refractive index for a refractive index detector or carbon content for an FID detector. The detector response (R_C) can be determined by injecting a known mass of the chosen solute (m) onto the column and measuring the response from the dimensions of the peak. Again assuming the concentration of the solute at the peak maximum is twice the average peak concentration, then

$$R_c = \frac{hwQ}{sm}$$

where (h) is the peak height and the other symbols have the meaning previously attributed to them.

Inevitably the response of a detector will differ between different solutes as well as between different detectors. *Ipso facto* the response of two detectors of the same type and geometry can only be compared using the same solute and the same mobile phase. Even so, if the response is specified in units of the physical property being measured, using the same solute and the same mobile phase, accurate comparisons may still not be possible. For example, in comparing the response of two UV detectors using the same wavelength, the same solute and mobile phase, the path length of the respective cells must still be taken into account to arrive at a rational comparison of response. This problem will be discussed in detail when light absorption detectors are discussed and the chromatographic apparatus is described.

Detector Noise

Detector noise is the term given to any perturbation on the detector output that is not related to an eluted solute. It is a fundamental property of the detecting system and determines the ultimate sensitivity or minimum detectable concentration that can be achieved. Detector noise has been arbitrarily divided into three types, '*short term noise*', '*long term noise*' and '*drift*' all three of which are depicted in figure 3.

Short Term Noise

Short term noise consists of baseline perturbations that have a frequency that is significantly higher than the eluted peak. Short term detector noise is not often a serious problem as it can be easily removed by appropriate noise filters without significantly affecting the profiles of the peaks. Its source is usually electronic, originating from either the detector sensor system or the amplifier.

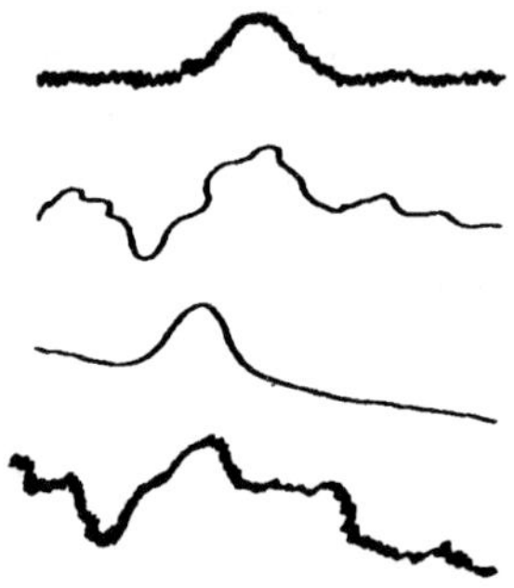

Figure 3 Different Types of Detector Noise

Long Term Noise

Long term noise consists of baseline perturbations that have a frequency that is similar to that of the eluted peak. This type of detector noise is the most significant and damaging as it is indiscernible from very small peaks in the chromatogram. Long term noise cannot be removed by electronic filtering without affecting the profiles of the eluted peaks. In figure 3, it is clear that the peak profile can easily be discerned above the high frequency noise, but is lost in the long term noise. Long term noise usually arises from temperature, pressure or flow rate changes in the sensing cell. Long term noise is largely controlled by detector cell design and ultimately determines the limits of detector *sensitivity* or the *minimum detectable concentration.*

Drift

Drift results from baseline perturbations that have a frequency that is significantly larger than that of the eluted peak. Drift is almost always due to either changes in ambient temperature, changes in mobile flow rate, or column bleed in GC; in LC drift can be due to pressure changes, flow rate changes or variations in solvent composition. As a consequence, with certain detectors baseline drift can become very significant in GC at high column temperatures, and in LC at high column flow rates. Drift is easily constrained by choosing operating parameters that are within detector and column specifications. A combination of all three sources of noise is shown by the lowest trace in figure 3. In general, the sensitivity of the detector should never be set above the level where the combined noise exceeds 2% of the F.S.D. (full scale deflection) of the recorder (if one is used), or appears as more than 2% F.S.D. of the computer simulation of the chromatogram.

Measurement of Detector Noise

The detector noise is defined as the maximum amplitude of the combined short and long term noise measured over a period of 10 minutes (the E19 committee recommends a period of 15 minutes). The detector must be connected to a column and mobile phase passed through it during measurement. The detector noise is obtained by constructing parallel lines embracing the maximum excursions of the recorder trace over the defined time period, as shown in figure 4. The distance between the parallel lines measured in millivolts is taken as the measured noise (v_n), and the noise level (N_D) is calculated in the following manner:

$$N_D = v_n A = \frac{v_n}{B}$$

where (v_n) is the noise measured in volts from the recorder trace,
(A) is the attenuation factor,
and (B) is the alternative amplification factor (see below).

It should be noted that attenuation is the reciprocal of amplification and manufacturers may use either function as a control of detector sensitivity. For example the sensitivity settings on a RI detector for LC may take either of the following forms.

Minimum	(Amplification)	1	2	4	8	16	32	Maximum
Sensitivity	(Attenuation)	32	16	8	4	2	1	Sensitivity

Figure 4 The Measurement of Detector Noise

The noise level of detectors that are particularly susceptible to variations in column pressure or flow rate (*e.g.* the katharometer in GC and the refractive index detector in LC) are often measured under static conditions (*i.e.* no flow of mobile phase). Such specifications are not really useful, as the analyst can never use the detector without a column flow.

It should be noted that at the high sensitivity range settings of some detectors, filter circuits are automatically introduced to reduce the noise. Under such circumstances the noise level should be determined at the lowest attenuation (or highest amplification) that does not include noise-filtering devices (or at best the lowest attenuation with the fastest response time) and subsequently corrected to an attenuation of unity.

Detector Sensitivity or the Minimum Detectable Concentration

Detector sensitivity or minimum detectable concentration (MDC), is defined as the minimum concentration of solute passing through the detector that can be unambiguously discriminated from noise. The size of the signal that will make it distinctly apparent from the noise (the signal–to–noise ratio) is a somewhat arbitrary choice. It is generally accepted that a signal can be differentiated from the noise in electronic measurements when the signal to noise ratio is *two* and this criteria has been adopted for defining detector sensitivity.

Thus for a concentration sensitive detector, the detector sensitivity (X_D) is given by

$$X_D = \frac{2N_D}{R_c} \quad (g / ml)$$

(R_c) and (N_D) being determined in the manner previously described.

The sensitivity of a detector is *not* the minimum mass that can be detected. This would be the *system mass sensitivity,* which would also depend on the characteristics of the apparatus as well as the detector and, in particular, the type of column employed. During the development of a separation, the peaks become broader as the retention increases. Consequently, a given mass may be detected if eluted as a narrow peak early in the chromatogram, but if eluted later, its peak height may be reduced to such an extent that it is impossible to discern it from the noise. Thus detector sensitivity quoted as the 'minimum mass detectable' must be carefully examined and related to the chromatographic system and particularly the column with which it was used. If the data to do this are not available, then the sensitivity must be calculated from the detector response and the noise level in the manner described above.

Some manufacturers have taken the minimum detectable concentration and multiplied it by the sensor volume and defined the product as the minimum detectable mass. This way of defining detector sensitivity is particularly misleading. For example a detector having a true sensitivity of 10^{-6} g/ml and a sensor volume of 10 μl would be attributed a sensitivity of 10^{-8} g. This value is, of course, grossly incorrect as it is the peak volume that controls the mass sensitivity, not the sensor volume. Conversely, if the peak volume does approach that of the sensor, then very serious peak distortion and loss of resolution occurs, and this effect will be discussed in detail later. Detector specifications given by manufacturers are becoming more rational and helpful to the analyst, but misleading values are still given and any specifications not described here, or not defined in the units that are recommended here, need to be careful examined to assess their true significance.

The Mass Sensitivity of a Chromatographic System

The mass sensitivity of a chromatographic system (including column, sample valve or injection system, connecting tubes and detector) is defined as the mass of solute (m_D) that will provide a peak with a height equivalent to twice the noise level.

Consider a chromatographic peak with height equivalent to twice the noise level, being sensed by a detector with a maximum sensitivity of (X_D) g/ml. The peak volume can be taken as ($4\sigma_c$) [5], where (σ_c) is the volume standard deviation of the peak as it is eluted from the column and the concentration at the peak maximum as twice the mean concentration of the solute in the peak.

Then,
$$\frac{2m_D}{4\sigma_c} = X_D$$

or
$$m_D = 2\sigma_c X_D \qquad (2)$$

Now from the Plate Theory [6],

$$\sigma_c = \frac{V_r}{\sqrt{n}}$$

where (V_r) is the retention volume of the solute,
and (n) is the column efficiency.

Thus,
$$m_D = \frac{2V_r X_D}{\sqrt{n}} \qquad (3)$$

Now, also from the Plate Theory [6],

$$V_r = V_o(1 + k') \qquad (4)$$

where (V_o) is the column dead volume,

and (k') is the capacity ratio of the solute.

and $$V_o = \varepsilon \pi r^2 l \tag{5}$$

where (l) is the column length,
(r) is the column radius,
and (ε) is the fraction of the column volume that is considered filled with mobile phase (usually taken as 0.65).

Consequently, substituting for (V_r) in equation (3) from equation (4) and for (V_o) from equation (5),

$$m_D = \frac{2\varepsilon \pi r^2 l(1+k')X_D}{\sqrt{n}} \tag{6}$$

Thus (m_D), the mass sensitivity of the chromatographic system depends on the detector sensitivity, column dimensions, column efficiency and the capacity factor of the eluted solute. However, irrespective of the column properties, the mass sensitivity is still directly related to the detector sensitivity. It will also be seen that the column radius will depend on the extra-column dispersion, much of which arises from the detector connecting tubes and sensor. It follows that the design of the detector and its sensitivity has a major influence on the mass sensitivity of the overall chromatographic system. It is also clear from equation (6) that the mass sensitivity decreases as the square of the column radius. This means that if the quantity of sample is limited, then the analysis should be carried out on a capillary column with on column injection if GC is the technique being used, or microbore columns if the analysis is to be carried our by LC.

The Concentration Sensitivity of a Chromatographic System

The concentration sensitivity of a *chromatographic system* (X_C) (as opposed to the concentration sensitivity of the detector alone) is defined as that concentration which will provide a peak with a height equivalent to twice the noise level. An expression for concentration sensitivity of the system can be obtained directly from the system mass sensitivity. If the minimum detectable mass is dissolved in the maximum permissible sample volume [7] (that sample volume that will limit the increase in sample variance to 10% of the column variance), then this solute concentration will constitute the minimum detectable sample concentration.

Now the maximum sample volume (V_S) that can be used while maintaining the limited peak dispersion is $1.1\sigma_C$ and the mass sensitivity given by equation (2) is $2\sigma_C X_D$. Thus the concentration sensitivity (X_C) is given by

$$X_c = \frac{2\sigma_c X_D}{1.1\sigma_c} = 1.8X_D \qquad (7)$$

It is seen that the system concentration sensitivity is also directly (and solely) dependent on the detector sensitivity and, in contrast to the system mass sensitivity, is independent of column or solute properties. It should be emphasized, however, that in order to realize the minimum concentration sensitivity, the maximum permissible sample volume must be used.

The Maximum Capacity Factor of an Eluted Peak

The maximum (k') at which a solute can be eluted is also determined by the detector sensitivity. As the (k') of the eluted solute increases, the peak becomes more disperse, the peak height is reduced and eventually the peak will disappear into the detector noise. In a general chromatographic analysis, if it were merely necessary to unambiguously identify the existence of a peak, then the specified signal–to–noise ratio might not be acceptable and the peak maximum would probably need to be about 5 times the noise level. This is also an arbitrary assumption based on experience, but in fact any desired signal-to-noise level can be chosen and the mathematical arguments will remain the same.

Now if the detector sensitivity as defined is (X_D), at the maximum (k') the concentration will be $2.5X_D$. The concentration at the peak maximum can be taken as twice the average concentration so if there is a mass (m) contained in a peak of base width $4\sigma_c$ then from the Plate Theory [6],

$$4\sigma_c = 4\sqrt{n}(v_m + Kv_s)$$

where (v_m) is the volume of mobile phase in a theoretical plate,
(v_s) is the volume of stationary phase in a theoretical plate,
and (K) is the distribution coefficient of the solute between the two phases.

Then,
$$\frac{2m}{4\sqrt{n}(v_m + Kv_s)} = 2.5X_D \qquad (8)$$

or
$$m = 5X_D\sqrt{n}(v_m + Kv_s)$$

Now as stated previously, the maximum permissible sample volume (V_s) is given by

$$V_s = 1.1\sigma_c$$

and from the Plate Theory [7],

$$\sigma_c = \sqrt{n}(v_m + Kv_s)$$

Thus,
$$V_s = 1.1\sqrt{n}(v_m + Kv_s) \qquad (9)$$

The early peaks, being the narrowest, are the most affected by sample volume. Furthermore the resolution of both early and late peaks must be given the same priority. Consequently, the sample volume should be chosen so that it does not increase the dead volume peak width (the narrowest peak) by more than 5%. Thus, (K) in equation (8) must be made zero and thus,

$$V_s = 1.1\sqrt{n}(v_m)$$

Consequently, if the sample concentration is (X_s),

$$m = X_s V_s = 1.1X_s\sqrt{n}(v_m) \qquad (10)$$

Equating equations (8) and (10),

$$1.1X_s\sqrt{n}(v_m) = 5X_D\sqrt{n}(v_m + Kv_s)$$

Now, bearing mind that from the Plate Theory [8], $k' = \frac{Kv_s}{v_m}$.

Then $$\frac{1.1X_s}{5X_D} = (1+k') \quad \text{or} \quad k' = \frac{0.22X_s}{X_D} - 1 \qquad (11)$$

It is seen that the maximum (k') depends on both the detector sensitivity and the sample concentration. Consequently, the detector sensitivity determines the greatest value for (k) that can be obtained from a column.

Table 3 Limiting Values for (k') for Detectors of Different Sensitivities and a Sample Concentration of 0.1% w/v

Detector Sensitivity	Maximum (k')
10^{-5} g/ml	21
10^{-6} g/ml	219
10^{-7} g/ml	2199
10^{-8} g/ml	22,000

Providing the detector has a reasonable sensitivity, the limit for (k'), however, is fairly high as shown in table 3 where the sample concentration is assumed to be 0.1 % w/v. The data given above is subject to the caveat that the sample is placed on the column in the maximum permissible volume, but not overloaded. The maximum (k') of a sample is rarely important in quantitative chromatographic analysis except, perhaps in GC, when the solute has an extremely low volatility and is extensively retained even at the maximum operating temperature. This situation might occur when there is a need for very high sensitivity and for some other reason can not be separated by LC.

Practical Considerations

Specific operating details for different detectors will be given when a chromatographic apparatus is described, but there is some useful advice that can be given here.

In general a detector with the best linearity and widest linear dynamic range should be chosen for quantitative analysis. If adequate sample is available, and column performance is not compromised, the detector should be operated at the higher end of its linear dynamic range by employing as large a sample size as possible. The detector will function in a more stable manner at lower sensitivities and provide better accuracy and precision.

Do not choose a high sensitivity, specific detector with a limited linear dynamic range over a less sensitive more linear detector to merely cope with low concentration samples. It is better to concentrate the sample and use the less sensitive more linear detector. If the specific detector is used, however, to selectively monitor certain compounds to obviate the need for higher resolution columns and ensuing longer analysis times, then the sample size must be carefully chosen so that the detector is clearly operating within its limited linear dynamic range. *Ipso facto* the linear dynamic range of the detector must be known or experimentally identified.

Do not introduce response time constants or signal smoothing with fast eluting peaks (as with capillary columns in GC and microbore columns in LC) or the peaks will be distorted and quantitative accuracy seriously impaired.

Synopsis

There are two essential elements to a successful quantitative chromatographic analysis, the solutes of interest must be adequately resolved by the column and they must be measured by a linear and sufficiently sensitive detector. There are twelve generally important specifications of a detector, but six specifically apply to quantitative analysis, *viz,* dynamic range, linear dynamic range, response index, response, noise level and sensitivity. The dynamic range is that concentration range over which the detector will respond and the linear dynamic range is that range of concentration over which the response is linear. The response index is a measure of the detector linearity and the response is the detector output per unit change in concentration. The noise level is the perturbation of the detector output when no solute is eluted and the sensitivity the concentration of solute that gives a signal equivalent to twice the noise level. There are three types of noise. High frequency noise is large compared with the frequency of the chromatographic peaks. Low frequency noise is about the same as that of the peak and has the greatest effect on sensitivity and chromatographic performance. The third source of noise is called drift which has a frequency much greater than that of the peak. The response index and the noise are the two specifications that are particularly important to the analyst. The response index can be measured experimentally by successive injections of solute at different concentrations onto the chromatograph and relating the peak heights or peak areas to the mass injected. The response index can also be determined by a logarithmic dilution method using specific apparatus for the purpose. The detector response can be determined from the output provided by a given concentration of solute and the sensitivity or minimum detectable concentration taken as the ratio of twice the noise level to the detector response. The mass sensitivity and concentration sensitivity of the overall chromatograph although related directly to the detector sensitivity is also a function of the column properties. The detector sensitivity also limits the maximum capacity of an eluted peak that can be satisfactorily used for quantitative analysis. This limitation, however, is rarely met in practice. From a practical point of view, the detector with the best linearity and widest linear dynamic range should be chosen for quantitative analysis. The detector will function in a more stable manner at lower sensitivities and provide better accuracy and precision, so these conditions should be chosen where possible. When a specific detector is used to selectively monitor certain compounds to obviate the use of high resolution columns and ensuing long analysis times, the sample size must be carefully chosen so that the detector is clearly operating within its limited linear dynamic range.

References

1. R. P. W. Scott, *Chromatography Detectors*, Marcel Dekker, Inc., New York (1996)17.
2. A. Fowlis and R. P. W. Scott, *J. Chromatogr.*, **11**(1963)1.
3. C. G. Scott, ASTM E19 No. E689–79.
4. J. E. Lovelock, *Gas Chromatography 1960* (Ed., R. P. W. Scott), Butterworths, London (1960)26.
5. R. P. W. Scott, *Techniques and Practice of Chromatography*, Marcel Dekker, Inc., New York (1995)23.
6. R. P. W. Scott, *Introduction to Analytical Gas Chromatography*, Marcel Dekker, Inc., New York (1998)21.
7. R. P. W. Scott, *Liquid Chromatography for the Analyst,* Marcel Dekker, Inc., New York (1994)95.

Chapter 5

Processing Chromatographic Data

Introduction

A quantitative analysis can only be successfully carried out if the substances of interest are adequately resolved. That is, the peaks must be separated sufficiently to allow the unique area or height of each peak to be accurately measured or calculated. This is necessary irrespective of whether the measurements are to be made manually on the chart paper or be calculated from acquired data by a computer. It follows that the degree of resolution that is necessary to achieve adequate quantitative accuracy needs to be addressed. In addition, despite chromatographic optimization, some components may still be only poorly resolved and, as a consequence, some action must be taken to access the peak areas and heights by special techniques. First, the conditions necessary to achieve adequate resolution will be considered.

Chromatographic Resolution

The resolution of a pair of closely eluted solutes is usually defined as the distance between the peaks measured in units of the average standard deviation of the two peaks. In practice, the standard deviation (σ) of two peaks that are eluted close together are virtually identical. However, it is necessary to decide the degree of separation that constitutes adequate resolution for accurate quantitative analysis. In figure 1, five pairs of peaks are shown, separated by 2σ, 3σ, 4σ, 5σ

and 6σ, the area of the smaller peak being half that of the larger peak. However, it should be noted that the relative size of the peaks will affect the minimum separation that will be suitable for accurate quantitative analysis and thus, must be taken into account.

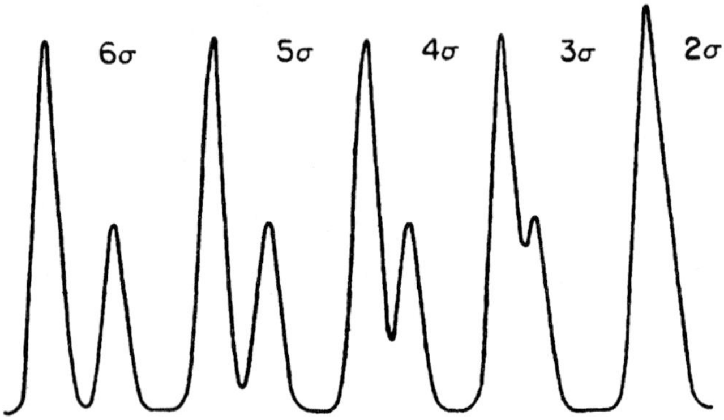

Figure 1 Peaks Showing Different Degrees of Resolution

It is clear that a separation of 6σ would be ideal for accurate quantitative results. Unfortunately, such resolution will often demand very high efficiencies, which may also entail long columns and consequently, very long analysis times. Furthermore, a separation of 6σ is *far more than necessary* for accurate quantitative analysis and this would be true even if manual measurements of peak area and peak height are employed. In fact, accurate quantitative results can usually be obtained with a separation of only 4σ and, as will be seen later, if peak heights are employed, then even less resolution may be needed. Duplicate measurements of peak area or peak height on peaks separated by 4σ should not differ by more than 2%. If the chromatographic data are acquired and processed by a computer, then with modern software, even a separation of 4σ could be more than adequate. In the following mathematical argument it will be assumed that a resolution of 4σ is sufficient, but should greater resolution be thought more appropriate, the equations below can be easily modified and the same arguments used. Defining the resolution required for accurate quantitative analysis as 4σ, it is now possible to calculate the number of theoretical plates necessary for the separation of a specific pair of solutes.

The Efficiency Required to Achieve a Specific Resolution

Consider the two peaks shown in figure 2. The various parameters of the chromatogram are labeled according to the expressions derived from the Plate Theory. A discussion of the Plate Theory is not germane to the subject of this book, but certain relationships will be used that are derived from it and so those readers interested in their origin are recommended to another book in this series, *Introduction to Analytical Gas Chromatography* [1].

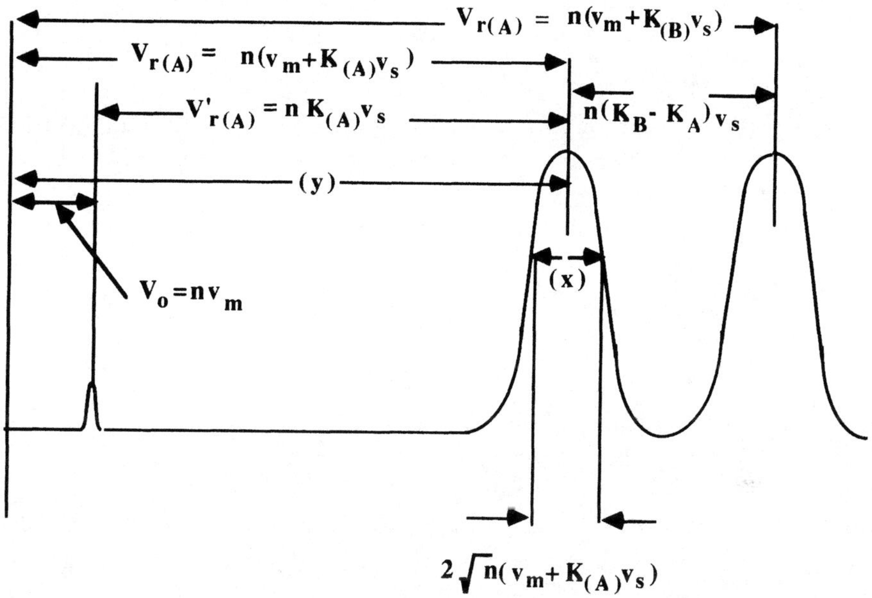

Figure 2 The Separation of Two Solutes

The difference between the peaks is given by (ΔV), where

$$\Delta V = n(v_m + K_B v_s) - n(v_m + K_A v_s) = n(K_B - K_A)v_v \qquad (1)$$

where (n) is the column efficiency in theoretical plates,
(v_m) is the volume of mobile phase per plate,
(v_s) is the volume of stationary phase per plate,
and (K_A),(K_B) are the distribution coefficients of solutes (A) and (B).

Bearing in mind that the separation ratio (α) is defined as

$$\alpha = \frac{K_B}{K_A}, \quad \text{then} \quad \Delta V = nK_A(\alpha - 1)v_s \qquad (2)$$

Now, by definition for adequate separation, $\Delta V = 4\sigma$

$$\text{or} \qquad \Delta V = 4\sqrt{n}(v_m + K_A v_s) \qquad (3)$$

Equating equations (2) and (3)

$$4\sqrt{n}(v_m + K_A v_s) = nK_A(\alpha - 1)v_s$$

Dividing throughout by (v_m) and remembering that the separation ratio (k') is given by $k'_A = \frac{K_A v_s}{v_m}$

$$\text{Then} \qquad 4\sqrt{n}(1 + k'_A) = n k'_A (\alpha - 1)$$

Solving for (n),

$$n = \left(\frac{4(1 + k'_A)}{k'_A(\alpha - 1)}\right)^2 \qquad (4)$$

Equation (4) was first developed by Purnell in 1959 [2] and is one of the more important equations used in column design and one that can be a great aid in column selection for the practicing analyst. In a given analysis, if the separation ratio of the 'critical pair' is measured (the critical pair comprise the two solutes that are eluted closest together), the column efficiency required (for a separation of 4σ) can be calculated from equation (4). If the actual efficiency is found to be inadequate, the flow rate can be reduced until sufficient efficiency is attained or until the maximum efficiency is realized at the optimum velocity. If, at the optimum flow rate, even more efficiency is necessary, then a longer column is required or a column packed with smaller particles. The latter is to be preferred, if possible, as the increased efficiency will be obtained at reduced analysis times (assuming adequate inlet pressure is available). As most modern LC columns are homogeneously packed, or, in the case of the GC capillary column, homogeneously coated, then for capillary columns of the same diameter and columns packed with the same diameter particles, the column efficiency can be assumed to increase linearly with the column length. Employing a similar argument, an expression for the efficiency required from a TLC plate to effect a given separation can also be developed. The (Rf) value of a solute is given by

$$R_f = \frac{\text{Distance Migrated by Solute}}{\text{Distance Migrated by Solvent}} = \frac{Z_s}{Z_m}$$

Now the distance (D) between two solutes (A) and (B) will be

$$D = Z_{s(B)} - Z_{s(A)}$$

$$= Z_m(R_{f(B)} - R_{f(A)}) \qquad (5)$$

Now if the diameter of a spot is (d) then, [3]

$$\frac{Z_{s(A)}}{d} = \frac{\sqrt{n}}{4} \quad \text{or} \quad d = \frac{4Z_{s(A)}}{\sqrt{n}} \qquad (6)$$

Now the diameter of the spot can be considered to be equivalent to (4σ) and thus to achieve resolution, the spots must be separated by a distance equivalent to their diameter, then from equations (5) and (6)

$$\frac{4Z_{s(A)}}{\sqrt{n}} = Z_m\left(R_{f(B)} - R_{f(A)}\right)$$

Rearranging, and bearing in mind that $R_f = \frac{Z_s}{Z_m}$,

$$\sqrt{n} = \frac{4Z_{s(A)}}{Z_m\left(R_{f(B)} - R_{f(A)}\right)} = \frac{4}{\left(\alpha_{B/A} - 1\right)}$$

where $\alpha_{B/A} = R_{f(B)}/R_{f(A)}$ is the TLC equivalent to the separation ratio used in GC and LC.

Thus $$n = \frac{16}{(\alpha_{B/A} - 1)^2} \qquad (7)$$

Equation (7) gives the efficiency required from a TLC plate to separate a particular pair of substances employing a specific phase system. To increase the plate efficiency the particle size of the adsorbent must be reduced, which, in most cases, is difficult to change. In practice, the same plate adsorbent will probably be employed, but the phase system will be adjusted to increase the value of (α_{AB}) so that the separation is achieved with the efficiency available.

Peak Deconvolution

Unfortunately there will be certain samples, usually complex in nature, where adequate resolution is not obtainable and, as a consequence, it is necessary to resort to other procedures to determine the respective areas of the merged peaks. This situation can arise when there is a large number of peaks in the chromatogram and attempts to improve the resolution of one pair results in the fusion of another pair. One such technique that may be used under such circumstances is *peak deconvolution.* Many data acquisition and processing systems include software that mathematically analyze convoluted (unresolved) peaks, identify the individual peaks that make up the composite envelope, and then determine the area of the individual peaks. The algorithms in the software must embody certain tentative assumptions in order to analyze the peak envelope, such as the peaks are Gaussian in form and are symmetric or that there are explicit equations available that describe the specific asymmetry of each peak. Furthermore, the software must assume that *all* the peaks can be described by the *same* function (*i.e.*, the efficiency of all the peaks are the same), which is also not always true. Nevertheless, providing the composite peak is not too complex, deconvolution can, under some circumstances, be reasonably successful.

However it must be emphasized that the use of deconvoluting software has limited capabilities and must be used with considerable caution.

Every effort should be made to separate all the components of a mixture by chromatography and only employ peak deconvolution as a last resort. In general, it should be remembered that,

"Clever algorithms are not an acceptable substitute for good chromatography."

Having made the point, if resolution is partial, and the components are present in equal quantities, then the deconvolution approach can be quite successful. An example of the convolution of two partially resolved peaks of equal size is shown in figure 3.

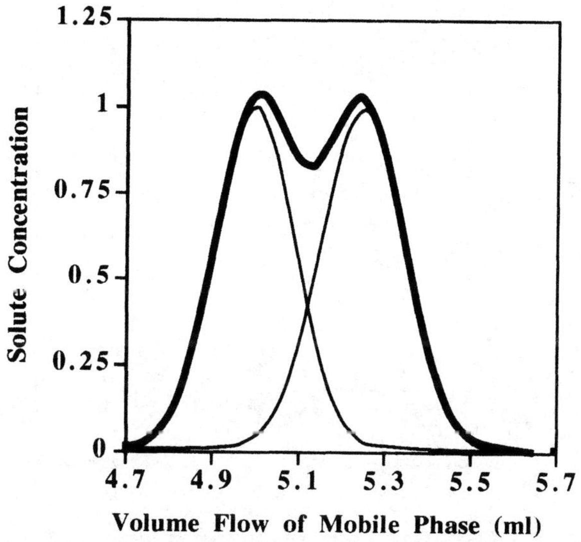

Figure 3 The Deconvolution of Two Partially Resolved Peaks Representing Solutes Present in Equal Quantities

It is seen that two Gaussian shaped peaks can be easily extracted from the composite envelope and the software could also supply values for either the heights of the deconvoluted peaks and/or their areas. It should be noted, however, that the two peaks are clearly discernible and the deconvoluting software can easily identify the approximate positions of the peak maxima and assess the peak widths. Such information allows the software to arrive at a valid analysis quickly and with reasonable accuracy. Nevertheless, most software would construct a perpendicular from the valley to the baseline, thus bisecting the combined peak envelope. The area of each half would then be taken as the respective area of each peak. If the peaks are of significantly different size the problem becomes more difficult but, nevertheless, solvable. An example of the analysis of a composite envelope for two peaks of different size is shown in figure 4. It is clear that the resolution will not permit peak skimming (a technique that will be discussed in detail later) with any hope of a reasonable degree of accuracy. It seen that the position of the peak maximum, and the peak width, of the major component is easily identifiable.

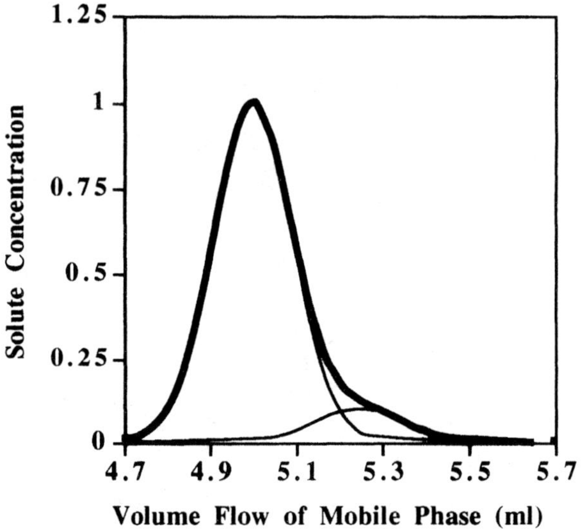

Figure 4 The Deconvolution of Two Partially Resolved Peaks Representing Solutes Present in Unequal Quantities

The software can accurately determine the Gaussian function of the major component and the reconstructed profile of the major component would then be subtracted from the total composite peak leaving the small peak as difference value. However, in the examples given, the solutes were at least partly resolved. Unfortunately, as the resolution becomes less and less, and the need for an accurate deconvolution technique becomes even greater, the value of the software presently available appears to become minimal. A typical example at the other extreme, where a deconvolution technique would be useless, is given in figure 5. It is seen that however sophisticated the software might be, it would be virtually impossible to deconvolute the peak into the three components. The peaks shown in the diagram are discernible because the peaks themselves were assumed and the composite envelope calculated.

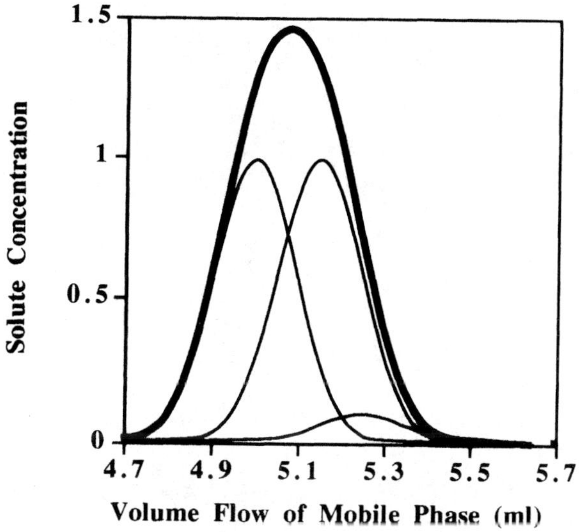

Figure 5 The Deconvolution of Three Unresolved Peaks

The envelope, however, would provide no basic information; there is no hint of an approximate position for any peak maximum and absolutely no indication of the peak width of any of the components or even how many peaks are present.

The Detector Response

One further form of peak deconvolution needs to be mentioned and that is the practical use of the detector response index. The detector response index has been presented as a means of defining linearity, but, as mentioned in an earlier chapter, it can also be used to take into account any non-linearity that is present and appropriately modify the peak height or peak area calculations and thus improve quantitative accuracy. An example of two peaks constructed from Gaussian functions using response factors of 0.95 and 1.05 are shown in figure 6. Such values would be considered outside those which would be acceptable for a detector to be defined as linear.

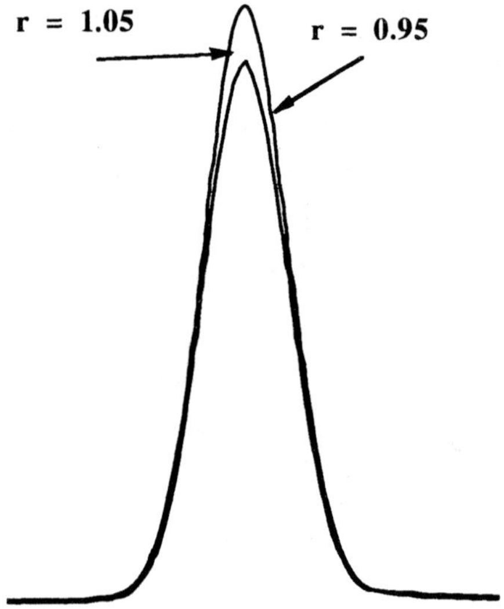

Figure 6 Two Identical Gaussian Curves Drawn with Response Indices of 0.95 and 1.05

It is seen that there is a clear difference between the peaks, and the error involved is quite considerable, as discussed in detail in Chapter 4. Thus, in order to obtain a value that truly reflects the solute concentration being measured, the detector output must not only be corrected for the linearity constant, but also the numerical value of the response factor. This is not a difficult calculation and should be automatically incorporated into any data processing software that is not dealing with the *ideal* detector. The equation given for detector response in Chapter 4 is

$$V = Ac^r$$

where (V) is the voltage output from the detector,
(A) is the linearity constant,

and (r) is the numerical value of the response index. Consequently the output that would truly represent the solute concentration would be

$$c = \sqrt[r]{\frac{V}{A}} \qquad (8)$$

or

$$c = A'\sqrt[r]{V} \quad \text{where } A' = \frac{1}{\sqrt[r]{A}}$$

If (r) is unity, then the expression reduces to the simple relationship that is normally used in quantitative analysis,

$$c = \frac{V}{A} \qquad (9)$$

Unfortunately, (r) is not usually unity and may differ from unity sufficiently to require equation (8) to be used as opposed to equation (9) in order to achieve the necessary accuracy. Equation (8) can be employed to calculate the effect of the response index on the level of the component as calculated assuming true linearity. The results are shown in table 1. It is seen that depending on the relative peak heights of the standard and the solute of interest the result may be greater or less than the true values.

Table 1 Effect of Response Index on Apparent Component Level

% of Component as Measured Assuming Linearity

% 1st Peak	r = 0.92	r = 0.94	r = 0.96	r = 0.98	r = 1.00	r = 1.02	r = 1.04	r = 1.06	r = 1.08
90	88.3	88.75	89.18	89.60	**90.0**	90.38	90.76	91.13	91.47
80	78.16	78.64	79.10	79.55	**80.0**	80.44	80.87	81.30	81.72
70	68.56	68.92	69.28	69.64	**70.0**	70.35	70.71	71.06	71.40
60	59.21	59.41	59.61	59.81	**60.0**	60.19	60.39	60.58	60.78
50	**50.0**	**50.00**	**50.0**	**50.00**	**50.0**	**50.00**	**50.0**	**50.00**	**50.0**
40	40.78	40.59	40.39	40.19	**40.0**	39.81	39.61	39.42	39.22
30	31.44	31.08	30.71	30.36	**30.0**	29.64	29.29	28.94	28.60
20	21.83	21.36	20.90	20.45	**20.0**	19.56	19.13	18.70	18.28
10	11.70	11.25	10.82	10.90	**10.0**	9.94	9.24	8.87	8.53

The data in table 1 can also be presented in the form of % error and these are shown in table 2.

Table 2 Effect of Response Index on Component Error

% 1st Peak	% Error of Component Assuming Linearity								
	r = 0.92	r = 0.94	r = 0.96	r = 0.98	r = 1.00	r = 1.02	r = 1.04	r = 1.06	r = 1.08
90	-1.70	-1.25	-0.82	-0.4	0	0.38	0.76	1.13	1.47
80	-1.83	-1.36	-0.90	-0.45	0	0.44	0.87	1.30	1.72
70	-1.44	-1.08	-0.72	-0.36	0	0.35	0.71	1.06	1.40
60	-0.79	-0.59	-0.39	-0.19	0	0.19	0.39	0.58	0.78
50	0	0	0	0	0	0	0	0	0
40	0.78	0.59	0.39	0.19	0	-0.19	-0.39	-0.58	-0.78
30	1.44	1.08	0.71	0.36	0	-0.36	-0.71	-1.06	-1.40
20	1.83	1.36	0.90	0.45	0	-0.44	-0.87	-1.30	-1.72
10	1.70	1.25	0.82	0.90	0	-0.36	-0.76	-1.13	-1.47

It is seen that absolute errors of 1.7% at the 10% level can be realized and in relative terms this would be a 17% error. It is clear that if the response index is outside the range 0.98.< r <1.02, then appropriate corrections should be applied to the results.

Chromatographic Data Processing

Chromatographic data can be processed either manually or by computer and, although virtually all modern chromatographs are provided with data acquisition and processing, there are still a considerable number of instruments in use that require manual processing. Consequently, both computer and manual processing will be discussed. As the logic of the computer software is based on manual measurement techniques, the manual procedure will be discussed first.

Manual Data Processing

The information required from the chromatogram is basically an accurate measure of, 1/ the retention time, 2/ the peak height, 3/ the peak width. The retention time is used for peak identification and the peak height and peak width for quantitative evaluation. In fact, it is the peak area that is often required, but manual measurement of peak area, although possible, is difficult and a compromise is usually reached by taking the peak area as proportional to the product of the peak height and the peak width measured at the points of inflexion. This assumption is theoretically sound provided the peak is Gaussian or near Gaussian in shape. A typical peak on which measurements might be made is shown in figure 7. Measurements, other than the peak width, are normally made directly on the chart using a good quality steel rule and should be estimated to the nearest 0.2 mm (if necessary using a lens or loupe). If capacity ratios ((k') values) are used for identification purposes, then the chart speed or flow rate should be adjusted so that the distance between the injection point and the dead point is at least 2 cm. This is necessary to ensure that the measurement error will be no greater than 1% (*i.e.* 0.02/2 x 100).

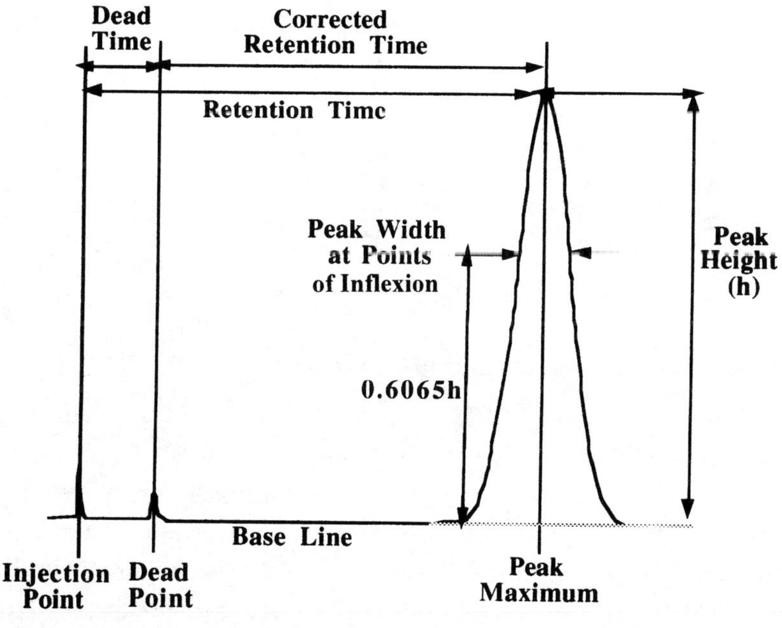

Figure 7 Peak Measurements

The most precise measurements must be made on the peak width and, as the peak width is frequently only a fraction of a centimeter wide, such measurements can not be made sufficiently accurately by means of steel rule. In some cases measurements can be made more precise by increasing the chart speed.

However, when capillary columns are employed the peak widths can be so narrow that even at high chart speed the peaks are still less than a centimeter in width. Under such circumstances it is advisable to use a comparator with a graticule that is calibrated in at least 0.1 mm. Comparators are simple to use and give accurate results, but care must be taken over the measurement procedure. A diagram depicting the measurement of peak width is shown in figure 8.

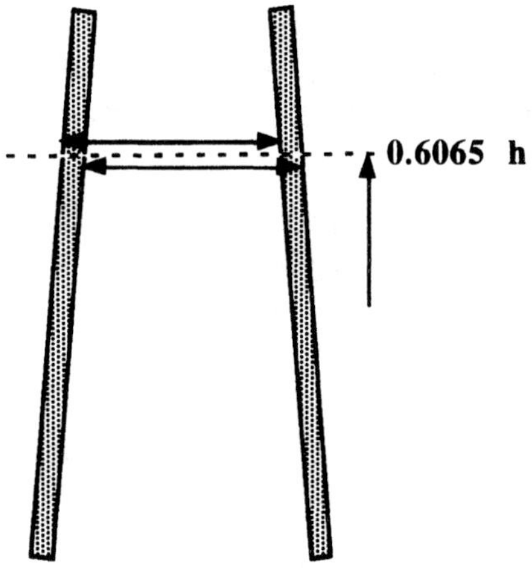

Figure 8 Measurement of Peak Width

The measurements should be taken from the inside of one line to the outside of the adjacent line, in order to eliminate errors resulting from the finite width of the ink line drawn on the chart. The measurement should be repeated using the alternate edges of the line and an average taken of the two readings to avoid errors arising from any variation in line thickness. It is recommended that three replicate runs should be made and the three values should not differ by more than 3%. Most data acquisition systems have software that will also measure peak widths but this will be discussed later.

Peak Area Measurements

The area of a peak is the integration of the mass per unit volume (concentration) of solute eluted from the column with respect to time. If the flow rate is constant, and the detector is a concentration sensitive device, the integration will also be with respect to volume *flow* of mobile phase through the column. It follows that the area of the peak is proportional to the total mass of solute. Measurement of peak area accommodates peak asymmetry and even peak tailing without compromising the simple relationship between peak area and mass. Consequently, peak area measurements give more accurate results under conditions where the chromatography is not ideal, the peak profiles are not truly Gaussian or Poisson, or described by some other specific function.

Unfortunately, whether a chart recorder is used and measurements are made directly on the chart, or computer data acquisition and processing are available, both systems are based on time as the variable and not volume of mobile phase flowing through the column. It follows that when using peak area measurements, a high quality flow controller (in the case of GC) or a high quality high pressure pump (in the case of LC) must be used to ensure that all *time* measurements are *linearly related* to volume *flow of mobile phase*. It is interesting to note that the flame ionization detector (FID) (the most commonly used detector for quantitative analysis in GC) responds to *mass* of solute entering it per unit time and thus the peak area integrated with respect to time is truly related to solute mass. This highly desirable property makes the accuracy of the quantitative analysis relatively indifferent to small changes in flow rate. Unfortunately, there is no equivalent detector in LC or, at least, not one that is commercially available.

Peak areas can be measured manually in a number of ways, the simplest being the product of the peak height and the peak width at 0.6065 of the peak height (2σ). This does not give the true peak area, but providing the peak is Poisson, Gaussian or close to Gaussian it will always give accurately the same proportion of the peak area. One of the older methods of measuring peak area was to take the product of the peak height and the peak width at the base (4σ). The peak width at the base is the distance between the points of intersection of the tangents drawn to the sides of the peak with the baseline produced beneath the peak. This technique has the disadvantage that the tangents to the elution curve must also be constructed manually, which introduces another source of error into the measurement. Another method involves the use of a planimeter (an instrument that provides a numerical value for the area contained within a perimeter traced out by a stylus). This is a very tedious method of measuring peak area and (partly for the same reason) is also not very accurate. The most accurate manual method of measuring peak area (which, unfortunately is also a little tedious) is to cut the peak out and weigh it. A copy of the chromatogram should be taken and the peaks cut out of the copy and weighed. This last procedure can accommodate any type of peak malformation and still provide an accurate measurement of the peak area. It is particularly effective for skewed or malformed peaks, where other methods of manual peak area measurement (with the exception possibly of the planimeter) fail dismally and give very inaccurate results. Cutting the peaks out and weighing them is, without doubt, the most accurate manual method for measuring peak areas. For general use, the recommended method is to use the product of the peak height and the peak width at 0.6065 of the peak height, but this does require adequate resolution of the components of the mixture. Depending somewhat on the chromatographic resolution, peak area masurements can be expected to give a precision of about ±0.5%.

Peak Height Measurements

As the peak height is inversely related to the peak width, then, if peak heights are to be used for analytical purposes, all parameters that can affect the peak width must be held constant. This means that the capacity factor of the solute (k') must remain constant and, consequently, the solvent composition is held stable. The temperature must also be held steady and a highly repeatable method of sample injection must be used. If computer data acquisition and processing are employed, then a direct printout of the peak heights is obtained and, with most systems, the analysis is calculated and also presented. If the peak heights are to be measured manually, which even today is the procedure used in many chromatography analyses, then the baseline is produced beneath the peak and the height between the extended baseline and the peak maximum measured. If peak heights are being measured, then a minimum resolution must be achieved and mathematical adjustments are not an alternative to chromatographic resolution. In theory there should be no part of an adjacent peak situated under the maximum of the peak height to be measured. However, this is impossible due to the nature of the Gaussian function. In practice the peaks should be separated sufficiently to allow a maximum error of 1.0% in peak height measurement.

It follows that the extent of overlap will affect the relative magnitude of the height of a small peak to a greater extent than that of a larger. If the peaks are of equal height, then the criteria of 1% error is met by a separation of about three standard deviations.

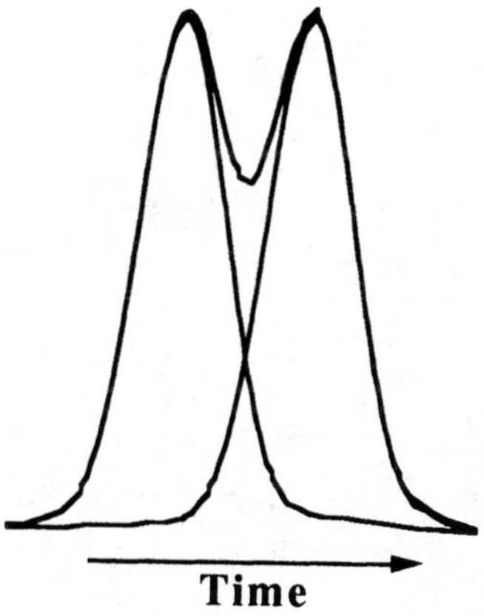

Figure 9 Two Peaks Separated by Three Standard Deviations

This situation is depicted in figure 9. It is seen that the overlap of the two peaks, at the point of the peak maximum of the first peak, is very slight. In fact, if the height of the first peak is taken as unity, the height of the envelope of the two peaks at the point of the maximum of the first peak is still only 1.011. It is interesting to determine the resolution required to achieve a minimum error of 1% in the peak height measurement, for a range of mixtures of differing composition. In figure 10 the minimum separation in terms of the standard deviation of the first peak is plotted against the percentage error for different binary mixtures.

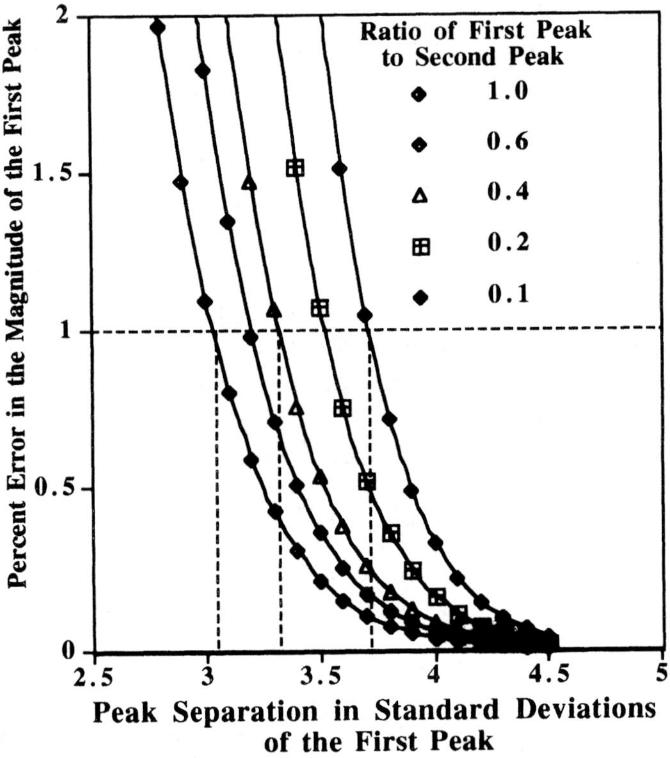

Figure 10 Resolution Required to Maintain a Maximum Error of 1% Employing Peak Height Measurements

It is seen that, as one might expect, the smaller the minor peak, the greater must be the resolution to maintain a maximum error of 1%. When the two peaks are equal in height, they need only be separated by 3.1σ; however when the smaller peak is only about 60% of the second larger peak, the necessary separation increases to about 3.35σ. At the extreme where the minor component is only about 10% of the larger peak, the resolution must be about 3.7σ. Nevertheless the degree of resolution would appear to be significantly less than that required for the accurate measurement of peak areas. This is particularly true if the sizes of the two peaks are significantly different. It follows that peak heights are still frequently employed in quantitative analysis, particularly where a chromatograph is being used that does not include data acquisition and processing equipment. Depending somewhat on the chromatographic resolution, peak height masurements can be expected to give a precision of about ±1%.

Computer Data Processing

Contemporary chromatographs will usually include a dedicated computer which will also control the operation of the chromatograph. In large laboratories, chromatographic data acquisition and processing for a number of instruments are sometimes handled by a central computer. Central computers can time share with many different chromatographs and can also provide laboratory housekeeping functions, such as monitoring the progress of samples through the laboratory from admission to the final report. The analyst does not need to know the details of the necessary electronic circuitry, nor the format of the software that is used in data processing. Notwithstanding, some perception of the logic behind the acquisition process, and the mathematical algorithms that are used, may help to identify the accuracy limits and the precision that can be expected from analysis.

Data processing can be logically divided into two stages; first, the analog data is converted into digital form, acquired and stored in an appropriate memory; second, the digital data is recalled, processed and the results reported in a chosen format.

The Acquisition of Chromatography Data

A diagram showing the individual steps involved in acquiring and processing chromatography data is shown in figure 11. After processing the output from the sensor, the detector provides an analog signal that is passed to the scaling amplifier and then to an analog-to-digital converter (A/D converter). The signal from the A/D converter, now the digital equivalent of the analog signal, then passes via an appropriate interface to the computer and is stored in memory. Either in real time, or at the end of the analysis, the computer recalls the data, carries out the necessary calculations and displays the result on a monitor and/or printer.

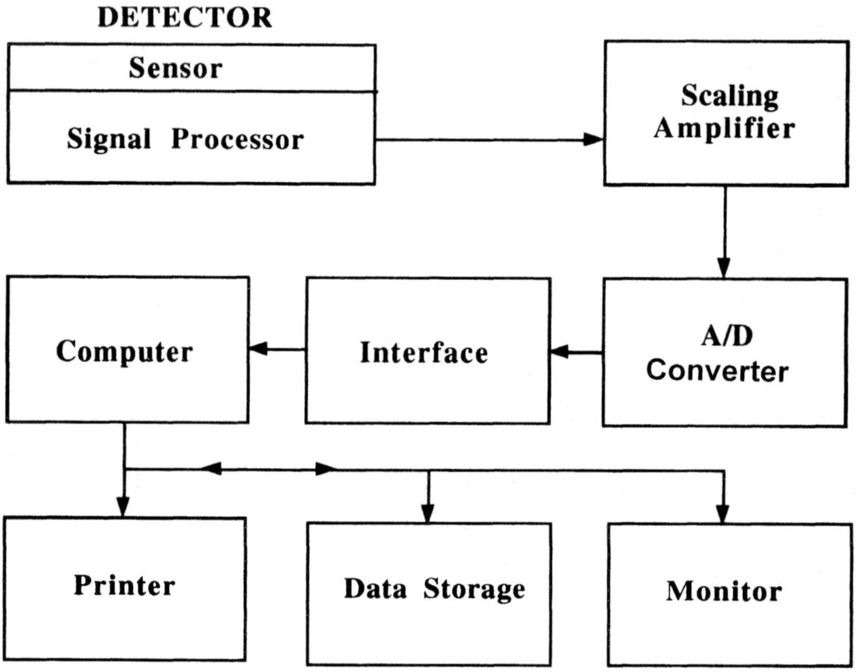

Figure 11 Block Diagram of a Typical Chromatography Data Acquisition System

The output from most detectors ranges from 0 to 10 mV, whereas the input to most A/D converters is considerably greater, *e.g.*, 0 to 1.0 V. Thus, the instantaneous measurement of 2 mV assumed from the detector must be scaled up to 0.2 V, which is carried out by a simple linear scaling amplifier.

After scaling, the signal must be digitized. There are a number of ways to digitize an analog signal and among others these will include sample and hold successive approximation conversion, the dual slope integrating converter, the single slope integrating converter, and the voltage to frequency (V/F) converter. A detailed discussion of the advantages and disadvantages of different A/D converters is not germane to the subject of this book and consequently only the simple V/F converter will be discussed as an example. Readers wishing to know more about the subject of A/D converters are referred to the *Data Acquisition and Conversion Handbook* [4]. A diagram showing the operating principle of an V/F type A/D converter is shown in figure 12. The converter consists of an integrator that can be constructed from an operational amplifier with a feedback capacitor. The capacitor is charged by the voltage from the scaling amplifier through the operational amplifier.

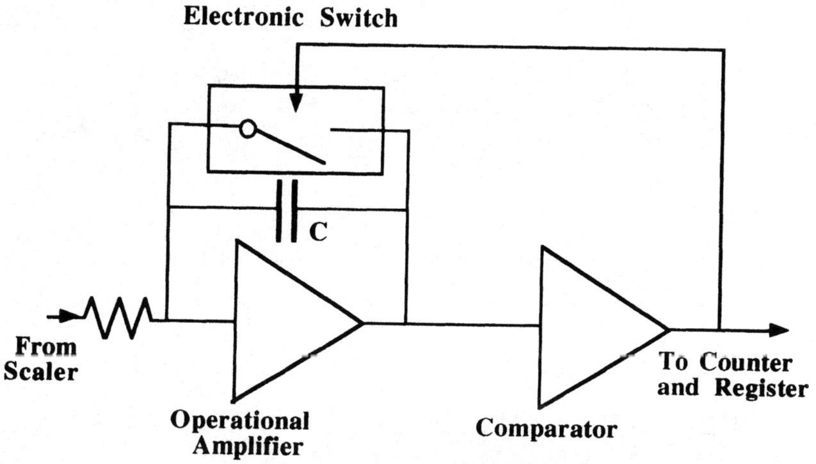

Figure 12 The Basic V/F Analog–to–Digital Converter

The output from the integrator is sensed by a comparator which activates the electronic switch when the potential across the capacitor reaches a preset voltage. The activation of the comparator also causes a pulse to be passed to a counter and at the same time the capacitor is discharged by the electronic switch. The process then starts again. The time taken to charge the capacitor to the prescribed voltage will be inversely proportional to the applied voltage and consequently the frequency of the pulses from the comparator will also be directly proportional to the applied voltage. The frequency of the pulses generated by the voltage controlled oscillator is sampled at regular intervals by a counter which then transfers the count in binary form to a register. The overall system is shown diagramatically in figure 13.

As already stated, the output from most detectors ranges from zero to ten millivolts and the input range of many A/D converters is usually from zero to one volt. Thus, the instantaneous measurement of 0.2 mV from the detector must be scaled up by a factor of 100 to 0.2 volts, which is carried out by the scaling amplifier. Now the A/D converter changes the analog voltage to a digital number, the magnitude of which is determined by the number of "bits" that the computer employs in its calculations.

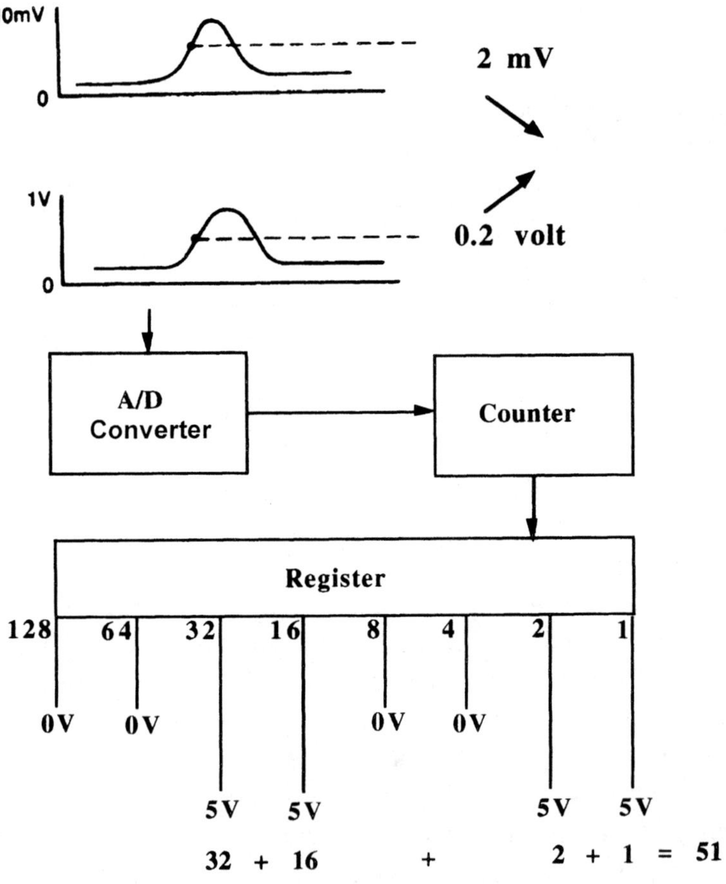

Figure 13 Stages of Data Acquisition

If, for example, eight bits are used, the largest decimal number will be 255. The digital data shown in figure 13 can be processed backward to demonstrate A/D procedure. It is seen that the third and fourth most significant "bits" (which are counted from the far left) and the two least significant "bits" (which are counted from the far right) are at the five volt level (high), which as shown in figure 13 is equivalent to 51 in decimal notation (32+16+2+1). It follows that the voltage that was converted must be $\frac{51}{255}$ x 1 volt = 0.2 volt. It should also be noted that because of the limitation of 8 "bits", the minimum discrimination that can be made between any two numbers is $\frac{1}{255}$ x 100 ≈ 0.4%. It follows that 8 bit systems are rarely used today and contemporary A/D converters usually have at least 12 or 18 bit outputs. Nevertheless, as most analysts are aware, the precision of any measurement, at whatever level, 50%, 5% or 0.5% is rarely better than 0.4%, *e.g.* 50 ± 0.2%, 5.0 ± 0.02% or 0.5 ± 0.002%, so in realistic terms, 8 bits of data is usually quite adequate. As a matter of interest 8 bit systems can be very inexpensive (similar to the cost of a hand calculator) and can be completely satisfactory for the vast majority of chromatographic analyses.

The rate of data acquisition is also important and a minimum of 20 data points per peak needs to be taken to ensure a satisfactory peak shape. Peaks containing less than 20 data points will be seriously distorted and give poor accuracy. There is an approximate rule of thumb for assessing the minimum data acquisition rate (f_A) given by the following expression:

$$f_A = \frac{20}{w_B}$$

where (w_B) is the width at the base, in seconds, of the narrowest peak.

Transmission of the Data to the Computer

After the analog signal has been converted to digital form, the next step is to make the digital data available to the computer. There are two major modes of signal transmission: serial transmission and parallel transmission.

In the serial mode, the digital word (number) is sent to the computer one bit at a time. Now a binary counter provides a parallel output since each of the output bits has its own data output channel and the value of each output bit is simultaneously available. To use a serial transmission scheme, this parallel output must be put into serial form. One way to accomplish this is to use a Universal Asynchronous Receiver/Transmitter (UAR/T). The detailed operation of the UAR/T will not be given here as it is not germane to the subject of this book. It is sufficient to say that the heart of the UAR/T is a shift register and the shift register is strobed by a signal from the computer that displaces the binary number, bit by bit, sequentially from the register to the computer.

The serial data transmission mode finds its greatest use in multiple detector/converter systems where data must be sent over moderate to long distances to the computer. The system is easy to implement, has a good noise immunity, and is reliable. The main disadvantage to the serial system is its moderate speed of transmission, which for most presently operating chromatographs was about 3000 bytes/sec when manufactured. Transfer speeds, however, have been rapidly increasing with the introduction of modern high speed solid state devices and today this is not a problem. In any event, the relatively low transfer speeds of earlier instruments should still be satisfactory for the vast majority of chromatography applications.

In parallel transmission the outputs of the counter are connected directly to a peripheral interface adapter (PIA) and thence to the data bus of the computer. The computer data bus is a parallel system of conductors by which the binary data is transferred between the central processor, memory and peripheral circuits. As the data bus is used for all data transfer, and each transfer involves different data levels, the data bus can not be continuously connected to the register of the A/D output. The isolation is achieved by means of a series of dual input *and*-gates with tri-state outputs, one gate for each bit. The *and*-gate only allows the data on the input to appear at the output on reception of a signal from the computer. An *and*-gate is a device, the output of which will be identical with the logic state of the input on applying a logic 1 voltage to the second input (*i.e.* the output of an and-gate will only go high when both inputs are high). The operation of the and-gate system is shown in figure 14. In the upper diagram the digital output from the A/D converter appears on one input of each and-gate and, as the second inputs are inactive, the output of all gates are logical zero. On activation of the second input of each gate with a logical 1 voltage from the computer, the logic state of the output of the gates are exactly the same as their first inputs. As soon as the data have been read into the computer, the activating voltage (the enabling pulse) is removed. The output of the gates is zero and the digital output from the A/D converter is again isolated from the computer. The data now located in the accumulator of the computer are then transferred either to the computer random access memory (RAM), to a buffer store or to disk.

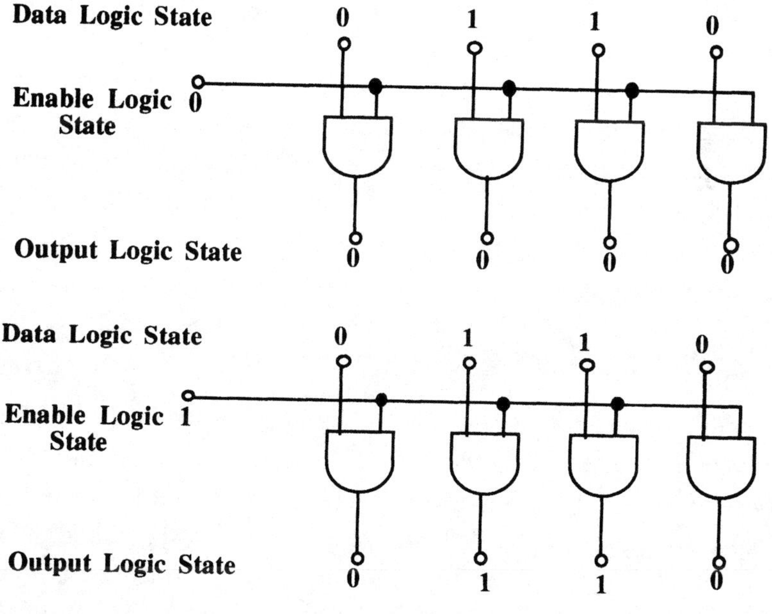

Figure 14 The Function of the *And*-Gates in the PIA

The parallel data transfer mode is simple to use and allows for greater flexibility than the serial mode. However, it requires numerous connections between the PIA and the A/D Converter, and is therefore limited to those cases where the converter and computer are relatively close. This type of transmission is common in chromatographs which have a built-in dedicated computer.

When the separation is complete, all the chromatographic information is stored in the computer memory as a series of binary words. Each binary word represents the detector output at a given time and the period between each consecutive word represents the same interval of time. Consequently, the computer records the concentration of solute passing through the detector at regular intervals of time throughout the separation. Thus, the chromatogram can be reconstructed by simply plotting detector reading against time (which in fact is the reconstructed chromatogram). If the chromatogram is registered on a potentiometric recorder, as opposed to a printer, the output must be passed through a D/A converter to recover the original analog signal. Typically, one chromatographic data point will occupy 2 words in memory (which for an 8 bit machine would require 16 bits). Consequently, a chromatogram 20 minutes long acquired at 5 data points per second would occupy 12,000 words. Part or all of the chromatogram can be reconstructed or, if so desired, certain portions can be expanded to permit more detailed examination of those areas of particular interest. Once the data is in memory, many different types of calculations can be carried out to help identify the components or carry out quantitative analysis.

Data Processing and Reporting

Data from the chromatograph can be handled in two ways. The data can either be processed in real time as it arrives at the computer, and only the results of the calculations, such as retention time, peak height and peak area, etc., are stored, the raw data being discarded. This is called "on the fly" processing and has the dubious advantage of providing results immediately after the peak is eluted. This procedure requires much less memory and, in the days when memory was limited and expensive, it was an attractive alternative. Today when many megabytes of RAM are available together with thousands of megabytes of disk storage, this procedure is rarely used. The big disadvantage is that the deletion of the raw data prevents further or alternative data processing if, subsequently, it is shown that the results require it. The only application where this procedure might still be useful is in process control where results are required rapidly in order to control a production process to ensure optimum yield and maximum purity.

The alternative, and far more advantageous, form of data processing is to store each detector output value as it is received (usually on a hard disk) and then process the results after the analysis has been completed. This method gives complete flexibility in data reduction, allows the chromatogram to be reconstructed in a number of alternative forms and provides different algorithms selected for baseline correction or peak area measurement, etc. It also allows specific portions of the chromatogram to be amplified to show the presence of trace materials.

Data Processing

The first function the software must achieve is peak detection. The start of the peak can be identified by a significant change in detector output (the level of the change being defined by the user in the initial set up for the analysis) or by the rate of change of the detector output (the rate also being defined prior to the analysis).

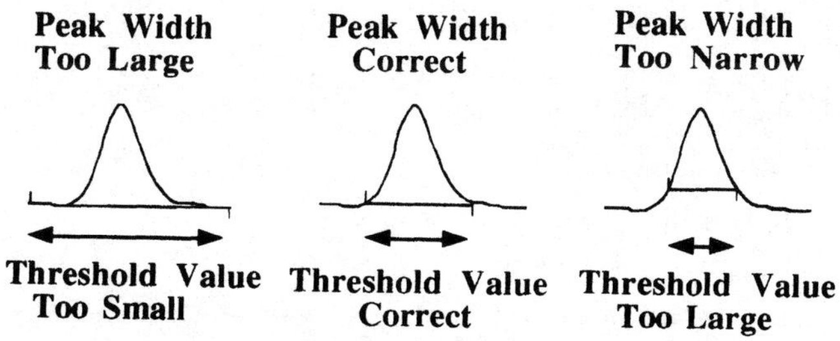

Figure 15 The Effect of Threshold Value on Apparent Peak Width

One of the essential parameters selected for data processing is the *threshold value,* which is the signal level above the baseline noise that the data processing system must recognize to define the start of the peak. This choice is critical and determines the accuracy of the analysis. The effect of the choice of threshold value is demonstrated in figure 15. It is clear from figure 15 that the threshold value must be very carefully set. if the computer is to identify the correct peak width and give the correct value for the peak height. Unfortunately the choice is often a trial and error iterative procedure, based on the results obtained from processing a synthetic mixture of known composition.

If peak areas are being measured, the start of the peak defines the point where integration must commence. In the simplest form integration involves summing the detector output from the peak start to the peak end. As the data are sampled at regular intervals, the sum will give a true value that is proportional to the peak area. The end of the peak is identified in the same way, but if the baseline is sloping, then the computer will extrapolate the baseline across the peak and calculate the level at which the peak can be assumed to have reached the baseline.

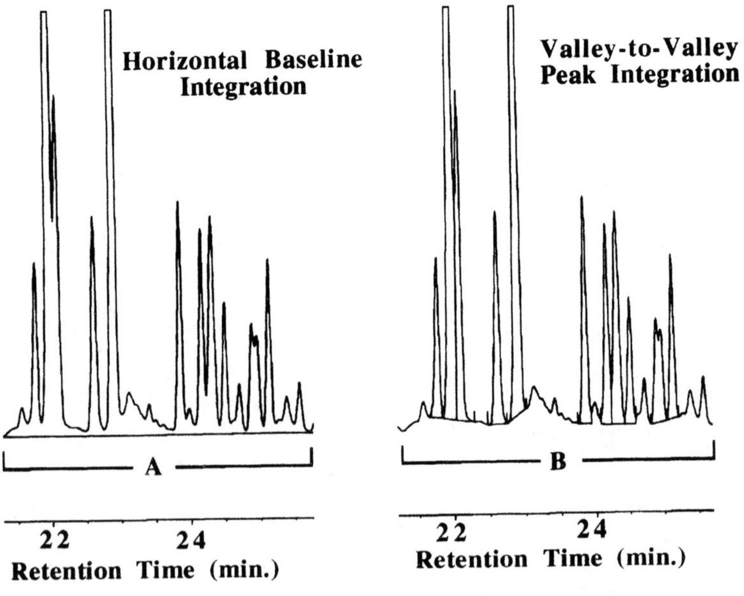

Figure 16 Horizontal and Valley-to-Valley Integration Protocols

This procedure will be discussed shortly. However, the type of baseline that is used is also important and two types are often made available for processing software. These are horizontal baseline integration and valley-to-valley integration. Both types are shown in figure 16. Wherever possible the horizontal baseline integration should be chosen as it obviously provides the more accurate results.

The next important parameter to identify is the peak maximum. There are a number of ways of doing this, but the simplest is to subtract consecutive values at (n) and (n – 1) between the peak start and the peak finish and determine when

$$S_{(n)} - S_{(n-1)} \text{ becomes negative.}$$

The is a very simple example and would only work if there was no noise on the signal, which in practice is never the case. Often, some smoothing routine is carried out on the points proximate to the peak maximum and then peak identification is repeated. In this way an accurate measure of the position of the peak maximum can be obtained. The certain identity of a component depends on the accuracy of the retention time measurement and thus the correct identification of the position of the peak maximum is critical. The peak height is taken as the difference between the signal at the point of the peak maximum and that directly beneath the peak maximum on the baseline produced under the peak. The projected baseline under the peak is calculated using a procedure similar to that described below. Thus the data processing has provided the position of the start of the peak, the position of the peak maximum, the position of the end of the peak, the peak area and the peak height. One method for constructing the baseline under a peak when the peak is eluted on a sloping baseline is as follows. In figure 17 a peak is depicted eluted on a sloping baseline. The average value of the baseline (Y_H) is obtained by the following simple relationship,

$$Y_H = \frac{Y_1 + Y_n}{2}$$

where (Y_1) is the detector signal at the start of the peak,
and (Y_n) is the detector signal at the end of the peak.

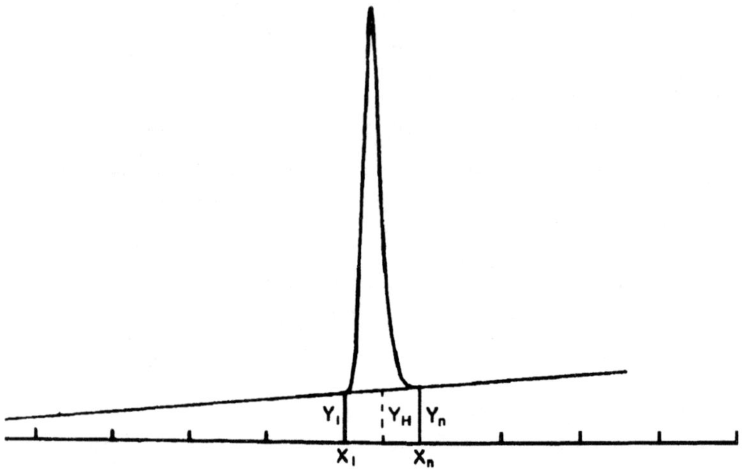

Figure 17 **Baseline Correction**

In general, the slope is usually considered linear under the peak and, in fact, if it is not, the chromatographic operating conditions probably need adjustment. Under some circumstances (*e.g.* at the end of a high-temperature program) the slope may not be linear and adjusting the operating conditions will not help. Thus, the curve of the baseline under the peak can be approximated to a second order polynomial function. If the slope of the baseline under the peak is linear, then the extrapolated baseline under the peak will obviously be given by

$$Y_{n'} = Y_1 + \frac{Y_n - Y_1}{n_n - n_1}(n_{n'} - n_1)$$

where (Y_1) is the data point at the peak start,
(n) is the data point at the peak end,
(n') is the data point between the peak start and peak end,
and $Y_{(n')}$ is the baseline at the data point (n').

Consequently the peak area (A) will be calculated as

$$A = \sum_{p=1}^{p=n'} Y_p - Y_1 + \frac{Y_n - Y_1}{n_n - n_1}(n_p - n_1)$$

The data processing techniques so far discussed are basic and relatively simple. A more detailed treatment is not appropriate here but alternative procedures such as peak skimming will be discussed later. However, there is one important point that should be made. In whatever way the software may manipulate the data, the results can never be more precise or more accurate than the basic data that have been produced by the chromatograph. In turn, this can only be as good as the sample that has been taken and prepared.

Some Simple Computing Techniques

There are a number of techniques employed in the computer calculation of peak areas that should be mentioned. These techniques are used for partially unresolved peaks and are not deconvoluting routines but approximation procedures. The technique used for unresolved peaks of approximately equal heights is illustrated in figure 18. In practice, the area of the first peak is obtained by integrating from the extreme of the envelope to the minimum between the peaks. The area of the second peak is obtained similarly by integrating between the minimum between the peaks to the end of the envelope.

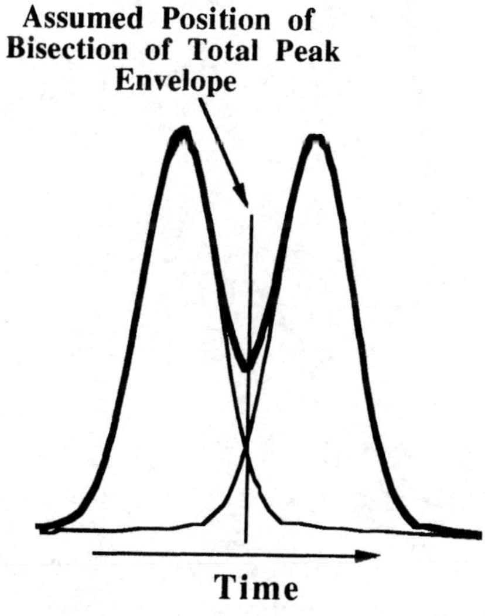

Figure 18 Area Assessment of Two Unresolved Peaks of Equal Height

Manual measurements are taken in the same manner by dropping a perpendicular from the minimum between the peaks to the baseline. The area on either side of the minimum is taken as that for each peak. It is seen that the procedure would give a fairly accurate value for each peak area. In fact the method often gives area measurements that are more accurate than employing the *skimming* procedure.

Peak *skimming* is carried out when resolution is incomplete and it is necessary to estimate the boundary between two peaks or a sloping baseline. A simple example of the comparative use of peak cutting and peak skimming is given in figure 19. The ratio of the area of the small peak to that of the larger is five. In the first instance the area of each peak is taken by integrating between the start of the envelope to the minimum between the peaks and from the minimum to the end of the envelope.

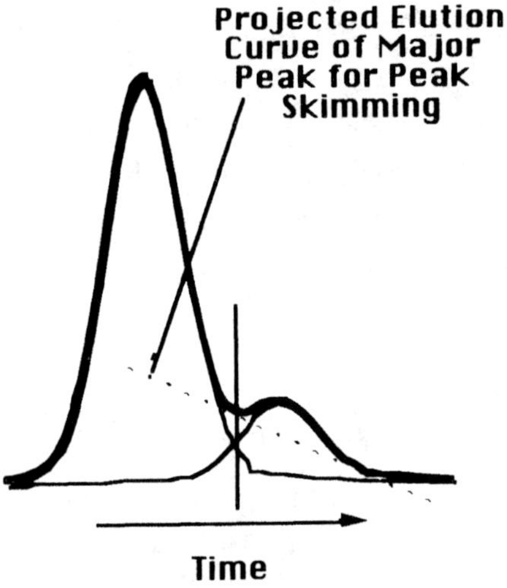

Figure 19 Measurement of the Area of Two Unresolved Peaks by Peak Cutting and Peak Skimming

The same can be achieved manually by dropping a perpendicular from the minimum to the baseline as shown and the area of each peak taken as that on either side of the bisector. It is seen that the division is fairly accurate, the first peak including a small portion of the area of the second peak and *vice versa*. The diagram also illustrates the process of peak skimming. Depending on the software available, the contour of the larger peak is projected under the smaller peak. The area of the smaller peak is taken as the difference between the area of the total envelope and the area of the first peak taken beneath the projected contour. The accuracy of this procedure depends entirely on the method used for projecting the baseline. The method used in the diagram utilizes a linear extrapolation under the second peak and it can be seen that this leads to *extensive* error. Employing a second order polynomial or an exponential extrapolation would probably provide more accurate results. The weakness of this procedure lies in the fact that each unresolved pair of peaks provides a unique contour and thus any standard function that is employed for extrapolation purposes can only be an approximation.

Another problem that often arises is the area measurement of a peak superimposed on a sloping baseline or on the tail of a large overloaded peak. The elution of a small peak on the tail of a large peak is depicted in figure 20. It is seen that the shape of the small peak is extensively distorted and that the use of a linear function to project the curve of the major peak below that of the smaller is virtually useless.

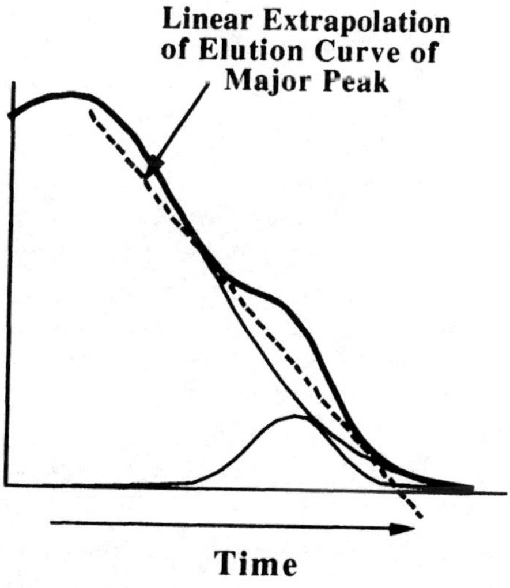

Figure 20 The Elution of a Small Peak on the Tail of a Large Peak

Again a second or third order polynomial or exponential function would certainly provide improved results. However, in this case, accurate values for the peak areas would probably only be obtained by peak deconvolution. The first peak would be reconstructed using parameters obtained from the ascending portion of the peak that is free from contamination from the smaller peak. The area of the first peak would be obtained by direct integration and that of the smaller peak by the difference between the area of the total envelope and that of the first peak. A skimming procedure can often be used when measuring the area of peaks on a sloping baseline and this is probably the most worthwhile utilization of the skimming technique. An example is depicted in figure 21. It is seen that if the slope is small, a linear function might be used to extrapolate the baseline beneath the peak. For baselines with greater slopes (*e.g.* that shown in the diagram) a second order polynomial will usually provide an accurate contour of the baseline below the peak and allow an accurate assessment of the peak area. In the case of manual measurement, the baseline beneath the peak can usually be quite accurately constructed with the aid of a set of French curves.

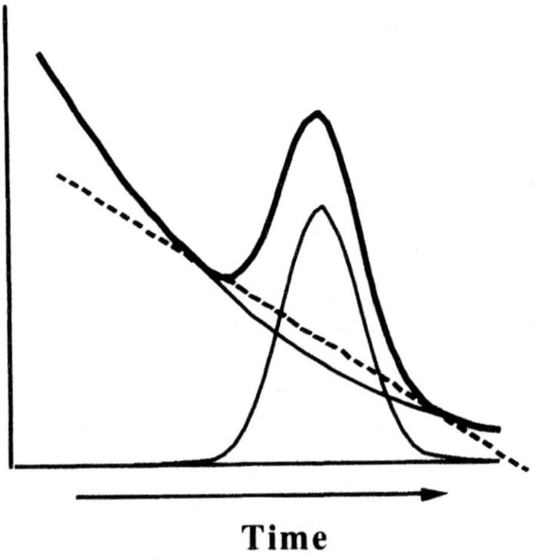

Figure 21 Peak Area Measurement on a Sloping Baseline

It should be pointed out that these subterfuges are almost always taken because the quality of the chromatography is inadequate. Although many and varied correction procedures can be devised to help improve the quantitative accuracy that is obtained from poorly shaped or poorly resolved peaks, this is not the best solution. It is advised that the effort should be put into improving the chromatographic conditions to achieve the necessary resolution and, in doing so, provide not only more accurate data, but also more reliable results.

Chromatographic Control

The computer that handles the output from the detector also provides other information in the analytical report. Today, most chromatographic systems, gas and liquid, have a dedicated computer associated with them which, as well as processing the information provided by the detector, will also control and record the operating conditions of the chromatograph. Temperatures and flow rates programs, etc., will be entered via the keyboard of the computer and the information stored for reporting purposes when required. An example of the print-out from a Perkin Elmer, LC140 liquid chromatograph giving the operating conditions and analytical results is shown in figure 22. Very sophisticated data processing programs are continually being developed providing, apparently, more information or more precise quantitative data from the sample.

Nevertheless, it must again be stressed that the quality of the information provided by the data processing system can only be as good as that which the chromatographic system provides and, specifically, the accuracy and precision of the measurements made by the detector and the integrity of the sample that was initially taken.

Quantitative Analytical Methods for GC and LC

Essentially, there are two procedures used in GC and LC for quantitative analysis: the first employs a reference standard to which the peak areas or peak heights of the other solutes in the sample are compared; the second is a normalization procedure where the area (height) of any one peak is expressed as a percentage of the total area (heights) of all the peaks.

Software Version :4.1(build 143)
Date : 6/7/95 12:44 PM

Sample Name	Tabasco	
Data File	C/TC4/Data/Pepper/Tobasd. Raw	Date 6/4/95 08:54 PM
Sequence File	C/TC4/Data/Pepper/Capsai.Seq.	Cycle: 7 Channel :A
Instrument	LC140 – 4:A Rack File 0/10	Operator
Sample Amount	1.000 Dilution Factor 1.00	

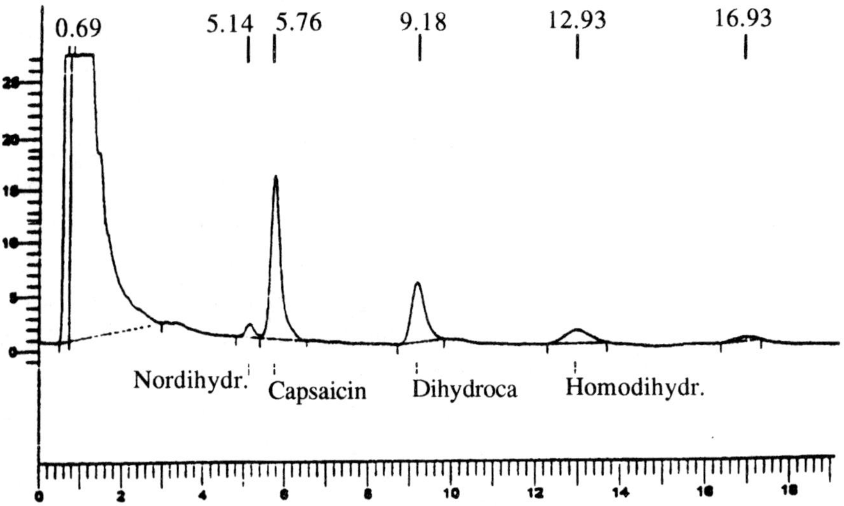

SCOVILLE "HEAT" IN PEPPER REPORT

Peak	Time (min)	Area (µVs	Component Name	Raw NG	λ max	Peak Purity	Capsaicin	Scov val.	Scov. Heat
1	5.138	17342	Nordihydrocapsaicin	3.374	279	1.3	0.001	9.3	78
2	5.759	257942	Capsaicin	50.183	280	1.2	0.013	16.1	2020
3	9.179	126620	Dihydrocapsaicin	24.440	280	1.2	0.006	16.1	984
4	12.931	51739	Homohydrocapsaicin	10.066	280	1.5	0.003	8.1	204
		452645		88.063		.5.1	0.022		3286

Courtesy of the Perkin Elmer Corporation

Figure 2 Analytical Report from Computer Data Processing

The most common method used in both GC and LC is that employing reference standards. However, there are certain circumstances, *e.g.* where the detector employed has the same response to all the solutes in the mixture, where the normalization method may be advantageous.

Quantitative Analysis Using Reference Standards

There are two procedures for using reference standards: in the first method a weighed amount of the standard can be added directly to the sample and the area or heights of the peaks of interest compared with that of the standard. This procedure is called the *internal standard method.* In the second procedure, a weighed amount of the standard can be made up in a known volume of solvent and chromatographed under exactly the same conditions as the unknown sample, but as a separate chromatographic run. In this case the peak area or height of the standard solute in the reference chromatogram is compared to the peak areas or heights of the solutes of interest in the sample chromatogram. This procedure is called the *external standard method.* The details of the two methods will now be described separately.

The Internal Standard Method

The most accurate quantitative results in GC and LC are probably obtained by the use of an internal standard. However, it can sometimes be difficult to employ, as the procedure depends upon finding an appropriate substance that will elute in a position on the chromatogram where it will not interfere or merge with any of the natural components of the mixture. Unfortunately, for a multi-component sample, this can become difficult to the extent of being virtually impossible, under which circumstances the external standard method must be used. However, for simple mixtures, where a standard can easily be introduced into a gap existing in the chromatogram, the internal standard method should be used as it will give the most accurate results. Having identified an appropriate reference standard, the response factors for each component of interest in the mixture to be analyzed must be determined. It should be noted that this usually does not include *all* the components. In many instances only certain components of the mixture are required to be assayed. A synthetic mixture is then made up containing known concentrations of each of the components of interest together with the standard.

Now, if there are (n) components, and the (p) component is present at concentration (c_p), the standard at a concentration (c_{st}) and the response index of the detector can be considered to be unity. Then

$$\frac{c_p}{c_{st}} = \frac{a_p}{a_{st}} \psi_p \tag{10}$$

where (a_p) is the area of the peak for component (p),
(a_{st}) is the area of the peak for the standard,
and (ψ_p) is the response factor for component (p).

Consequently, the response factor (ψ_p) for the component (p) can be calculated as follows:

$$\psi_p = \frac{c_p a_{st}}{c_{st} a_p} \tag{11}$$

If peak heights are used instead of peak areas, then by simple replacement,

$$\frac{c_p}{c_{st}} = \frac{h_p}{h_{st}} \psi_p \tag{12}$$

where (h_p) is the height of the peak for component (p),
(h_{st}) is the height of the peak for the standard,
and (ψ_p) is the response factor for component (p).

Thus, by rearrangement, the response factor (ψ_p) for the component (p) is given by

$$\psi_p = \frac{c_p h_{st}}{c_{st} h_p} \tag{13}$$

A weighed amount of standard is now added to the sample and the sample is chromatographed. Let the concentration of the standard be (C_{st}), the concentration of any component (p) be (C_p) and the peak areas of the component (p) and that of the standard (A_p) and (A_{st}) respectively; then

$$C_p = \frac{A_p C_{st}}{A_{St}} \psi_p \tag{14}$$

If peak heights are employed instead of peak areas,

$$C_p = \frac{H_p C_{st}}{H_{st}} \psi_p \tag{15}$$

where (H_p) is the peak height of component (p) in the sample, and (H_{st}) is the peak height of the standard in the sample.

Now if the response index (r) is not unity, which will be the case for virtually all practical detectors, then the value of (r) should be taken into account in the expressions used to obtain quantitative results. Consider first the computer integration of the area. The area is calculated by summing all the signals from the detector over the period of the peak. Now, if ($r \neq 1$), then the area that will incorporate the non-linearity will be

$$a_{(r)} = \sum \sqrt[r]{V}$$

where ($a_{(r)}$) is the corrected area of the peak for a non-linear detector with a response index of (r).

It should be noted that the areas, as supplied by normal data acquisition and processing software, will not produce this corrected area without program modification.

Thus, equation (10) must be modified to

$$\frac{c_p}{c_{st}} = \frac{a_{p(r)}}{a_{st(r)}} \psi_p \tag{16}$$

and equation (11) must be modified to

$$\psi_p = \frac{c_p a_{st(r)}}{c_{st} a_{p(r)}} \tag{17}$$

Now equation (12) employs peak heights and so the values for the peak height can be taken as those provided by the computer but must be taken to the (r)th root; thus,

$$\frac{c_p}{c_{st}} = \frac{\sqrt[r]{h_p}}{\sqrt[r]{h_{St}}}\psi_p \qquad (18)$$

and equation (13) becomes

$$\psi_p = \frac{c_p \sqrt[r]{h_{St}}}{c_{St} \sqrt[r]{h_p}} \qquad (19)$$

Again, a weighed amount of standard is now added to the sample and the sample is chromatographed. Let the concentration of the standard be (C_{St}), the concentration of any component (p) be (C_p) and the peak areas *corrected for the response index* (r) of the component (p) and that of the standard ($A_{p(r)}$) and ($A_{St(r)}$), respectively; then

$$C_p = \frac{A_{p(r)} C_{st}}{A_{St(r)}}\psi_p \qquad (20)$$

If peak heights are employed instead of peak areas,

$$C_p = \frac{\sqrt[r]{H_p C_{st}}}{\sqrt[r]{H_{St}}}\psi_p \qquad (21)$$

Thus, the concentration of any (or all) of the components present in the mixture can be determined, providing they are adequately separated from one another. It is interesting to note that if the maximum accuracy and precision is required, and the data are to be corrected for a response index that is *other than unity*, then if the computer program has not been written to accommodate detector response factors (although most modern data systems do), then peak heights must be used and the chromatographic data must be processed manually. For repeat analyses of the same type of mixture, the operating conditions can be maintained constant and, as there is no extreme change in sample composition, the response factors will usually need to be determined only once a day.

The External Standard Method

When using the external standard method, the solute chosen as the reference is chromatographed separately from the sample. However, as the results from two chromatograms, albeit run consecutively, are to be compared, the chromatographic conditions must be maintained extremely constant. In fact, to reduce the effect of slight changes in the operating conditions of the chromatograph, the sample and reference solutions are often chromatographed alternately. The data from the reference chromatograms run before and after the sample are then used for calculating the results of the assay. The external standard method has a particular advantage in that the reference standard (or standards) that are chosen can be identical to the solute (or solutes) of interest in the sample. This also means that the relative *response factors* between the reference solute and those of the components of interest are no longer required to be determined. In addition, using the external standard method, the reference mixture can be made to have standard concentrations that are closely similar to the components of the sample. As a consequence, errors that might arise from any slight non-linearity of the detector response are significantly reduced. Assuming, in the first instance that the response index of the detector is unity, if the (p)th solute in the mixture is at a concentration of ($c_{p(s)}$) in the sample and ($c_{p(st)}$) in the standard solution, then

$$c_{p(s)} = \frac{a_{p(s)}}{a_{p(st)}} c_{p(st)} \qquad (22)$$

where ($a_{p(s)}$) is the area of the peak for solute (p) in the sample chromatogram,
($a_{p(st)}$) is the area of the peak for solute (p) in the reference chromatogram,
and ($c_{p(st)}$) is the concentration of the standard in the reference solution.

If peak heights are employed, then $c_{p(s)} = \frac{h_{p(s)}}{h_{p(st)}} c_{p(st)}$ (23)

where ($h_{p(s)}$) is the area for solute (p) in the sample chromatogram,
($h_{p(st)}$) is the area solute (p) in the reference chromatogram,
and ($c_{p(st)}$) is the concentration of the standard in the reference solution.

If the value or the response index, (r), is not unity, then the corrected area and height must be used. Thus again assuming

$$a_{(r)} = \sum \sqrt[r]{V}$$

then equation (15) becomes

$$c_{p(s)} = \frac{a_{p(s)(r)}}{a_{p(st)(r)}} c_{p(st)} \qquad (24)$$

and equation (16) becomes

$$c_{p(s)} = \frac{\sqrt[r]{h_{p(s)}}}{\sqrt[r]{h_{p(st)}}} c_{p(st)} \qquad (25)$$

Theoretically, providing the chromatographic conditions are kept constant, the reference chromatogram need only be run once a day. However, in practice, it is advisable to run the reference chromatogram at least every two hours and, as suggested above, many analysts run a reference chromatogram immediately before and after each sample. It is a common task in many control laboratories to analyze a large number of repeat samples of a very similar nature for long periods of time, often with automatic sampling. Under such circumstances, reference samples and assay samples are often run alternately throughout the whole batch of analyses. Not only will this give more accurate results, but it will also alert the operator to any change in operating conditions, which can then be immediately rectified.

The Normalization Method

The normalization method is the easiest and most straightforward to use and requires no reference standards or calibration solutions to be prepared. Unfortunately, to be applicable, the detector must have the same response to all the components of the sample. Thus, it will depend on the detector that is employed and the sample being analyzed. Unhappily, it is the least likely method to be appropriate for most LC analyses. In GC the response of the flame ionization detector

(FID) depends largely on the carbon content of the solute. Thus the technique can be used in GC when employing the FID for compounds of similar types such as a mixture of high molecular weight paraffins. An exceptional example in LC, where the normalization procedure is frequently used, is in the analysis of polymers by exclusion chromatography using the *refractive index detector*. The refractive index of a specific polymer is a constant for all polymers of that type having more than 6 monomer units. Under these conditions normalization is the obvious quantitative method to use. The method of calculation is also very simple but again will depend on whether the value of (r) is assumed to be unity or not. Assuming, first, that (r) is unity, the percentage $x_{(p)}\%$ of any specific polymer (p) in a given polymer mixture can be expressed by

$$x_p = \frac{a_p}{a_1 + a_2 + a_3 + \ldots\ldots + a_n} 100 \qquad (26)$$

where (a_p) is the peak area of polymer (p),

or

$$x_p = \frac{a_p}{\sum_{p=1}^{p=n} a_p} 100 \qquad (27)$$

If peak heights are used, the percentage $x_{(p)}\%$ of any specific polymer (p) in a given polymer mixture can be expressed by similar equations,

$$x_p = \frac{h_p}{h_1 + h_2 + h_3 + \ldots\ldots + h_n} 100 \qquad (28)$$

where (h_p) is the peak area of polymer (p),

or

$$x_p = \frac{h_p}{\sum_{p=1}^{p=n} h_p} 100 \qquad (29)$$

In most computer data acquisition systems there is an iterative program that allows the user to identify the reference and sample chromatograms and the pertinent peaks for processing. The software will then carry out the necessary calculations and provide the output in the form of a printed analytical report.

However, assuming that ($r \neq 1$) then equation (26) becomes

$$x_p = \frac{a_{p(r)}}{a_{1(r)} + a_{2(r)} + a_{3(r)} + \ldots + a_{n(r)}} 100 \qquad (30)$$

equation (27) becomes

$$x_p = \frac{a_{p(r)}}{\sum_{p=1}^{p=n} a_{p(r)}} 100 \qquad (31)$$

equation (28) becomes

$$x_p = \frac{\sqrt[r]{h_p}}{\sqrt[r]{h_1} + \sqrt[r]{h_2} + \sqrt[r]{h_3} + \ldots + \sqrt[r]{h_n}} 100 \qquad (32)$$

and equation (29) becomes

$$x_p = \frac{\sqrt[r]{h_p}}{\sum_{p=1}^{p=n} \sqrt[r]{h_p}} 100 \qquad (33)$$

For the most part, contemporary computer chromatography software does not provide for situations where ($r \neq 1$) so the analyst can only use peak height data from the program and make the corrections manually. Alternatively, the chromatogram can be processed completely manually in which case peak areas can be used. The magnitude of any error that can arise from not correcting for (r) when it is other than unity can only be accessed if and when the actual value of (r) is determined in the manner discussed previously. Unfortunately, as stated previously, very few instrument manufacturers provide a value for (r) and thus the response index will usually need to be determined experimentally by the user.

Quantitative Analysis by TLC

TLC is not the best chromatographic technique for quantitative analysis and, although it can provide quantitative results, the necessary procedure tends to be more cumbersome and tedious compared with other chromatographic methods. Furthermore, for accurate work, expensive scanning equipment is required, which, as already discussed, rather reduces the cost advantage of the technique.

In principle there are three approaches used in quantitative TLC: extraction of the spot and separate measurement by spectroscopic or other techniques, comparative techniques employing visual assessment and finally optical scanning. The first, for obvious reasons, is rarely used and, due to the difficulties involved in the extraction of the material from the spot, gives relatively poor accuracy.

Comparative Spot Assessment by Visual Estimation

The quantity of material contained in a TLC spot can not be estimated visually, as the retinic response of the human eye is not linear and varies widely with the wavelength of the reflected light. Visual estimation is further complicated by the variation of the iris with light intensity. The eye, however, is very sensitive to slight *differences* in light intensity and therefore can be used very effectively for *comparing* light intensities. In fact, in the early twentieth century, the majority of trace analyses (*e.g.* oxygen in boiler water, heavy metal contaminants such as chromium, manganese and cadmium, alkaloids, etc.) were carried out by forming colored derivatives and visually comparing the color with that of standards.

In TLC, a similar procedure is employed. The sample is run in parallel with a series of calibration solutions each containing the solute of interest at different concentrations. The first set of calibration standards can have relatively large concentration intervals, which allows the approximate concentration of the solute of interest to be identified. A second set of standards is then made up that embraces the suspected concentration in smaller intervals and the separation is repeated. By matching the intensity of the spots, the concentration of the sample can be estimated. The TLC spots can not be rendered visible by a destructive process (*e.g.* concentrated sulfuric acid) as such processes are not quantitative. If the solute of interest is colorless, a suitable colored derivative must be formed either before separation (in which case a special eluting solvent must be identified) or more commonly after the separation by treating the plate with an appropriate reagent. Providing the separation is clean and a suitable colored derivative is employed, the visual comparative procedure can give quantitative results with an accuracy of about +/-10%. Today, such accuracy is generally considered inadequate and as a result, the visual comparative technique has been largely replaced by instrumental methods of spot density measurement. Scanning densitometry will be discussed in another chapter but it was clear that this method of quantitative measurement of a TLC plate would be far superior to that of color matching. Nevertheless, the necessary expense involved in the scanning instrument, together with its limited accuracy, makes liquid chromatography a very strong competitor.

Synopsis

Quantitative analysis can only be successful if the peaks of interest are adequately resolved. In general peaks that are separated by four standard deviations can be quantitatively assessed. However, if the peaks are very different in size, then a greater resolution may be necessary. Using the Purnell equation the actual efficiency needed for adequate resolution can be calculated. In a similar way the number of theoretical plates required by a thin layer plate can also be estimated. If peaks are unresolved, deconvoluting software is available that will allow the retention times and areas of unresolved peaks to be calculated. However, such techniques are not an alternative to good chromatography and adequate resolution. If the response index of the detector is significantly greater than unity, it should be used to correct peak heights and peak areas, otherwise significant errors may arise. Chromatograms can be manually processed using retention distances, peak heights and the peak width at 0.6975 of the peak height. In most modern chromatographs the data are acquired by a computer and processed using suitable software. If peak height measurements are used for quantitative analysis, then often the resolution necessary will be less than that required for peak area measurements, particularly if the separation ratios are small. In computer data processing the detector output is amplified by a scaling amplifier, digitized and passed either serially or by parallel transmission to the computer where it is stored on disk. 256 bit acquisition will provide a resolution of 0.4% but in modern equipment 12 and 16 bit processing is more common although often not necessary. Data are usually reported by the computer which prints out a summary and will include operating variables as well as the actual quantitative data. Various general techniques are used in computer data processing, which will include baseline extrapolation, composite peak bisection, peak skimming, etc. The two most common analytical methods in chromatography use either internal or external standards. When using an internal standard a reference substance is added in a known amount to the sample and the areas or heights of the peaks of interest are compared with that of the added standard. If an external standard is used, the standard is chromatographed separately and the area of the standard peak on the reference chromatogram compared with the areas of the peaks of interest on the sample chromatogram. If an external standard(s) is used then the same substances can be used as those being measured and thus there is no need to determine the relative response factors. The other method is the normalization method and can only be successful if all the peaks in the chromatogram have the same response factors. Then the percentage of any specific substance is taken as the peak area or height of that substance divided by the sum of all the peak areas or heights multiplied by 100. The quantitative assessment of a thin layer plate is more difficult and can only be achieved by either comparing spot intensities to those on a series of reference plates or by the use of expensive scanning devices. In general, normalization procedures are not satisfactory for quantitative TLC analyses.

References

1. R. P. W. Scott, *Introduction to Analytical Gas Chromatography*, Marcel Dekker, Inc., New York (1998)21.
2. J. H. Purnell, *Nature (London)*, **183**(1959)2009.
3. R. P. W. Scott, *Techniques and Practice of Chromatography*, Marcel Dekker, Inc., New York (1995)23.
4. *Data Acquisition and Conversion Handbook* (Ed., E. L. Zuch), Mansfield, MA (1979).

Part 2

Quantitative Gas Chromatographic Analysis

Chapter 6

Gas Chromatographic Apparatus for Quantitative Analysis

Virtually all contemporary gas chromatographs are designed primarily for quantitative analysis. However, the analyst needs to be aware of those parts of the equipment that are critical from the point of view of accuracy and precision and those procedures that can significantly affect the reliability of an assay. In particular, the analyst should be aware of any specific development techniques that can influence the quality of an analysis.

The modern gas chromatograph is a highly complex, well designed and well presented instrument. Despite its somewhat glamorous appearance, however, it is well to remember that the impressive separations it provides are performed in a simple tube through which passes a stream of gas. The ancillary apparatus, though impressive in appearance, merely supports this simple, but essential, device. All modern gas chromatographs are closely associated with a computer and in some cases there are no manual overrides for any operating variable, and all running conditions can only be set up, or changed, *via* the computer keyboard. The general layout of a contemporary gas chromatograph, together with the separate flow patterns of the gas supply and analog and digital electrical signals are shown in figure 1. Although all the parts are controlled directly by the computer, each functions discretely and thus the individual components will be briefly described separately.

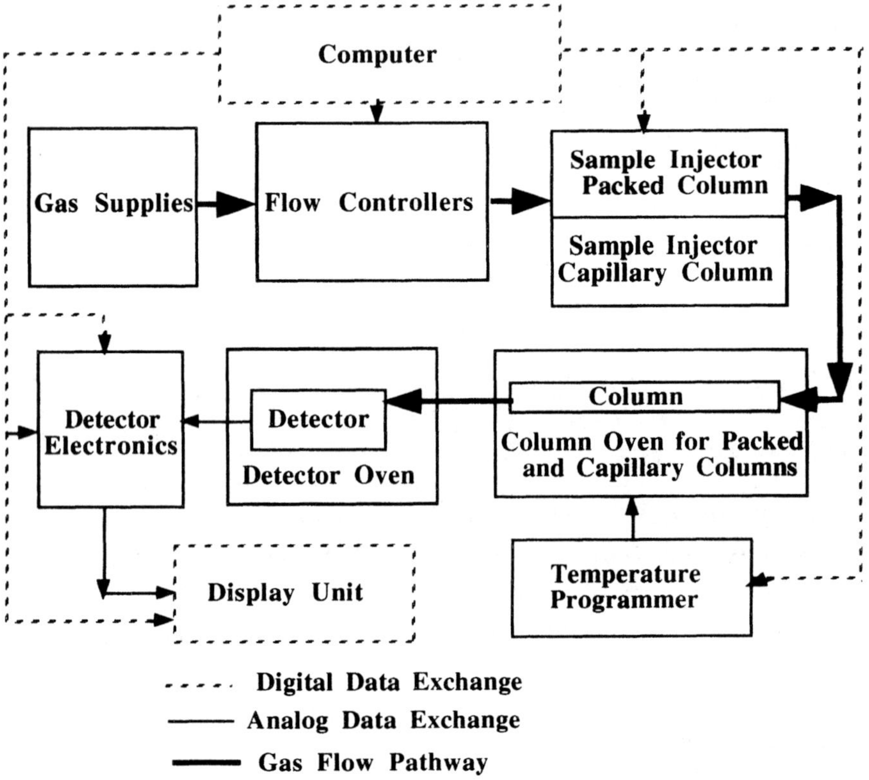

Figure 1 The Layout of a Contemporary Gas Chromatograph

Gas Supplies

There are a number of different gasses that may be required by the gas chromatograph, the nature of which will be determined partly by the type of detector that is selected and partly by the type of column that is used. As far as the column is concerned, low density gases provide a lower resistance to mass transfer in the gas phase and thus their choice determines the optimum gas velocity that should be used. The optimum velocities for a packed and capillary column can be derived from the Van Deemter equation and the Golay equation, respectively [1].

For a packed column,

$$u_{(o)opt} = \frac{D_{m(o)}}{d_p}\sqrt{\frac{2\gamma}{f_1(k')}}$$

where($u_{(o)opt}$) is the optimum exit velocity,
($D_{m(o)}$) is the solute diffusivity at atmospheric pressure,
(d_p) is the particle diameter of the packing,
(k') is the capacity ratio of the solute,
and (γ) is a packing factor,

and for a capillary column,

$$u_{o(opt)} = 2\frac{D_{m(o)}}{r}\left(\frac{12(1+k')^2}{\left(1+6k'+11k'^2\right)}\right)$$

where (r) is the radius of the capillary column.

It is clear that the higher the diffusivity of the solute in the mobile phase, the higher the optimum velocity and consequently the shorter the analysis time. Consequently, if a large sample throughput is important, then low density, high diffusivity gases would be preferred.

The minimum HETP (the plate height at the optimum velocity where the maximum efficiency is obtained) for packed and capillary columns respectively is given by the following equations [1]:

$$H_{min} = 2\lambda d_p + 2\sqrt{2\gamma D_{m(o)}\left(\frac{f_1(k')d_p^2}{D_{m(o)}} + \frac{f_2(k')d_f^2}{D_S\left(\gamma_{i/o}+1\right)}\right)}$$

where (H_{min}) is the minimum plate height,
(λ) is another packing constant,
(D_S) is the solute diffusivity in the stationary phase,
and ($\gamma_{(i/o)}$) is the inlet/outlet pressure ratio,

$$\text{and} \quad H_{min} = r\left(\frac{\left(1+6k'+11k'^2\right)}{3(1+k')^2}\right)$$

It is clear that the diffusivity of the solute in the mobile phase has no effect on the maximum column efficiency. It follows that the use of low density gasses such as hydrogen or helium, which exhibit high solute diffusivities, will not provide high column efficiencies, but can permit the use of high gas velocities to provide high speed separations and reduced analysis times. Examples of HETP curves obtained for different gases are shown in figure 2. The curve for hydrogen was obtained from a different column and thus the minimum differs from that for the other gasses.

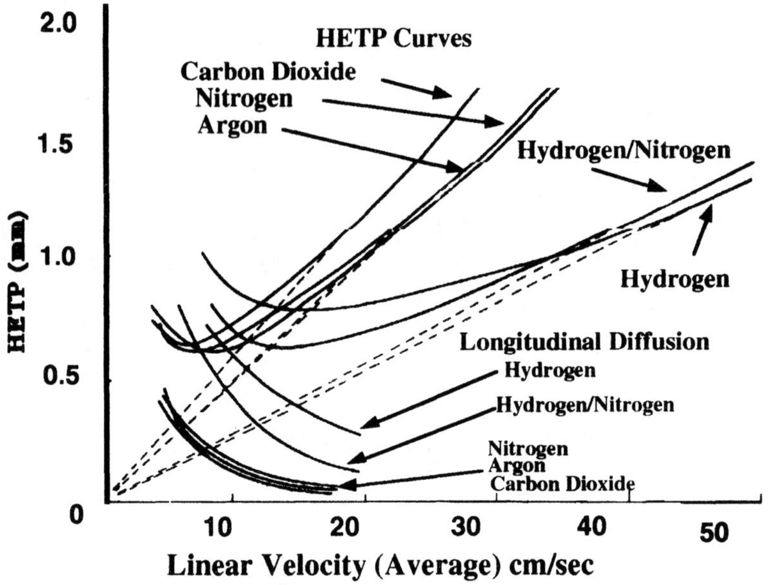

Figure 2 HETP Curves for Different Carrier Gases

Short analysis times provide high sample throughput and are thus very important in routine assays for quality control. Unfortunately, the detector may restrict the choice of mobile phase, *e.g.*, the argon detector will require argon to be used as the carrier gas. In addition, the electron capture detector will usually require a 10% methane in argon mixture for optimum performance. Nevertheless the electron capture detector can operate with nitrogen albeit somewhat less efficiently. Detectors based on argon or helium plasma formation obviously define the gas to be employed and the flame ionization detector (FID) plainly needs hydrogen and an oxidizing gas such as air or oxygen to function correctly. It is clear that a number of different gases may be required for a specific column–detector combination. As the FID is the most common detector used in GC for quantitative analysis, at least three gas supplies are needed, one for the mobile phase and two for the FID. Providing the mobile phase gas supply is not designed exclusively for a particular gas, it can be used with other gases to meet specific chromatographic needs. This versatility, of course, excludes inflammable gases such as hydrogen which utilize left-hand threaded unions and must not be used with other gases.

Gas Supplies from Tanks or Cylinders

Gases are commonly stored in large cylindrical tanks at high pressure fitted with reducing valves that supply the gas to the instrument at the specified pressure. The cylinders are usually situated outside and away from the chromatograph for safety purposes and the gases are passed to the chromatograph through copper or stainless steel conduits. However, as a result of the perceived danger from large volumes of gases stored under pressure, hydrogen, oxygen and nitrogen generators have been developed and their use as an alternative to gas cylinders is steadily increasing.

Pure Air Generators

These instruments can utilize either compressed air from tanks or air supplied directly from the laboratory compressor pumps. The Packard Zero Air Generator passes the gas through a 0.5 μ filter that removes oil and water and then over a catalyst to remove hydrocarbons. Finally the hydrocarbon-free air is passed through a 0.01 μ cellulose fiber filter to remove any residual particulate matter in the gas stream. The product contains less than 0.1 ppm total hydrocarbons and can deliver the air at a pressure of 125 psi, and flow rates up to 2,500 cc per min.

Pure Nitrogen Generators

Nitrogen generators also use the laboratory compressed air supply. The general contaminants are first removed and the purified air passes over layers of polymeric hollow fiber membranes through which nitrogen selectively permeates. The residual nitrogen depleted air that carries about 30% oxygen is vented to atmosphere at the back of the instrument. The nitrogen produced by the Air Products nitrogen generator contains less than 0.5 ppm of oxygen, less than 0.5 ppm of water vapor and less than 2.0 ppb of halocarbons or hydrocarbons. It can supply up to 1 l/min at pressures up to 100 psi.

Hydrogen Generators

In the Packard Hydrogen Generator, Hydrogen is generated electrolytically from pure deionized water. Much of the technology used in hydrogen generators is proprietary and technical details are often difficult to obtain. The electrolysis system uses a solid polymer electrolyte and thus does not need to be supplied with replacement electrolytes, but only with deionized water. The device generates 99.999% pure hydrogen with a reservoir capacity of 4 liters, and a delivery pressure ranging from 2 to 100 psi. Different models produce flow rates ranging from 0 to 125 ml/min to 0 to 1200 ml/min. The oxygen, produced simultaneously with hydrogen, but at half the flow rate, is vented as pure oxygen at the back of the instrument.

Pressure Controllers

There are a number of pressure controllers associated with a gas chromatograph. The reducing valves on the gas tanks are in fact simple *pressure* controllers and the *flow* controllers that are used for detector and column flow control consist of servo controlled *pressure* controllers. A diagram of a pressure controller is shown in figure 3. Essentially the device contains two gas chambers separated by a diaphragm, in the center of which is a needle valve that is actuated by the diaphragm. The diaphragm is held down by a spring which is adjustable so that the pressure in the second chamber and thus the outlet flow can be set at a chosen value. Gas enters the lower chamber and the pressure on the lower part of the diaphragm acts against the spring setting and opens the valve. Gas then passes into the upper chamber and out to the chromatograph.

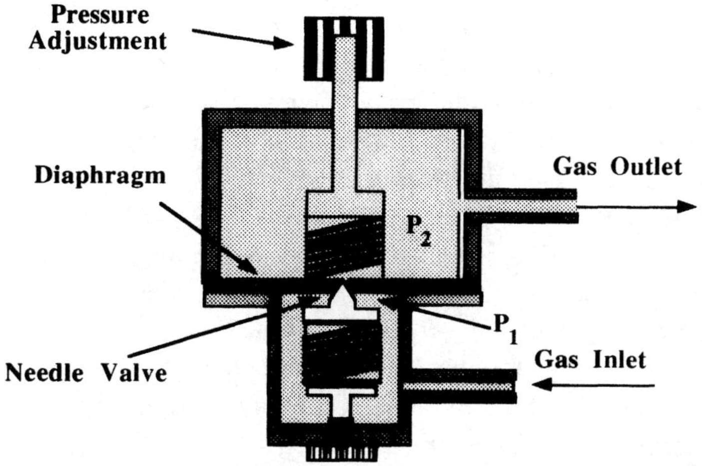

Figure 3 The Gas Pressure Controller

The pressure builds up in the upper chamber to the set value at which time the diaphragm moves downward closing the valve. If the pressure falls in the upper cylinder, the diaphragm again moves upward due to the pressure in the lower chamber, which opens the valve and the pressure in the upper chamber is brought back to its set value.

Flow Controllers

Applying constant pressure to a column does not ensure a constant flow of mobile phase though the chromatographic system, particularly when temperature programming is employed. As the temperature of a gas is raised, so the viscosity increases, and thus at a constant inlet pressure the flow rate will fall. The reduction in flow rate will depend on the temperature program limits and to a certain extent on the temperature gradient. To obviate the flow rate change, mass controllers are used which ensure a constant mass of mobile passes through the column in unit time irrespective of the system temperature.

The sensing system utilizes a bypass tube with a heater situated at the center. Precision temperature sensors are placed equidistant up stream and down stream of the heater. A proprietary baffle system in the main conduit creates a pressure drop that causes a fixed proportion of the flow to be diverted through the sensor tube. At zero flow both sensors are at the same temperature. When there is flow, the down stream sensor is heated producing a differential temperature across the sensors. As the temperature of the gas will be proportional to the product of mass flowing and its specific heat, the differential temperature will be proportional to the mass flow rate. The voltage generated from the differential sensor signal is compared to that of a set voltage and the difference is used to generate a signal to actuate a valve controlling the flow. Thus a closed loop control system is formed that maintains the mass flow rate set by the reference voltage. The device can be made extremely compact, is highly reliable and affords accurate control of the carrier gas flow rate irrespective of gas viscosity changes in the column due to temperature programming.

Flow Programmers

There are three methods used to accelerate the elution of strongly retained peaks during chromatographic development. The first technique (used exclusively in liquid chromatography) is gradient elution, where the composition of the mobile phase is progressively changed during chromatographic development. This procedure is not practical in gas chromatography, because the interactions of the solute molecules with those of the carrier gas are relatively infrequent, and when they do occur are extremely weak. Consequently changing the nature of the gas will have little impact on the elution rate. The second technique is temperature programming which is most commonly used in gas chromatography. Temperature programming is, of course, also used in liquid chromatography, but usually in conjunction with gradient elution. The distribution coefficient of a solute is exponentially related to the reciprocal of the absolute temperature, and thus the peak is accelerated through the column as the temperature is increased. The third development procedure is that of flow programming, where the flowrate is increased progressively during chromatographic development. The theory of flow programming is as follows. If the mobile phase is incompressible, the retention volume of the solute (V_r) is related to the retention time (t_r) and the flow rate (Q) by the simple expression,

$$t_r = \frac{V_r}{Q}$$

Now from D'Arcy's Law for the flow of fluid through a packed bed (which is basically the same as Poiseuille's Law for the flow of a fluid through an open tube except the proportionality constants differ) at point (x) where the pressure is (p), (Q) will be given by

$$Q = \frac{Dd^4}{\eta}\frac{dp}{dx}$$

where (d) is either the particle diameter of the packing or the radius of the tube whichever being appropriate,
(η) is the viscosity of the fluid,
and (D) is the constant appropriate for a packed bed or tube.

If the fluid is assumed to be incompressible then the equation can be integrated directly. If the fluid is compressible, as in the case of a gas, then it is necessary to recognize that it is the *mass* of fluid passing at any point in the column per unit time that is constant and not the volume. Thus if (ρ) is the density of the gas at any point along the bed or tube, then (ρQ) is constant and as the density of a gas is directly proportional to the pressure, then PQ is constant at any point along the tube or column. Consequently, if (P_1) and (Q_1) are the pressure and volume flow entering the tube or bed respectively, then

$$P_1 Q_1 = PQ = \frac{Dd^4 P}{\eta}\frac{dp}{dx}$$

or

$$\int_0^L P_1 Q_1 dx = \frac{Dd^4}{\eta}\int_{P_1}^{P_2} P\,dp$$

Integrating,

$$P_1 Q_1 = \frac{Dd^4}{2L\eta}\left(P_1^2 - P_2^2\right) \qquad (1)$$

or

$$P_1 Q_1 = \frac{Dd^4}{2L\eta} P_1^2\left(\frac{\gamma_{(i/o)}^2 - 1}{\gamma_{(i/o)}^2}\right) = \varphi P_1^2\left(\frac{\gamma_{(i/o)}^2 - 1}{\gamma_{(i/o)}^2}\right)$$

where ($\gamma_{(i/o)}$) is the inlet/outlet pressure ratio (P_1/P_2), and

$$\varphi = \frac{Dd^4}{2L\eta}$$

Thus, $$Q_1 = \varphi P_1 \left(\frac{\gamma_{(i/o)}^2 - 1}{\gamma_{(i/o)}^2} \right)$$

Now, $$P_1 Q_1 = P_2 Q_2 \quad \text{or} \quad Q_2 = \frac{P_1 Q_1}{P_2} = Q_1 \gamma_{(i/o)}$$

where (Q_2) is the flow rate at the column exit.

Thus, $$Q_2 = \varphi P_1 \left(\frac{\gamma_{(i/o)}^2 - 1}{\gamma_{(i/o)}} \right)$$

Consequently, for either a packed column or a capillary column,

$$t_r = \frac{V_r}{\varphi P_1 \left(\frac{\gamma_{(i/o)}^2 - 1}{\gamma_{(i/o)}} \right)} \tag{2}$$

Again it should be noted that (φ) is a constant the magnitude of which will depend on the type of column and mobile phase that is being employed.

Now employing the standard pressure correction factor,

$$V_r = \frac{V_{r(0)} 3\left(\gamma_{(i/o)}^2 - 1\right)}{2\left(\gamma_{(i/o)}^3 - 1\right)}$$

where (V_r) is the true retention volume of the solute,
and ($V_{r(0)}$) is the retention volume measured at the outlet.

Then,
$$V_{r(0)} = \frac{2V_r\left(\gamma^3_{(i/o)} - 1\right)}{3\left(\gamma^2_{(i/o)} - 1\right)} = t_{r(0)}Q_2$$

Now
$$Q_2 = \varphi P_1\left(\frac{\gamma^2_{(i/o)} - 1}{\gamma_{(i/o)}}\right)$$

$$t_0 = \frac{2V_r\left(\gamma^3_{(i/o)} - 1\right)\gamma}{3\varphi P_1\left(\gamma^2_{(i/o)} - 1\right)\left(\gamma^2_{(i/o)} - 1\right)} \qquad (3)$$

If ($\gamma_{(i/o)}$) is large compared with unity,

$$t_0 = \frac{2V_r\gamma^4_{(i/o)}}{3\varphi P_1\gamma^4_{(i/o)}} = \frac{2V_r}{3\varphi P_1} \qquad (4)$$

It is seen that at high values of (γ), the inlet/outlet pressure ratio loses its significance and the retention time (t) becomes inversely proportional to the pressure.

The relationship between $\frac{\left(\gamma^3_{(i/o)} - 1\right)\gamma_{(i/o)}}{P_1\left(\gamma^2_{(i/o)} - 1\right)\left(\gamma^2_{(i/o)} - 1\right)}$ and P_1 is shown in figure 4. It is seen that the function that controls the retention time falls quite dramatically as the inlet pressure is increased which, as might be expected, will result in reduced analysis times with an increase in pressure. The net effect of pressure programming on elution time can be evaluated as follows. Reiterating equation (3),

$$t_0 = \frac{2V_r\left(\gamma^3_{(i/o)} - 1\right)\gamma}{3\varphi P_1\left(\gamma^2_{(i/o)} - 1\right)\left(\gamma^2_{(i/o)} - 1\right)}$$

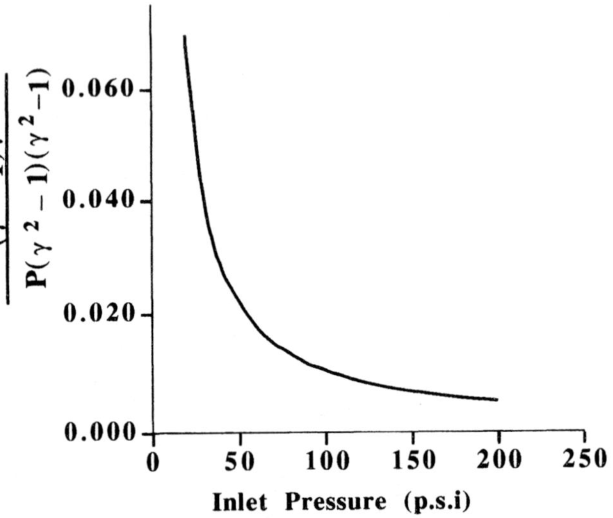

Figure 4 Graph of $\frac{\left(\left(\gamma_{(i/o)}^3-1\right)\right)\lambda}{P_1\left(\gamma_{(i/o)}^2-1\right)\left(\gamma_{(i/o)}^2-1\right)}$ **against Inlet Pressure P1**

Rearranging,
$$V_r = \frac{3\varphi P_1\left(\gamma_{(i/o)}^2-1\right)\left(\gamma_{(i/o)}^2-1\right)t_0}{2\left(\gamma_{(i/o)}^3-1\right)\gamma_{(i/o)}}$$

Therefore, over a range of changing values for (P_1) that takes place during a pressure or flow program, employing a time interval of (Δt), after (q) intervals of time, the contribution of the column flow to the retention volume ($\Delta V_{r(0)}$) will be given by

$$\Delta V_r = \frac{3\varphi P_1\left(\gamma_{(i/o)}^2-1\right)\left(\gamma_{(i/o)}^2-1\right)\Delta t}{2\left(\gamma_{(i/o)}^3-1\right)\gamma_{(i/o)}}$$

Taking (Δt) as unit time (1 second), then

$$V_r = \sum_{q=1}^{q=n}\Delta V_{r(q)} = \sum_{q=1}^{q=n}\frac{\varphi P_q\, 3\left(\gamma_{(i/o)}^2-1\right)\left(\gamma_{(i/o)}^2-1\right)}{2\left(\gamma_{(i/o)}^3-1\right)\gamma_{(i/o)}}$$

Now assuming a linear program of ΔP/sec. starting at a pressure (P_1), then at time (q), $P_q = P_1 + q\Delta P$. Substituting for (P_q),

$$V_r = \sum_{q=1}^{q=n} \Delta V_{r(q)} = \sum_{q=1}^{q=n} \frac{\varphi(P_o + \Delta Pq)3\left(\gamma_{(i/o)}^2 - 1\right)\left(\gamma_{(i/o)}^2 - 1\right)}{2\left(\gamma_{(i/o)}^3 - 1\right)\gamma_{(i/o)}} \qquad (5)$$

and $n = T_{t(0)}$ where ($T_{t(0)}$) will be the retention time of the solute under the defined pressure programming conditions.

To simplify the calculations, a practical situation will be taken where the retention volume is 1000 ml on a given column operating at a (P_o) value of 15 psi (that is 15 psi above atmospheric, *i.e.* $\gamma_{(i/o)}=2$) and the retention time of the solute is 10 minutes (600 seconds). This will define the flow properties of the column, *i.e.* $\frac{3\varepsilon}{2}$. Then from equation (3)

$$t_0 = \frac{2V_r\left(\gamma_{(i/o)}^3 - 1\right)\gamma_{(i/o)}}{3\varphi P_o\left(\gamma_{(i/o)}^2 - 1\right)\left(\gamma_{(i/o)}^2 - 1\right)}$$

or

$$600 = \frac{2x1000\left(2^2 - 1\right)\left(2^2 - 1\right)2}{3\varphi 15\left(2^3 - 1\right)} = \frac{2}{3\varphi}\frac{18000}{105}$$

$$\frac{3\varphi}{2} = 0.286$$

Substituting for $\frac{3\varphi}{2} = 0.286$ in equation (5),

$$V_r = \sum_{q=1}^{q=n} \Delta V_{r(p)} = \sum_{q=1}^{q=n} \frac{0.286(P_o + \Delta Pq)\left(\gamma_{(i/o)}^2 - 1\right)\left(\gamma_{(i/o)}^2 - 1\right)}{\left(\gamma_{(i/o)}^3 - 1\right)\gamma_{(i/o)}} \qquad (6)$$

It is now possible to calculate the change in retention time for a series of solutes separated under pressure programming conditions in the defined column. Consider five solutes having actual retention volumes of 400, 600, 800, 1000 and 1200 ml eluted under pressure programming conditions where (P_O) is 15 psi and increases at a rate ranging from 0 to 0.075 psi/sec. Employing equation (6) and with the aid of a simple computer program, the retention time of the solutes can be calculated for each of the different programming rates. The results are shown in figure 5.

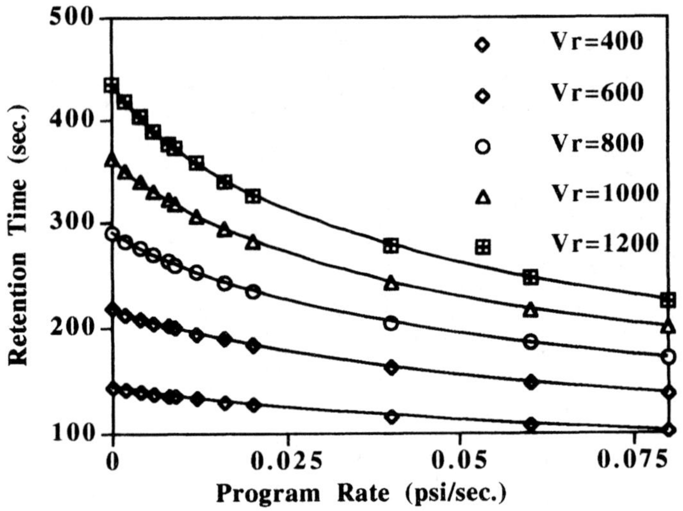

Figure 5 The Change in Elution Time of a Series of Solutes with Flow Programming Rate

It is seen that although the use of pressure programming does indeed reduce the retention time of all solutes, it is also seen that as the program rate is increased the reduction in analysis time becomes relatively less. The results shown relate to a practical situation and demonstrate the effect that could be expected when flow programming either a capillary or packed column. The program rates of 0.01 to 0.075 psi/sec. are readily available from most gas chromatographs that have electronic actuated flow controllers. There is an interesting effect that can be observed from figure 5, and that is the effect of flow programming on the separation ratio between two solutes.

It is clear that all the peaks become bunched together, as their retention time is decreased, by increasing the flow programming rate. This contraction of the chromatogram must inevitably reduce the separation ratios between the peaks and thus reduce the resolving power of the column. From a physical chemical point of view this reduction in the separation ratios would be expected. As the programming rate increases the peaks in the column are continually subjected to an acceleration and as a result those remaining in the column longer will suffer greater acceleration than those eluting initially at a greater rate. Consequently, the later peaks will tend to catch up with those solutes eluting in front of them. From the data given in figure 5, the separation ratios of each peak, with respect to the first eluted peak, can be calculated and the results are shown plotted against program rate in figure 6.

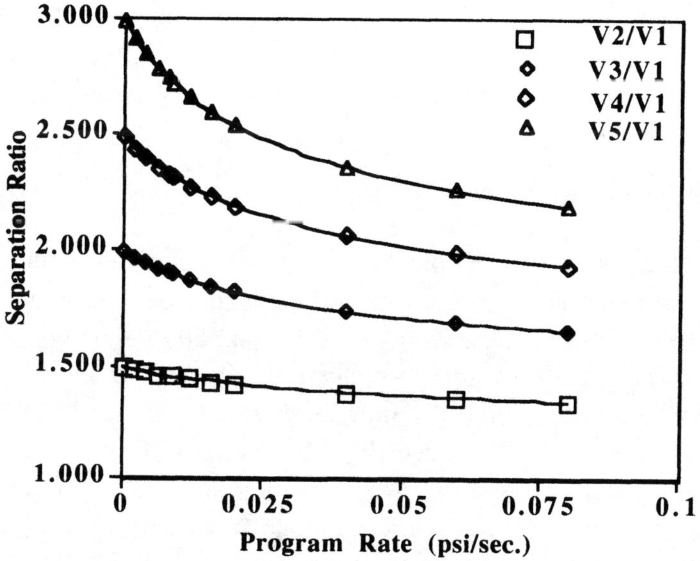

Figure 6 Curves Relating Separation Ratios with Flow Programming Rate

It is seen that the separation ratios between the solutes are greatly reduced with increased program rate and that the column resolution would be seriously affected. It should also be noted that the effect is much greater at the lower program rates. This reduction in separation ratio between the solutes will be accompanied by significant band dispersion (reduced column efficiency) and at the higher flow velocities, the resolving power of the column will be reduced to an even greater extent. Despite the disadvantages, however, there will be situations where it is the only alternative development technique that can be employed, for example where thermally labile materials are being analyzed.

Flow programming can be used in quantitative analysis providing the response of the detector is independent of the flow rate, which is true for a *very limited number* of detectors. Of those detectors that are used in GC for quantitative work, only the *flame ionization detector can tolerate flow programming* without the accuracy and precision of the analysis being effected. This is because the FID responds to the *mass* of solute passing through it per unit time and not the *concentration* of the solute in the eluent gas. This means that any progressive dilution that takes place from flow programming will not change the detector response, providing the mass of sample passing through the detector per unit time is unaffected.

Injection Devices

The injector places the sample to be separated onto the column and must do so in an accurate and reproducible manner. As there are a number of column configurations used in quantitative analysis, and a range of sample types, diverse designs of sample injectors are necessary to accommodate the unique character of each sample and operate satisfactorily with specific column types. For example, gas sampling devices are quite different from liquid injection systems and the capillary column requires an injector that is quite different from that used for a packed column. Irrespective of the design of the subsequent equipment the overall precision and accuracy of a GC analysis can be no better than that provided by the sample injector. It follows that the sample injector is a very critical part of the chromatographic equipment and needs to be well designed and well maintained. The precision provided by the sample injector is particularly important when *external standards* are used.

Gas Sampling Systems

Gas samples are usually placed on the column with an external loop sample valve, although as microbore columns and packed capillaries become available for gas analysis, the internal loop valve may become more appropriate for some samples. The external loop valves are used for packed columns and the larger diameter (PLOT) columns. Generally, in gas analysis the *accuracy* of the device need not be high but the sample volume delivered to the column must be extremely *precise*. A diagram of the external loop sample valve is shown in figure 7.

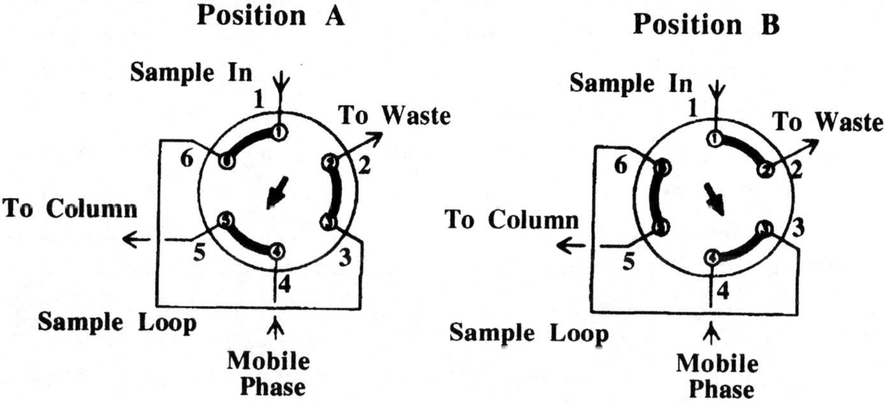

Courtesy of Valco Instruments, Inc.

Figure 7 The External Loop Gas Sample Valve

Basically, the valve consists of two plates: one plate with six holes in it oriented round the circumference in the way shown in figure 7; the other plate has slots milled in its surface such that when the plates are face to face, different pairs of holes are connected. In the loading position shown on the left, the carrier gas is connected by a rotor slot to port (4) and the column to port (5), thus allowing the carrier gas to flow directly through to the column. In this position the sample loop is connected to ports (3) and (6). Sample flows from a syringe into port (1) through the rotor slot to the sample loop at port (6), fills the sample loop, and passes out through port (3) into port (2) and to waste. On rotating the valve to place the sample on the column, the sample loop is interposed between the column and the carrier gas supply, by connecting ports (3) and (4) and ports (5) and (6). Thus, the sample is swept onto the column. In the sampling position, the third rotor slot connects the syringe port to the waste port. After the sample has been placed on the column, the rotor can be returned to the loading position, the system purged with a suitable gas and the sample loop loaded in readiness for the next injection. This type of sample valve functions well and the appropriate loop volume must be chosen on the basis of the dimensions of the column with which it is to be used. However, due to the parabolic velocity profile of the flow through the sample tube, significant band dispersion can occur. Nevertheless, if an internal standard or a normalization procedure is employed in the data processing, then the valve can be rotated back to its fill position immediately after injection. This cuts the tail of the sample from the main bulk and significantly reduces peak dispersion and its consequent loss of resolution. This procedure, however, is not very reproducible, and introduces poor sample–volume precision and therefore must *not be used* when *external standards* are employed.

The internal loop valve is used with very small diameter columns. A diagram of an internal loop sample valve is shown in figure 8.

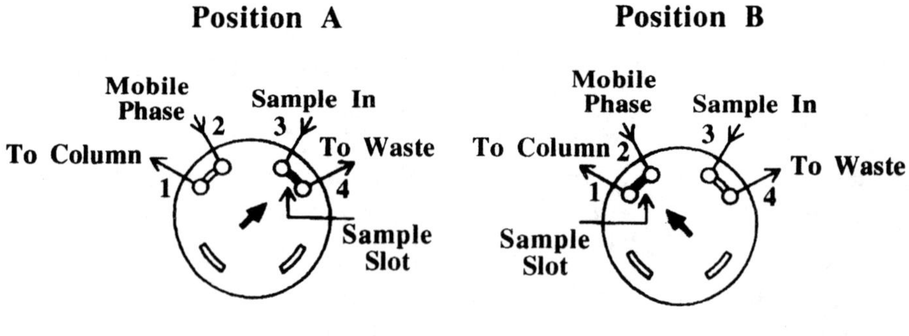

Courtesy of Valco Instruments, Inc.

Figure 8 The Internal Loop Gas Sample Valve

The internal loop sample valve is not often used in GC gas analysis and the valve usually has only four ports, one pair for the sample to be placed on the column, the other pair to connect the carrier gas to the column. The sample volume is contained in the connecting slot of the valve rotor and delivers sample volumes ranging from 0.1 µl to about 0.5 µl to the column. On the left-hand side of figure 8, the sample is shown being loaded into the valve, from port 3 through the sample slot to port 4 and out to waste. In this position, the carrier passes through the valve from port 2 directly to the column through port 1. On rotation (depicted on the right-hand side of figure 8), the valve slot containing the sample is now imposed between the carrier gas supply and the column by connecting ports 1 and 2, and the sample is swept onto the column. In practice, a few seconds is allowed to pass to permit the sample to be completely transferred to the column. The valve can then be returned to the load position. The external loop valve is the generally recognized sampling device for placing gas samples on a GSC column.

Packed Column Injection Systems

The sampling system used in modern GLC equipment injects a defined volume of solution, via a septum injector, directly onto the column or into a flash heater. There is some disagreement with respect to the use of the flash heater as opposed to on-column injection. The flash heater vaporizes the sample which is then absorbed on the stationary phase. This procedure inherently broadens the injection band but disperses the sample into the packing. Direct injection into the packing contains the sample in a small volume, but it is claimed cools the packing. Generally the heat adsorbed is very small, and in the authors' experience, on-column injection is to be preferred particularly for quantitative work. An example of a septum injection system used for packed columns is shown in figure 9. The silicone rubber septum is compressed between metal surfaces in such a manner that an hypodermic needle can pierce it and be withdrawn with no gas leak. The glass liner is there to prevent the sample coming in contact with the heated metal wall and possibly suffering thermal decomposition.

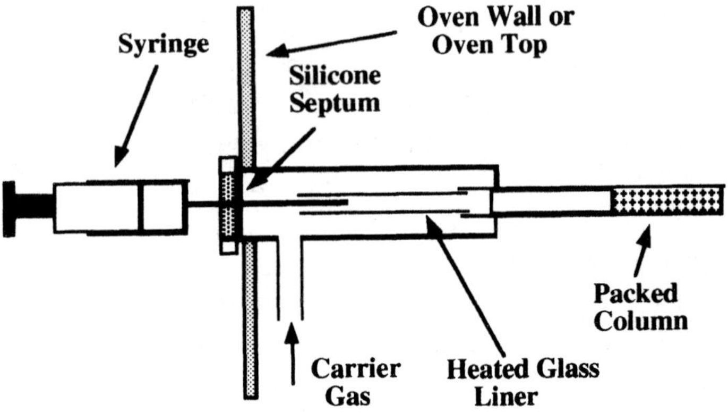

Figure 9 A Septum Injector

In some instruments the glass liner is provided with a separate heater so that the temperature at which the sample is volatilized can be controlled. Such a device has been termed the "flash heater". If a syringe with a long needle is used, it can penetrate past the liner and discharge its contents directly into the column packing. This procedure is called on-column injection and with many samples is the preferred procedure as it reduces peak dispersion on injection and thus often provides higher column efficiencies and is generally more reproducible. However, if on-column injection is carried out, the flash heater should be turned off, otherwise the sample may be boiled out of the syringe needle before it can be injected into the packing.

Open Tubular Column Injection Systems

Due to the very small sample size that must be placed on narrow bore capillary columns, a split injection system is necessary, a diagram of which is shown in figure 10. The basic difference between the two systems is that the capillary column now projects into the glass liner and a portion of the carrier gas sweeps past the column inlet and to waste. As the sample passes the column opening, a small fraction flows directly into the capillary column. The split ratio can be changed by adjusting the portion of the carrier gas that flows to waste. This device is only used for small diameter capillary columns where the charge size is critical.

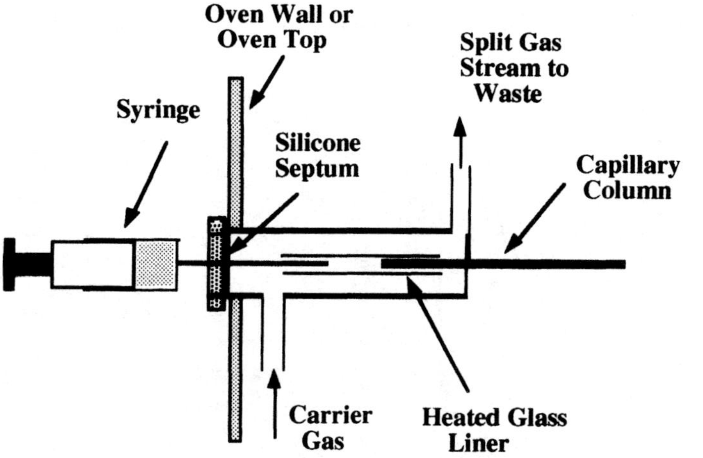

Figure 10 The Split Injection System

Unfortunately, the device is not ideal due to component differentiation and, as a result, the sample placed on the column may not be truly representative of the actual material being analyzed. In general the solutes with the higher diffusivities (low molecular weights) are lost preferentially to those with lower diffusivities (higher molecular weights). It follows that quantitative analyses carried out using the high efficiency small diameter capillary columns may have limited accuracy and precision, depending on the nature of the sample. This is an inherent problem with small bore capillary columns which is extremely difficult to obviate. Consequently, capillary columns that utilize split injection systems are not suitable for very accurate quantitative analysis particularly if the components of the sample have a wide molecular weight range. In an attempt to overcome this problem, larger diameter open tubular columns have been employed that would permit on-column injection. The columns have an I.D. of about 0.053 in., which is slightly greater than the diameter of a specific hypodermic needle. The injection system is shown in figure 11. Unfortunately, there are also difficulties with this type of injector, not so much with its effect on accuracy and precision but on column resolution. When the sample is injected into the column, it breaks up into separate parts, as bubbles form along the first part of the column.

This causes the sample to be deposited at two or more positions along the tube as the solvent evaporates.

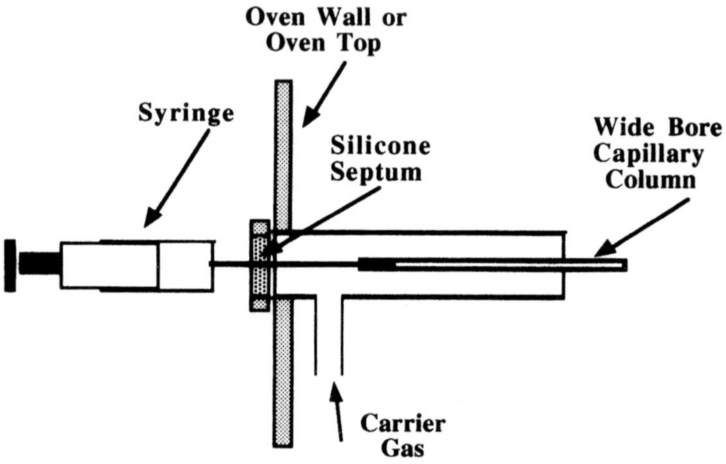

Figure 11 Injection System for Large Bore Capillary Columns

When the separation is developed, each local concentration of sample can act as a separate injection and, as a result, a chromatogram containing very wide or double peaks is produced. There have been a number of procedures introduced in an attempt to eliminate the sample splitting on the column. The first solution was the 'retention gap method' which is depicted in figure 12. This procedure involves removing the stationary phase from the first few centimeters of the column. The sample is injected into this section of the column and, if the sample is split, it will still vaporize in the normal way. As there is no stationary phase present, the solutes will all travel at the speed of the mobile phase until they reach a coated section of the column. At this point they will be absorbed into the stationary phase and all the sample will accumulate at that point. This technique is usually practiced in conjunction with temperature programming, starting the program at a fairly low temperature. This facilitates the accumulation of all the solutes at one point in the column, that is, where the stationary phase coating begins. The temperature program is then started and the solutes are eluted through the column in the normal way.

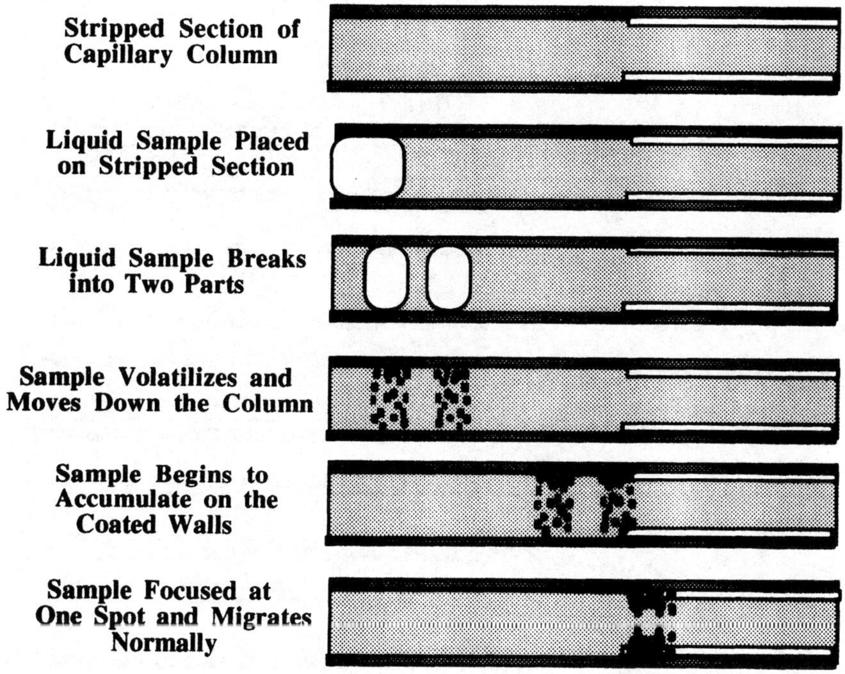

Figure 12 The Retention Gap Method of Injection

The success of this method depends on there being a significant difference between the boiling points of the sample solvent and those of the components of the sample.

An alternative procedure, called the "solute focusing method" is more effective, but requires more complicated equipment. The injector is designed so that there are two consecutive, independently heated and cooled zones located at the beginning of the column. A diagram of the solute focusing system is shown in figure 13.

Initially both zones are cooled and the sample is injected onto the first zone where sample splitting almost inevitably occurs. The carrier gas is allowed to remove the solvent and then the first zone is heated and the second zone kept cool. The solutes in the first zone are eluted through the zone at the higher temperature but all the components accumulate at the beginning of the cooled second zone.

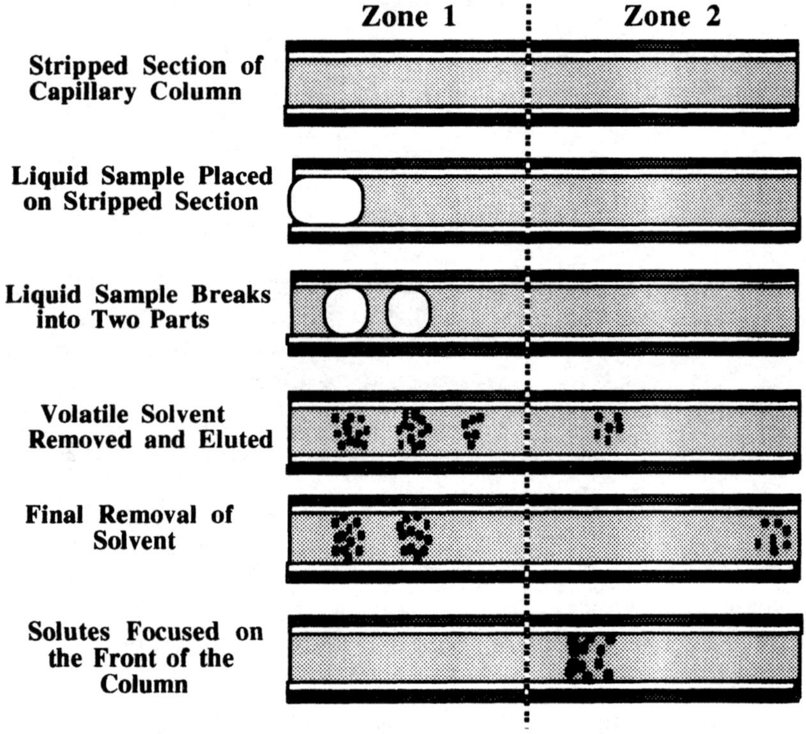

Figure 13 The Solute Focusing Method of Injection

All the sample has now been focused at the beginning of the column. The second zone is now heated and the separation developed in the usual manner. This technique has more flexibility than the "retention gap method" but the apparatus and the procedure is more complex. It should be pointed out that sample splitting does *not occur* in packed columns. It follows, that if the sample is amenable to separation on such columns, (*i.e.* they can resolve the components of interest), then the packed column may be the column of choice if the best accuracy and precision are required. In general, if very high efficiencies are not required, the packed column will give more precise and accurate quantitative results than the open tubular column. Unfortunately, despite their technical advantages in quantitative GC, packed columns are not so popular as open tubular columns as they are not seen as the 'state of the art' technology.

Automatic Injection Systems

Today, most gas chromatographs that are used for routine analysis also include an automatic sampling device. This involves the use of a transport mechanism that may take the form of a carousel or some type of conveyor system. The transporter carries a series of vials that may alternately contain sample and washing solvent. The sampling mechanism can be quite intricate, involving a complex sequence of operations that are automatically controlled by the device. First the syringe is washed with solvent, then it is rinsed with the sample, then reloaded with the sample and finally the contents are discharged into the column. After the analysis the next cycle commences with the next syringe washing procedure. In routine analytical laboratories, which often have very sophisticated GC assemblies, there may also be robotic sample preparation such as extraction, concentration, derivatization procedures, etc.

To simplify the operation of the automatic sampler, the injection system is usually arranged to be vertical with the septum at the top. This will allow the syringe to be lifted to the correct height, rotated so that it is over the septum of the injection device, then lowered to allow the needle to penetrate the septum and enter the vaporizer or column. The syringe is then mechanically discharged and the syringe reverses its three-dimensional program of movement until it is over the next sample vial.

The Column Oven

The operating temperature is an important variable in GC and must be carefully chosen to suit the polarity or boiling points of the constituents of the sample. The technique of temperature programming, like gradient elution in LC, is also used to accommodate mixtures that have a wide boiling range or broad polarity span. Consequently, not only must the temperature be accurately controlled, but it must also be possible to change the temperature at precisely controlled rates during chromatographic development.

The temperature range of the oven, and consequently the temperature programmer, varies a little from manufacturer to manufacturer but most modern equipment can operate between the limits of 5°C and 400°C. Operating below 5°C is possible, but problems can arise from condensation, and cooling coils must be fitted to reduce the temperature below that of ambient. Very few separations are carried out at 400°C or above and, in fact, 250°C is the more general upper temperature limit for many analyses (largely because of the limited thermal stability of many of the stationary phases that are available). The capacity of most GC ovens ranges between 1 and 2 cu ft, they are well insulated, and are fitted with devices that will allow rapid cooling after a temperature program. This usually involves the circulation of ambient air through the oven from outside, by opening external flaps or the oven lid to allow cool air entry. Efficient GC ovens will cool back to the starting temperature of a program in 10 minutes or less. The oven is fitted with a fairly powerful circulating fan to ensure even heating throughout and also aids in the rapid reduction of oven temperature after a program. This fan also provides rapid and efficient distribution of heat from the heater element to the oven and its contents during a program. The temperature control of the oven should be better than ± 0.2°C. The oven usually contains supporting structures or fittings, for holding packed and capillary columns, split injection systems, automatic sample valves for multidimensional analysis and other ancillary apparatus. The ovens of a gas chromatograph emit a considerable amount of heat during operation and it is highly desirable that the unit is installed in an air-conditioned environment.

The Temperature Programmer

Temperature programming was invented in the early days of GC and is now an essential accessory to all gas chromatographs. The technique is used for the same reasons as flow programming, that is to accelerate the elution rate of the late peaks that would otherwise take an inordinately long time to elute.

The distribution coefficient of a solute is exponentially related to the reciprocal of the absolute temperature and as the retention volume is directly related to the distribution coefficient, temperature will govern the elution rate of the solute.

The corrected retention volume (V'_r) is given by $V'_r = KV_S$ and by thermodynamic argument [2], it can be shown that

$$\ln K = -\frac{\Delta H_o}{RT} + \frac{\Delta S_o}{R}$$

Thus

$$V'_{r(T_o)} = V_S e^{-\frac{\Delta H_o}{RT_o} + \frac{\Delta S_o}{R}}$$

Thus for a period (Δt) at (T_o) the effective retention volume can be considered to be constant at

$$V^t_{r(T_o)} = V_S e^{-\frac{\Delta H_o}{RT_o} + \frac{\Delta S_o}{R}}$$

Furthermore after time (t_p) between (t_p) and ($t_p + \Delta T$) during a temperature program at a rate of α°C per unit time, the effective retention volume can be considered to be constant at

$$V'_{r(T_o+\alpha p)} = V_S e^{-\frac{\Delta H_o}{R(T_o+\alpha p)} + \frac{\Delta S_o}{R}}$$

It follows that if the retention time in seconds is (n) and (Δt) is conveniently taken as one second, then the mean value of the retention volume throughout the program will be

$$\overline{V}'_{r(T_0)to(T_0+\alpha p)} = \frac{\sum_{p=1}^{p=n} V_S e^{-\frac{\Delta H_0}{R(T_0+\alpha p)} + \frac{\Delta S_0}{R}}}{\sum_{p=1}^{p=n} \Delta n}$$

$$= \frac{1}{n} \sum_{p=1}^{p=n} V_S e^{-\frac{\Delta H_0}{R(T_0+\alpha p)} + \frac{\Delta S_0}{R}}$$

As in this case, pressure programming is not employed, then for simplicity, the inlet outlet pressure ratio is considered small and the pressure correction ignored. Then if the mean flow rate is ($\overline{Q}$),

$$\overline{V}'_{r(T_0)to(T_0+\alpha p)} = \frac{1}{n} \sum_{p=1}^{p=n} V_S e^{-\frac{\Delta H_0}{R(T_0+\alpha p)} + \frac{\Delta S_0}{R}} = \overline{Q}n \qquad (7)$$

Thus by allotting appropriate values to (ΔH_0), (ΔS_0) and (Q), and employing a simple computer program, the effect of the program rate (α) on retention time for temperature programmed separations can be evaluated. The program will be required to find that value of (n) that meets the conditions given in equation (7). Equation (7) can also be used to determine the effect of different relative values for the entropy (ΔS_0) and energy (ΔH_0) terms on the elution time at different program rates.

The value of (ΔH_0) for *n*-heptane eluted from a column carrying octadecane as the stationary phase is about –8.4 kcal/mol and the value of (ΔS_0) about –22 cals/mol (3). In the normal series of alkanes, the values of (ΔH_0) increases by about 1 kcal/mol and (ΔS_0) by about –1.5 cal/mol per unit increase in the number of carbon atoms in the chain. Employing equation (7) with a simple program the retention time of three solutes were calculated for a range of different program rates. The three solutes differed either by a small increment in (ΔH_0) or a small increment in (ΔS_0). The flow rate was taken as 60 ml/min at atmospheric pressure and ambient temperature and the initial temperature was taken as 47°C. The results obtained are shown in figure 14.

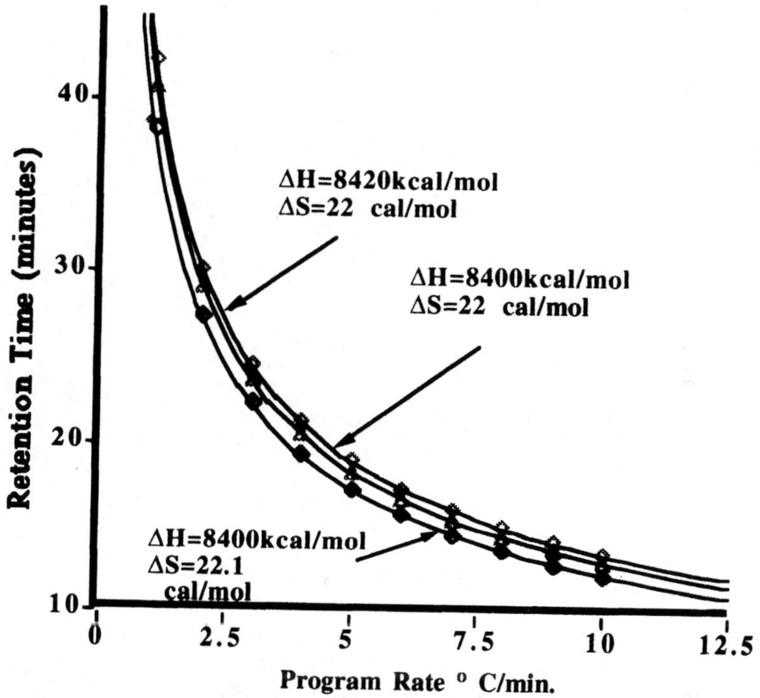

Figure 14 Graph of Retention Time against Program Rate for Solutes Having Different Enthalpy and Entropy Values

It is seen that all solutes behave very similarly and that the retention time can be easily calculated from equation (7). In practice the necessary constants can be evaluated for any solute, from the retention time taken isothermally at two different temperatures. It might be useful in certain difficult analyses if the elution time versus program rate curves were calculated from measured data for the last eluted solute and the closest eluted pair.

As the temperature $T_p = T_o + \alpha p$, the curves shown in figure 14 can be depicted in the more conventional form of curves relating the elution temperature against program rate. These curves are shown in figure 15.

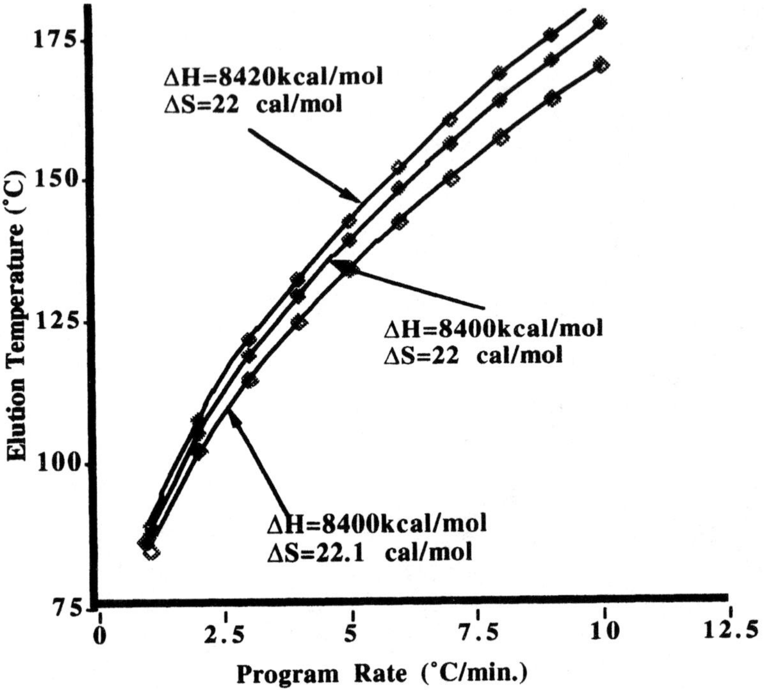

Figure 15 Graph of Elution Temperature against Program Rate for Solutes Having Different Enthalpy and Entropy Values

It is seen that, as would be expected, the elution temperature increases with program rate. However, the practical use of equation (7) arises from the fact that the elution temperature or retention time can be easily calculated for a range of different program rates from the data obtained from two initial isothermal experiments.

Finally the data used in figure 14 can be used to calculate any change in retention ratio between solutes having slightly different entropy values and slightly different enthalpy values but very close retention volumes. These are shown in figure 16. The reference solutes had (ΔH_O) values of 8.4 kcal/mol and 9.4 kcal/mol respectively and identical values for (ΔS_O) of -22 cal/mol. The values for the 4 closely eluting solutes are included in figure 16.

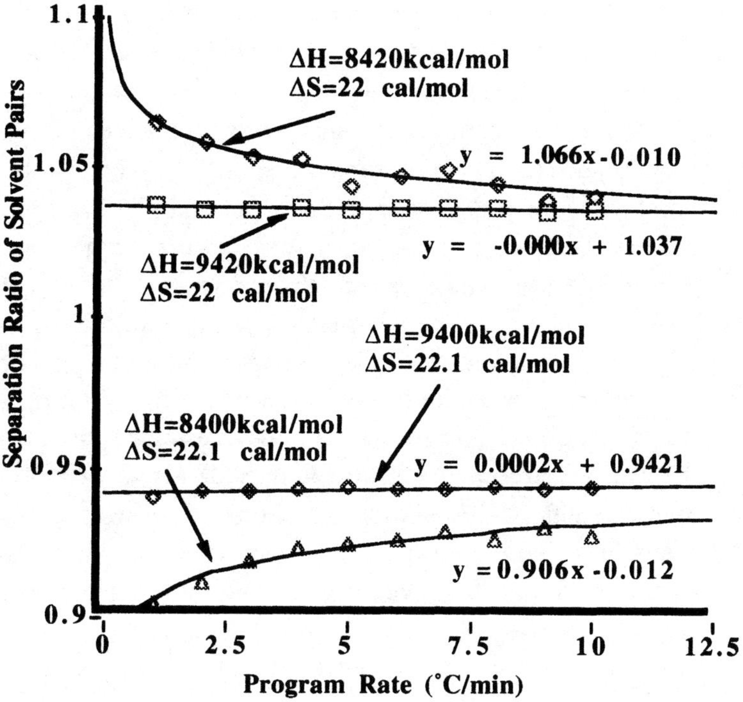

Figure 16 The Effect of Program Rate on Separation Ratio for Solutes with Small Enthalpy and Entropy Differences

It is seen that the effect of program rate on the retention ratio for solute pairs having small changes in either enthalpy or entropy is not significant for the more retained solute. For the less retained solute there appears to be a change in retention ratio with program rate but this is still relatively small. Furthermore as the ratio is smaller at the higher program rates, the reduction in peak dispersion resulting from the increase in solute diffusivity at the higher temperatures will probably compensate, and there will be little net change in resolution.

The separation ratios of solute pairs that may elute close together isothermally, but have significantly different enthalpy and entropy values, will change substantially as a result of temperature programming.

The temperature programming device is usually incorporated in the same unit as the oven temperature controller. The early programmers were electromechanical and thus rather clumsy machines, complicated to set up, and not highly reliable. Modern programmers are solid state electronic modules, compact, precise, dependable and easy to operate. Some can be manually operated, but many are controlled solely from the computer keyboard. The programmer should provide isothermal oven temperature control at any specified time and for any specified period. It should also provide temperature programs between prescribed temperature limits at chosen ramp rates that will be automatically initiated at defined times. All programmers provide linear gradients, but some provide exponential and logarithmic temperature programs as well. The temperature programming limits should range from about 5°C to 400°C although temperatures much above 300°C are not frequently used for reasons already mentioned. Linear program rates ranges from at least 0.5°C/min to 20°C/min should be available and optional non-linear functions can also be useful but are not essential. If operating below ambient temperature, oven cooling must be used.

Detector Ovens

The detector oven is physically distinct from the column oven and is controlled by a separate thermostat. It is usually relatively small, sufficient in size to enclose the detector only, and is operated isothermally and consequently is usually not programmable. As a result, it does not require a particularly rapid heat response and so does not usually contain a circulating fan. The detector oven temperature must, at all times, be held at least 25°C above the maximum column temperature used in the analysis to ensure no condensation takes place in the detector. It is also important to ensure that the conduit between the column and the detector, that often passes through a thermally insulating wall between the two ovens, does not fall below the detector oven temperature. If eluted solutes condense in the conduit between the column and detector, this will cause serious peak distortion, loss of trace components and denigrated quantitative accuracy. Some detectors have a maximum operating temperature so the detector oven should not be operated above this. It follows that the maximum column temperature will also need to be 25°C less than the detector temperature.

Column/Detector Connecting Conduits

As already discussed, any conduit between the column and the detector must be maintained at a temperature at least as high as that of the column oven and preferably equal to that of the detector oven. However, there are some other important aspects of column/detector conduits to be considered. Having resolved a pair of solutes in the column, if the integrity of the separation is to be maintained, then the peaks must not be allowed to disperse in the connecting conduits during passage to the detector otherwise they will merge together and the separation will be impaired. To a first approximation, the dispersion of a solute in a connecting conduit is inversely proportional to the diffusivity of the solute in the mobile phase, consequently the problem is far more serious in LC than GC, as the diffusivities are five orders of magnitude smaller. However, although there is little difficulty with packed GC columns because the peak volumes are relatively large, occasionally problems can arise when capillary columns are used. The peaks eluted from capillary columns will only be a few microliters in volume and thus any connecting conduits must have commensurate dimensions to that of the column. In general, when using a capillary column, it is wise to use the last few centimeters of the column as the connecting conduit and lead the column directly into the detector. This will virtually eliminate any extra-column dispersion that might take place between the column and the detector. If a packed column is used, then an appropriate length of quartz capillary tubing (about 500 μm I.D.) can be used as a connecting conduit. In general, the length of the connecting conduit should be kept as short as possible, irrespective of the type of column used. It should again be pointed out that extra-column dispersion is not nearly as serious in GC as it can be LC.

Gas Chromatography Detectors

There are a large number of GC detectors available from which the analyst can choose for any particular application. In general, the more specific the detector response, the higher its sensitivity and the more restricted its linear dynamic range. A specific detector may be selected for one of two reasons. First, the sample may be very dilute and thus the sample size small and the specificity of the detector is needed exclusively for its high sensitivity. Second, the specific detector may be used because the substance of interest is one of a multitude of different components and the specificity of the detector is used to selectively detect the substance of interest against a background of many unresolved peaks. In fact, the specificity of the detector is used as an alternative to a high efficiency column. Nevertheless, in quantitative analysis, the specific detector should only be chosen when a more general detector can not be used. Appropriate sample preparation techniques can often increase the sample concentration, or a longer open tubular column will often give the necessary resolution to handle multicomponent mixtures. It is recommended that the specific detector is used only as a last resort. The most effective general detector is, without doubt, the flame ionization detector (FID), which has a wide linear dynamic range, a more than adequate sensitivity, is robust, reliable and easy to operate. In addition, it can detect virtually all carbon-containing substances at very low concentrations. The FID is the detector of choice for quantitative gas chromatography.

The Flame Ionization Detector

The FID, invented by Harley and Pretorious [4], and separately by McWilliams and Dewer [5], evolved from the Heat of Combustion Detector developed by Scott [6]. The FID detector was originally used with hydrogen or a mixture of hydrogen and nitrogen as a carrier gas, which was burnt at a small jet situated inside a cylindrical electrode. A potential of a few hundred volts was applied between the jet and the electrode and when a carbon-containing solute was burnt in the jet, ions were formed that were collected at the cylindrical electrode. The resulting current was amplified and fed to a recorder or to the A/D converter of a computer data acquisition system. Later the hydrogen was passed into the detector eluent subsequent to the column, thus allowing other gases to be used as the carrier gas. A diagram of the basic FID is shown in figure 17.

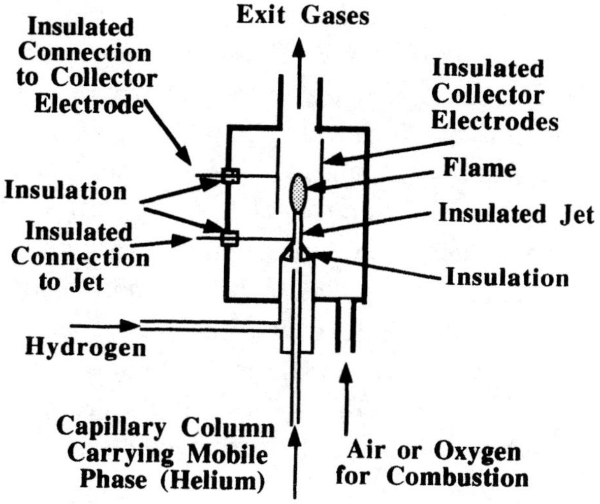

Figure 17 The Flame Ionization Detector

The properties of the FID have been examined in careful detail by Ongkiehong [5], Condon *et al.* [6] and Desty *et al.* [7]. The currently accepted simple explanation of the function of the FID is as follows.

During the process of oxidation, the solute molecules form oxidized or partially oxidized fragments in the flame that can emit electrons by thermionic emission, much like the heated filament of a thermionic "valve" or "tube". A potential of one or two hundred volts is applied between the jet and the cylindrical electrode that allows the electrons and any ions that are also formed to be collected and measured. The background current from the flame (ions and electrons formed by the combustion of hydrogen alone) is normally very small ($1\text{-}2 \times 10^{-12}$ amperes) and consequently the noise level is commensurably minuscule (about 10^{-14} amperes). The ionization process is relatively inefficient, only about 0.0018% of the solute molecules produce ions, that is about two ions or electrons per 10^5 molecules. However, due to the very low noise level, the minimum detectable mass of *n*-heptane is still as little as 2×10^{-12} g/sec. At a column flow rate of 20 ml/min this would be equivalent to a minimum detectable concentration of about 3 x 10^{-12} g/ml.

The detector response, as already stated, is almost independent of flow rate and thus can be used very easily with capillary columns. The eluent from the column is mixed with the hydrogen prior to entering the detector. The diluting effect of the hydrogen has no impact on the sensitivity, as the detector responds to *mass per unit time* entering the detector, not *mass per unit volume*. The FID detects very nearly all carbon-containing solutes, with the exception of a limited number of small molecular weight compounds such as carbon disulfide, carbon monoxide, etc. In fact, due to its diverse and comprehensive response, it could be classed as a universal detector, at least for organic materials. Due to the nature of the detecting system, the FID can operate at very high temperatures and still maintain its high sensitivity and wide linear dynamic range.

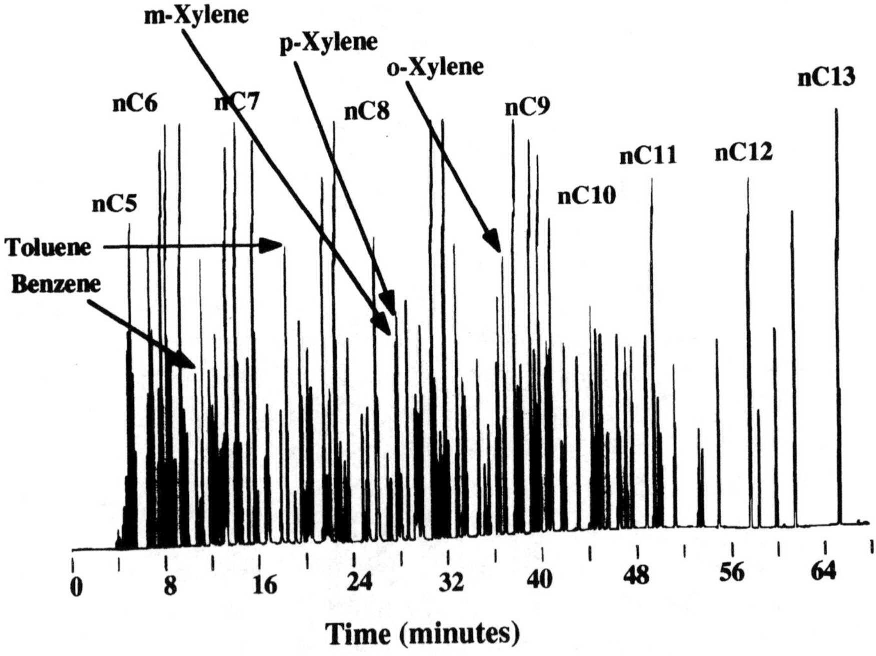

Courtesy of Supelco, Inc.

Figure 18 The Separation of a PIANO Standard Mixture

The FID is by far the most popular GC detector in common use today. It is used for essential oil analysis, solvent analysis and in the analysis of many pharmaceuticals that are either volatile or can be made into volatile derivatives. Its greatest area of application is, however, in hydrocarbon analysis. An example of the use of the FID in a *paraffin, isoparaffin, aromatic, naphthene and olefin* analysis of a hydrocarbon mixture (PIANO analysis) is shown in figure 18. The column was the Petrocol DH 50.2, 50 m long and 0.5 mm I.D. and made from fused silica. The column temperature was held a 35°C for 5 minutes and then programmed up to 200°C at 2°/min. The carrier gas was helium and the mobile phase velocity was 20 cm/sec.

The Nitrogen Phosphorus Detector (NPD)

The nitrogen phosphorus detector (NPD) is a specific detector with a high sensitivity that evolved from the FID. It gives a strong response to organic compounds containing nitrogen and/or phosphorus. It would appear to function in a very similar manner to the FID but, in fact, operates on an entirely different principle. A diagram of an NPD is shown in figure 19.

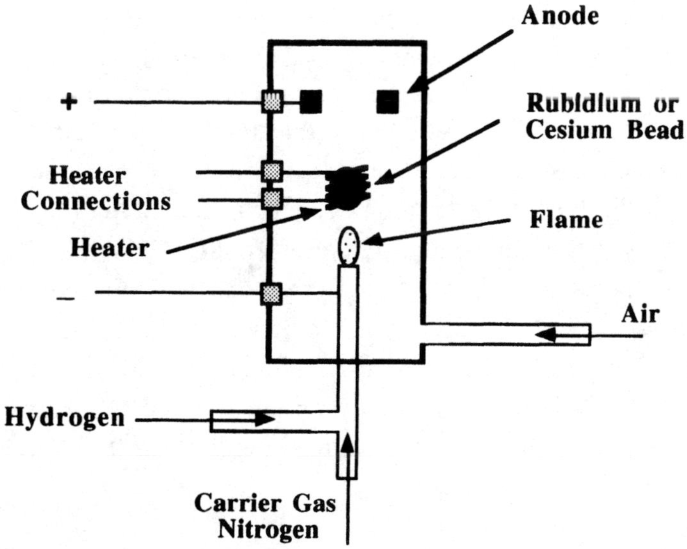

Figure 19 The Nitrogen Phosphorus Detector

The essential part of the NPD sensor is a rubidium or cesium bead contained inside a small heater coil. The nitrogen carrier gas from the column is mixed with hydrogen and flows into the detector through a small jet. The bead, heated by a current passing through the coil, is situated above the jet, and the nitrogen-hydrogen mixture passes over it. There are two basic modes of operation. If the detector is to respond to both nitrogen and phosphorus, then a minimum hydrogen flow is employed to ensure that the gas does not ignite at the jet. In contrast, if the detector is to respond to phosphorus only, a large flow of hydrogen can be used and the mixture burned at the jet.

An appropriate potential is applied between the bead and the anode. The heated alkali bead emits electrons by thermionic emission which are collected at the anode and thus produce an ion current. When a solute that contains nitrogen or phosphorus is eluted, the partially combusted nitrogen and phosphorus materials are adsorbed on the surface of the bead. This adsorbed material reduces the work function of the surface and, as a consequence, the emission of electrons is increased which raises the current collected at the anode. The sensitivity of the NPD is about 10^{-12} g/ml for phosphorus and 10^{-11} g/ml for nitrogen).

The main disadvantage of this detector is that its performance deteriorates with time. Reese [10] examined the function of the NPD in great detail. The alkali salt employed as the bead is usually a silicate and Reese demonstrated that the reduced response was due to water vapor from the burning hydrogen converting the alkali silicate to the hydroxide. At the operating temperature of the bead, the alkali hydroxide has a significant vapor pressure and consequently the rubidium or cesium is continually lost during the operation of the detector. Eventually all the alkali is evaporated, leaving a bead of inactive silica. This is an inherent problem with all NP detectors and as a result the bead needs to be replaced regularly if the detector is in continuous use.

The specific response of the NPD to nitrogen and phosphorus, coupled with its relatively high sensitivity, makes it especially useful for the analysis of many pharmaceuticals and in particular in environmental analyses involving herbicides. Employing appropriate column systems traces of herbicides at the 500 pg level can easily be determined. An example of the separation and identification of a series of herbicides employing the NPD is shown in figure 20.

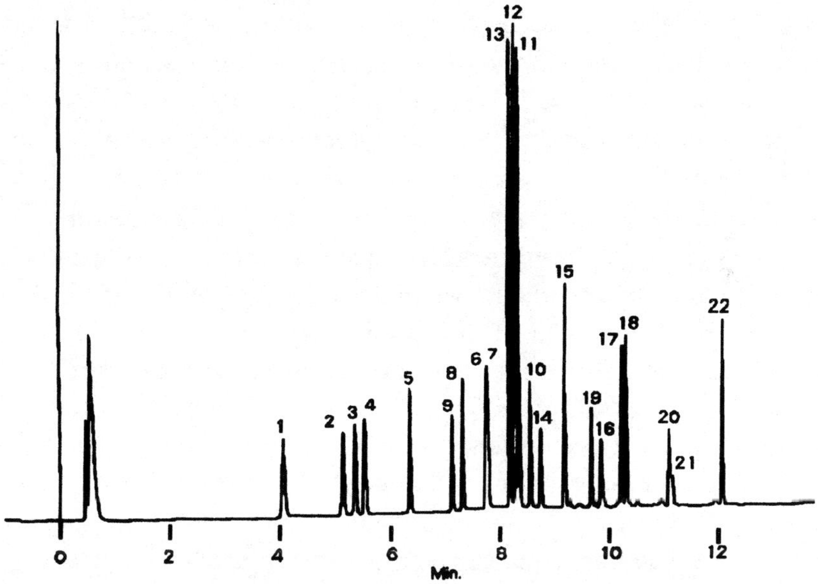

1/ Eptam®	2/ Sutan®	3/ Vernam®	4/ Tillam®
5/ Odram®	6/ Treflan®	7/ Balan®	8/ Ro-Neet®
9/ Propachlor	10/ Tolban®	11/ Propazine	12/ Atrazine
13/ Simazine	14/ Terbacil	15/ Sencor®	16/ Dual®
17/ Paarlan®	18/ Prowl®	19/ Bromacil	30/ Oxadiazon
21/ Goal®	22/ Hexazinone		

Courtesy of Supelco, Inc.

Figure 20 The Separation and Specific Detection of Some Herbicides Using the Nitrogen Phosphorus Detector

An SPB-5 column was used, 15 m long and 0.53 mm I.D. carrying a 0.5 µm film of stationary phase. The column temperature was held at 60°C for 1 minute and then programmed at 16°/min to 290°C and then held there for 5 minutes. The flow rate was 5 ml/min. and the carrier gas helium. The sample size was 1 μl of ethyl acetate containing 5 ng of each herbicide.

The Electron Capture Detector

The electron capture detector is one of a family of detectors invented by Lovelock [11] around the late 1950s and early 1960s. The first member of the family was the macro argon detector and although it is rarely used in GC today, it will be briefly described, as its function has a bearing on that of the electron capture detector. Due to the outer octet of electrons being complete in the noble gases, collisions between argon atoms and electrons are perfectly elastic. Consequently, if a potential is set up between two electrodes in argon, and ionization is initiated by a suitable radioactive source, electrons will drift towards the anode. If the potential of the anode is high enough, the electrons develop sufficient kinetic energy that on collision with an argon atom, a *metastable* atom can be produced. Consequently, a cloud of metastable atoms collect around the anode. A metastable atom carries *no* charge but has an electron in an outer orbit that gives it an energy of about 11.6 electron volts. Now 11.6 electron volts is sufficient to ionize most organic molecules. Hence, collision between a metastable argon atom and an organic molecule will result in the outer electron of the metastable atom collapsing back to its original orbit, followed by the expulsion of an electron from the organic molecule. The electrons produced by this process are collected at the anode generating a large increase in anode current. The argon detector is an extremely sensitive detector and can achieve ionization efficiencies of greater than 0.5%. The detector did not achieve popularity, however, largely because its response was neither linear nor predictable. Another deterrent arose due to the early detectors employing a "hot radioactive source" (strontium 90) to provide the ionization. The availability of less active sources and modern electronic linearizing circuits might well make it worthwhile to reexamine the argon detector for use in GC. A number of ionization detectors have evolved from the argon detector culminating in the development of the electron capture detector.

The electron capture detector (see figure 22) operates on a slightly different principle. A low energy β-ray source is used to produce electrons and ions. The first source to be used was tritium adsorbed into a silver foil but, due to its relative instability at high temperatures, was soon replaced by the Ni^{63} source. The detector can be used in two ways, either with a constant potential applied across the cell (the DC mode) or with a pulsed potential across the cell (the pulsed mode). In the DC mode, hydrogen or nitrogen can be used as the carrier gas and a small potential (usually only a few volts) is applied across the cell just sufficient to collect all the electrons available and provide a small standing current. If an electron capturing molecule (for example a molecule containing an halogen atom which has only seven electrons in its outer shell) enters the cell, the electrons are captured by the molecule and the molecules become charged. The mobility of the captured electrons are much reduced compared with the free electrons and thus the electrode current falls dramatically. There are, however, some distinct disadvantages to the use of the DC mode of detection. The most serious objection is the variation of electron energy with applied potential. As the electron capturing properties of molecules vary with the electron energy, the specific response of the detector to different molecules will depend on the applied potential

In the pulsed mode a mixture of 10% methane in argon is usually employed and the electron capturing environment is quite different. The electrons generated by the radioactive source rapidly assume only thermal energy and, in the absence of a collecting potential, exist at the source surface in an annular region about 2 mm deep at room temperature and about 4 mm deep at 400°C. A short period square wave pulse is applied to the electrode collecting the electrons and producing a base current. The standing current, using 10% methane in argon is about 10^{-8} amp with a noise level of about 5×10^{-12} amp. The pulse wave form is shown in figure 19. During the inactive period of the waveform, electrons having thermal energy only will become readily attached to any electron-capturing molecules present in the cell producing negatively charged ions. The negative ions quickly recombine with the positive ions and thus become unavailable for collection. Consequently the signal received will constitute a reduction in the standing current.

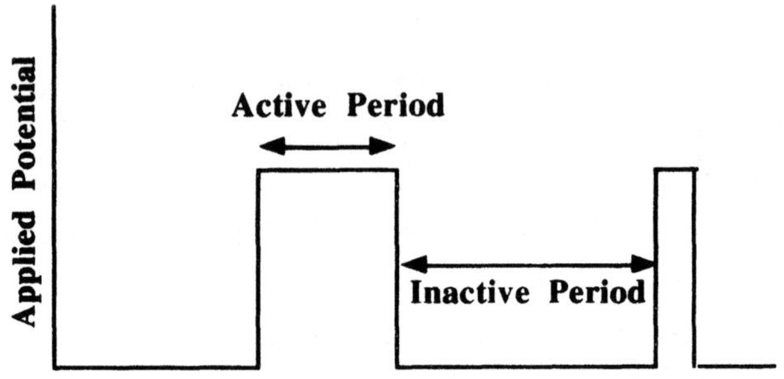

Figure 21 Wave Form of Electron Capture Detector Pulses

The period of the pulsed potential is adjusted such that relatively few of the slow negatively charged molecules have time to reach the anode, but the faster moving electrons are all collected. During the "off period" the electrons reestablish equilibrium with the gas. The three operating variables are the pulse duration, pulse frequency and pulse amplitude. By appropriate adjustment of these parameters the current can be made to reflect the relative mobilities of the different charged species in the cell and thus exercise some discrimination between different electron-capturing materials. A diagram of an electron capture detector is shown in figure 22.

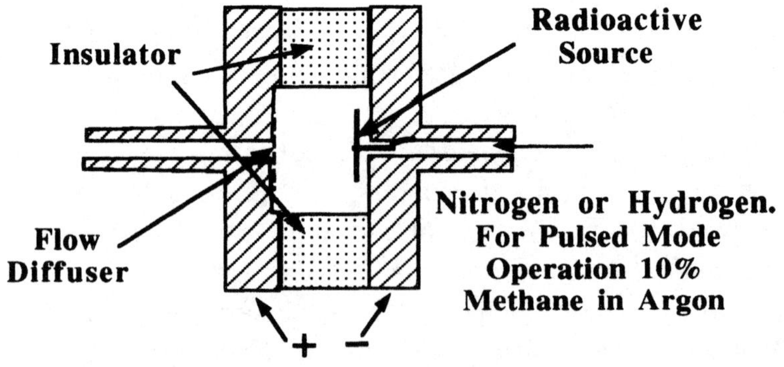

Figure 22 The Electron Capture Detector

There are a large number of different detector designs but the basic electron capture detector consists of a small chamber one or two ml in volume with two metal electrodes. The electrodes may be formed by concentric cylinders or by metal discs separated by a suitable insulator. The cell contains the radioactive source, usually electrically connected to the conduit through which the carrier gas enters and to the negative side of the power supply. A gauze "diffuser" is sometimes connected to the exit of the cell and to the positive side of the power supply. The electrode current is monitored by a suitable amplifier.

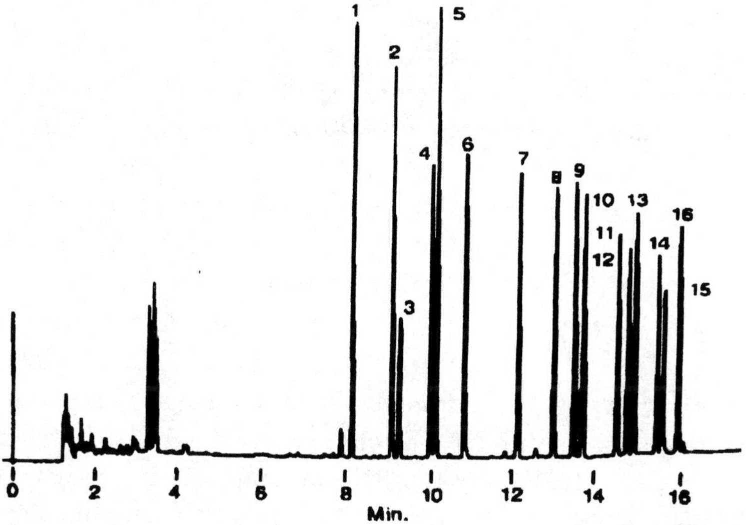

1 α–BHC	2 γ-BHC(Lindane)	3 β-BHC	4 Heptachlor
5 δ-BHC	6 Aldrin	7 Heptachlor Epox.	8 Endosulphan
9 4,4'-DDE	10 Dieldrin	11 Endrin	12 4,4'-DDD
13 Endosulphan 11	14 p,p'-DDt	15 Endin Aldehyde	16 Endosulp.Sulf.

Courtesy of Supelco, Inc.

Figure 23 The Analysis of Priority Pollutant Pesticides by Method 608

The electron capture detector is extremely sensitive, probably the most sensitive GC detector available (*ca.* 10^{-13} g/ml) and is widely used in the detection and analysis of halogenated compounds, in particular, pesticides. It will function using helium, argon or argon/methane mixtures as carrier gases. An example of a pesticide analysis employing an electron capture detector is shown in figure 23. The column used was a SPB-608 fused silica capillary column, 30 m x 0.53 mm I.D. with a 0.5 μ film of stationary phase. The column was programmed from 50°C at 1°/min to 150°C and then to 260°C at 8°/min. Helium was used as the carrier gas at a flow rate of 5 ml/min. The sample consisted of 0.6 μl of a solution of the pollutants in *n*-decane. The mass of each pollutant present was about 120 pg.

The Katharometer Detector

The katharometer detector (sometimes spelled catharometer and often referred to as the *thermal conductivity detector* or *hot wire detector*) is a relatively insensitive detector that has survived largely as a result of its catholic response and its sensitivity to the permanent gases. As a consequence, it is often used as the detector in gas analysis and in environmental testing. Its use in these types of applications has made it the fourth most commonly used GC detector. A filament carrying a current is situated in the column eluent and, under equilibrium conditions, the heat generated in the filament is equal to the heat lost by conduction and convection and consequently the filament assumes a constant temperature. At the equilibrium temperature, the resistance of the filament and thus the potential across it is also constant. The heat lost from the filament will depend on both the thermal conductivity of the gas and its specific heat and both these parameters will change in the presence of a foreign gas or solute vapor. The presence of a different gas entering the detector causes the equilibrium temperature to change, producing a change in potential across the filament. This potential change is amplified and fed to a suitable recorder. A diagram of the katharometer is shown in figure 24. There are two types of katharometer, the "in-line" cell where the column eluent actually passes directly over the filament and the "off-line" cell where the filaments are situated out of the main carrier gas stream and the gases or vapors only reach the sensing element by diffusion.

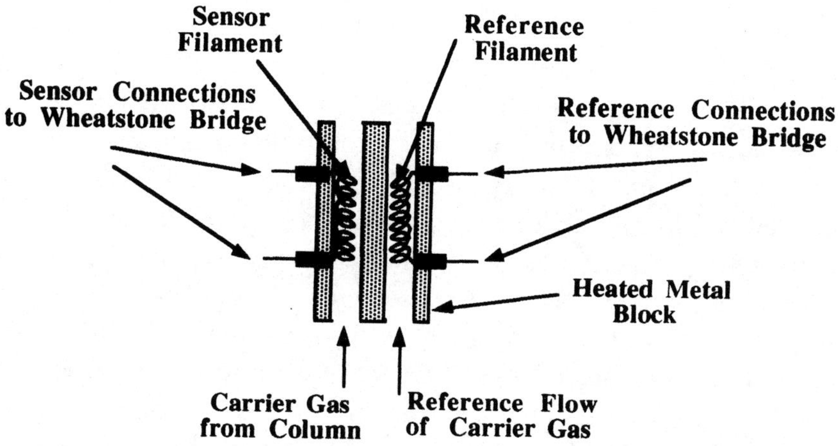

Figure 24 The Katharometer Detector "In-Line Cell"

Due to the high diffusivity of vapors in gases, the diffusion process can be considered as almost instantaneous. An example of the "off-line" cell is shown in figure 25.

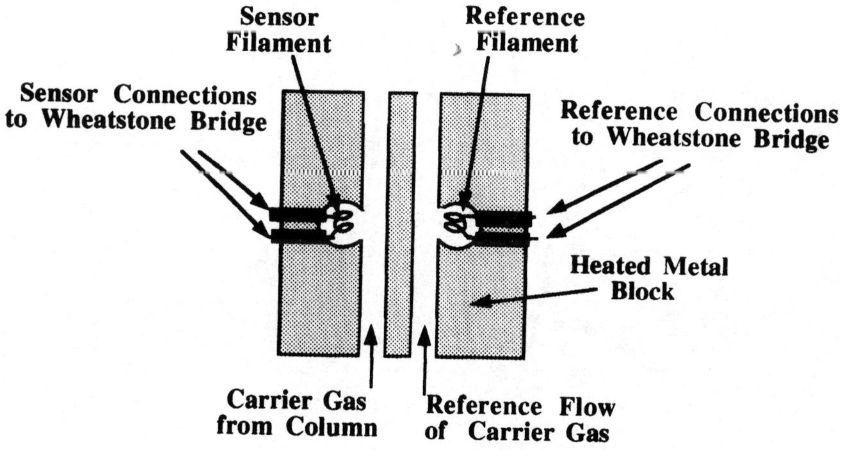

Figure 25 The Katharometer Detector ("Off-Line Cell")

Due to the principle of detection the katharometer detector is extremely *flow* and *pressure* sensitive. Consequently, all katharometer detectors are carefully thermostatted and fitted with reference cells to compensate for changes in pressure or flow rate. The filaments are made to be two of the arms of a Wheatstone bridge and the out-of-balance signal amplified and fed to a recorder or computer data acquisition system. For maximum sensitivity hydrogen should be used as the carrier gas, but to reduce fire hazards helium can be used with very little compromise in sensitivity. The sensitivity of the katharometer is only about 10^{-6} g/ml and has a linear dynamic range of about 500. Although the least glamorous, this detector can be used in most GC analyses that utilize packed columns and where there is no limitation in sample availability. The device is simple, reliable, and rugged and is particularly useful for those with limited experience in GC. An example of the separation of the various compounds of hydrogen, deuterium and tritium, employing gas solid chromatography and using a detector is shown in figure 26.

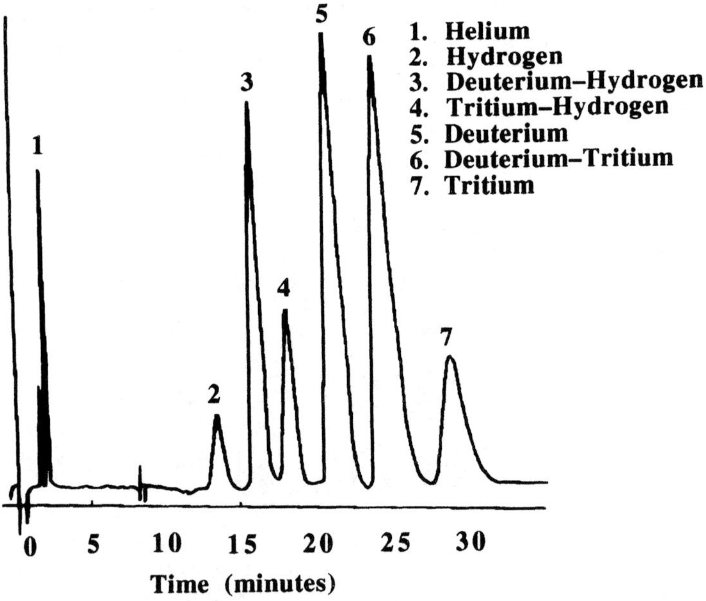

Courtesy of the Supelco Inc.

Figure 26 The Separation of the Compounds of Hydrogen, Deuterium and Tritium

The stationary phase was activated alumina (treated with $Fe(OH)_2$), and the column was 3 m long and 4 mm I.D. The carrier gas was neon, the flow rate 200 ml/min (at atmospheric pressure) and the column temperature was -196°C.

The four detectors described in detail so far are now well established and the most popular. In fact, as already stated the FID, ECD, NPD and the katharometer are employed in probably over 90% of all GC applications. Of the four, the FID is the most versatile, sensitive and linear, and the most likely to be found satisfactory for dealing with any analytical challenge that may arise. It is the detector most strongly recommended for quantitative analysis.

Synopsis

Modern gas chromatographs are designed primarily for quantitative analysis. The basic instrument consists of a controlling computer, gas supplies, flow controllers, appropriate sampling devices, a temperature programmable oven, detector and detector oven, a data acquisition system and a data display unit. The gasses are supplied from compressed gas cylinders or gas generators. Nitrogen and oxygen can be generated from the laboratory compressed air by selective diffusion and cleaning and hydrogen and oxygen generated by electrolysis. The flow of gas to the detector and/or column is managed by mass–flow controllers actuated by the computer. Thus, the column can be flow programmed if so desired. Care must be taken to employ an appropriate detector when using flow programming as the response of most detectors changes with the column flow rate. A sample valve is used to place gas samples on the column and a hypodermic syringe and silicone rubber diaphragms are used for placing liquid samples on the column. The sample can be discharged into a flash heater or directly into the actual column packing. Capillary column injectors employ similar devices but with a split-flow arrangement so that only a small portion of the sample passes onto the column. The use of a sample splitter reduces the precision and accuracy of the analysis and so large diameter tubular columns are often used so that the sample can be injected directly into the column. This procedure can introduce multiple peak formation due to droplet formation on the front of the column, and to prevent this happening techniques such as the "retention gap injection" and "solute–focusing" may need to be used. Column oven temperature should be controllable and programmable from 5°C to 400°C with a precision of ± 0.2°C and program rates provided that range from about 0.2°C per min to about 20°C per min. The column detector conduit must be designed to restrict peak dispersion to ensure no loss of resolution. The most common GC detectors used for quantitative work are the FID followed by the NPD and the ECD. The katharometer, another fairly popular detector, is used mostly for gas analysis. However, for quantitative work the FID is the most useful and effective; it is very sensitive (the minimum level of detection is about 2×10^{-12} g/s), linear, and has a linear dynamic range of over five orders of magnitude. In addition the response of the FID is not affected by changes in detector temperature or, within wide limits, by the column flow rate. In the FID carbon-containing molecules generate carbon fragments during combustion which emit electrons by thermal emission. These electrons are collected by an electrode system using a potential gradient ranging between 100–200 volts. The FID has a base current of about $1–2 \times 10^{-12}$ amperes and a noise level of less than 10^{-14} amperes. The NPD is similar to the FID except that the hydrogen containing the solute vapor passes over a heated cesium or rubidium silicate bead. The heated bead emits electrons by thermal emission which are collected by a suitably placed electrode. When a nitrogen or phosphorus compound is present in the eluent gas, the partially combusted material is adsorbed on the surface of the bead and reduces the work function of the surface. This results in a increase in electron emission and an increase in current. For the detector to respond to both nitrogen and phosphorus, a minimum flow of hydrogen that does not ignite must be used. When a large flow of hydrogen is used and is burned at a jet, then the detector only responds to phosphorus. The sensitivity is about 10^{-12} g/ml for phosphorus and 10^{-11} g/ml for nitrogen. During operation cesium or rubidium hydroxide is formed from the presence of water. As a result the alkali is continually lost at the bead temperature and consequently the bead has a limited life and must be occasionally replaced. The NPD is very sensitive, but has a much smaller linear dynamic range than the FID, and is not as linear (the response index being significantly greater than the FID over a much restricted concentration range). The ECD, another specific, high sensitivity detector, consists of a small volume cell containing a strong ionizing source of low energy β particles. The electron current is collected across two electrodes under a relatively small applied voltage. In the DC mode, hydrogen or nitrogen is used as the carrier gas and all the electrons are collected which provides a standing current. When an electron-capturing substance enters the cell, some electrons are captured, and the large charged molecules move slower than the electrons and consequently the current falls. In the pulsed mode, a 10% methane in argon mixture is used. A pulsed voltage is applied to the electrodes and the free electrons are collected at every pulse. When some electrons are removed by electron capturing molecules, then the quantity of mobile electrons falls, and the pulse of current is reduced. This detector is probably the most sensitive GC detector available (*viz.* 10^{-13} g/ml) but is selective for only those compounds that can capture electrons. The katharometer detector is a low sensitivity device and is not generally used with capillary columns. However, the katharometer is the detector of choice for packed columns particularly for gas analysis. It functions by measuring the change in resistance of an electrically heated wire situated in the column eluent. When a solute is eluted, the thermal conductivity and the specific heat of the carrier gas changes, which causes a change in the heat lost by the filament, resulting in a corresponding change in filament temperature and a consequent change in resistance. The katharometer has a sensitivity of about 1×10^{-6} g/ml, is very flow, pressure and temperature sensitive and has a restricted linear dynamic range of less than three orders of magnitude of concentration. Nevertheless, it has a very catholic response and thus is the most effective detector for the quantitative assay of permanent gas mixtures.

References

1. R. P. W. Scott, *Introduction to Analytical Gas Chromatography*, Marcel Dekker, Inc., New York (1998)65.
2. R. P. W. Scott, *Techniques and Practice of Chromatography,* Marcel Dekker,

Inc., New York (1995)27.

3. D. E. Martire, Private Communication.
4. J. Harley, W. Nel and V. Pretorious, *Nature, London*, **181**(1958)177.
5. G. McWilliams and R. A. Dewer, *Gas Chromatography 1958,* (Ed., D. H. Desty), Butterworths, London,(1958)
6. R. P. W. Scott, *Nature, London*, **175**(1955)422.
7. L.Onkiehong, *Gas Chromatography 1960* (Ed., R. P. W. Scott) Butterworths, London (1958)9.
8. R. D. Condon, P. R. Scholly and W. Averill, *Gas Chromatography 1960,* (Ed. R. P. W. Scott), Butterworths, London (1958)134.
9. D. H. Desty, A. Goldup and G. J. Leach, *Gas Chromatography 1960* (Ed. R. P. W. Scott) Butterworths, London (1958)156.
10. C. H. Reese, *Ph. D. Thesis,* University of London (Birkbeck College), 1992.
11. J. E. Lovelock and S. R. Lipsky, *J. Amer. Chem. Soc.*, **82**(1960)431.

Chapter 7

Gas Chromatography Applications

Gas chromatography is probably the most popular analytical technique available today and appears to be the technique of choice whenever and wherever it is appropriate and practical. There are a number of reasons for this. First, as a result of the high sensitivity of GC detectors, the technique can easily measure components in a mixture well below the nanogram level and is thus extremely important in environmental analyses. Second, GC columns can have exceedingly high efficiencies and thus resolve very complex mixtures or, alternatively, closely eluting solutes such as enantiomeric pairs. Third, the apparatus is relatively straightforward to operate, rugged and reliable and can provide results that are accurate and precise. The technique can, however, only be used for the separation of substances that are *volatile* (*i.e.* have a significant vapor pressure at the temperature of elution) and although derivatization can be used to increase the volatility of many compounds, the technique is naturally excluded from a large number of applications encompassing the analysis of involatile compounds.

A typical separation of a mixture of Arochlor 1016, 1232, 1248, 1260 (1:1:1:1) and monitored by an electron capture detector is shown in figure 1 and demonstrates the versatility and resolving power available from high efficiency open tubular GC columns.

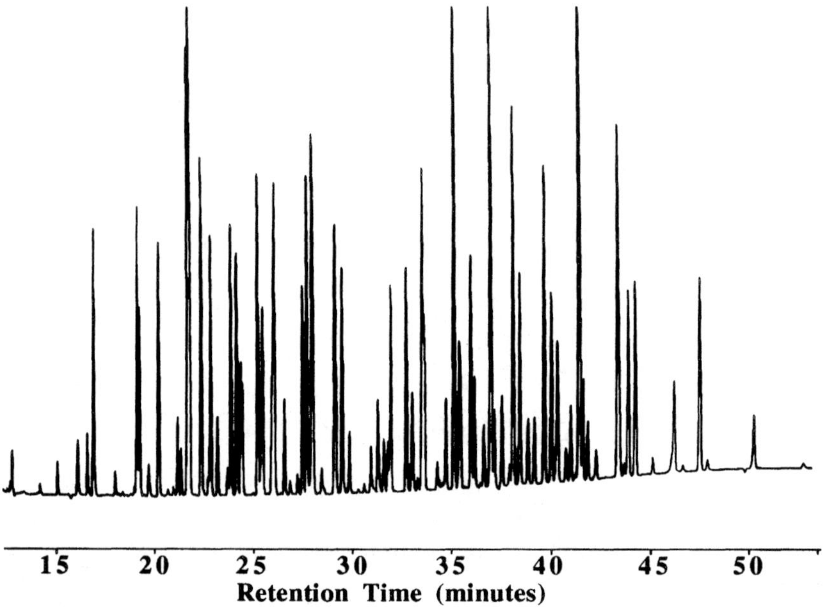

Courtesy of the International Journal of Environmental Analytical Chemistry [ref. 1]

Figure 1 The Separation of Arochlor 1016, 1232, 1248, 1260 (1:1:1:1) and Monitored by Electron Capture Detection

The column used was a fused silica tube, SIL-8, 50 m long, 0.22 mm I.D. carrying a film of diphenyl dimethyl siloxane 0.26 µm thick as the stationary phase. Helium was employed as the carrier gas at a velocity of 23 cm/sec and the column was temperature programmed from 90°C to 170°C at 20°/min, held at 170°C for 7.5 min, then programmed to 280° at 3°C/min and finally held at 280°C for 20 min.

The analysis of polychlorinated biphenyls (PCBs) in environmental samples usually requires that they be extracted from the sample matrix either by liquid-liquid extraction for aqueous samples or by Soxhlet extraction (or its equivalent) for solid samples. Interfering compounds, usually organopesticides, also need to be removed and this can often be achieved by acid treatment or by using column chromatography. Once isolated the PCBs can be separated either using a packed GC column or an open tubular (capillary) column. Both columns would normally be employed with an electron capture detector which would specifically monitor the elution of the PCBs. An alternative, more expensive option would be to use a mass spectrometer detector with electron impact ionization. The results from PCB analyses are usually reported in terms of their Aroclor content. Calibration is carried out employing one or more Aroclor standards for which the weight and relative retention is known and consequently the response factors can be calculated. This approach was developed and reported by Webb and McCall [2]. Whereas packed columns can be used for the quantitative analysis of congener mixtures, high resolution capillary columns can assay PCBs either as the individual congeners or as Aroclors. In the identification of Arochlors the chromatograms are compared with standard chromatograms by a pattern recognition procedure. The reference samples recommended by the EPA are Aroclors 1221, 1232, 1242, 1248, and 1254. Having identified the Arochlor, quantitative assays are obtained using calibration data. The procedure using high efficiency open tubular columns is briefly discussed in EPA method 8082 but details are somewhat lacking. If a mass spectrometric detection is employed using electron impact ionization either individual congeners or the Arochlors themselves can be quantitatively assessed at very high sensitivity. Further information on PCB analysis can be obtained from [3].

The GC applications that will be described and discussed in this chapter have been selected from a very large number of published reports. The examples that have been chosen (largely from more recent publications that utilize modern equipment and phases) will illustrate the practical range of GC for the analysis of volatile samples. For further examples, the reader is referred to the following two journals, *The Analyst,* published by the Royal Society of Chemistry, and *Analytical Chemistry*, published by the American Chemical Society. In addition, the *Journal of Chromatographic Science,* published by Preston Publications Inc., and the *Journal of Chromatography,* published by Elsevier, often include cases and reports on many of the more unusual and novel types of GC applications. It should be emphasized that quantitative applications of

GC have been published in the scientific literature for over 50 years, therefore it is likely that helpful information will be available in prior publications to assist the analyst faced with unfamiliar assays.

The Extraction and Analysis of Carbonyl Compounds from Some Natural Products [6]

Lipid hydroperoxides decompose to form aldehydes, the presence of which can be used to identify lipid peroxidation. The medium and short chain aldehydes are the cause of the unpleasant odor of rancid fat-rich foods [4]. In addition, the carbonyl compounds can be considered as the toxic messengers of the initial free-radical event of peroxidation [5] and their assay is often used to identify and measure the extent of peroxidation that is taking place. Estimation of aldehydes usually involves an extraction procedure that can be accomplished by liquid/liquid extraction or solid phase extraction. Stashenko *et al.* [6] used these two techniques to determine aldehydes in vegetable oil and also in human urine after the ingestion of beer over a period of two hours and at the same time evaluated the efficacy of the two extraction procedures.

Two milliliters of vegetable oil was placed in a capped tube and heated for one hour at 250°C in an oil bath. When cool, the sample was derivatized at room temperature with 2 ml of pentofluorophenyl-hydrazine solution (0.22 M in 0.1 M HCl) for three hours. The pentofluorophenyl-hydrazine derivatives were extracted from the aqueous layer three times with 2 ml of hexane and the bulked extracts concentrated to 1 ml by evaporation in a stream of nitrogen at room temperature. For solid phase extraction (SPE), the aqueous layer was passed through a SPE cartridge containing an octadecyl bonded packing and the derivatives eluted with 2 ml of hexane which was also concentrated to 1 ml by evaporation at room temperature in a stream of nitrogen. Two columns were used, one was 50 m long, 0.25 mm I.D. coated with a film of polydimethylsiloxane 0.25 μm thick and the other was 60 m long, 0.25 mm I.D. coated with a film of 5% phenyl-polydimethylsiloxane 0.33 μm thick. Both columns were held at 100°C for 5 min and then programmed up to 200°C at 2.5°C/min. They were then held at 200°C for 10 min and then programmed up to 250°C at 5°C/min. Helium was used as the carrier gas at 35.5 cm/s and argon-methane mixture was used as a makeup gas at 60 ml/min with the electron capture detector. Hydrogen at 5 ml/min and 30 ml/min was used as makeup gases for the nitrogen phosphorus detector and the flame ionization detector respectively. A chromatogram of the aldehydes in sunflower oil from a 5% phenylpolydimethylsiloxane column using the electron capture detector is shown in figure 2.

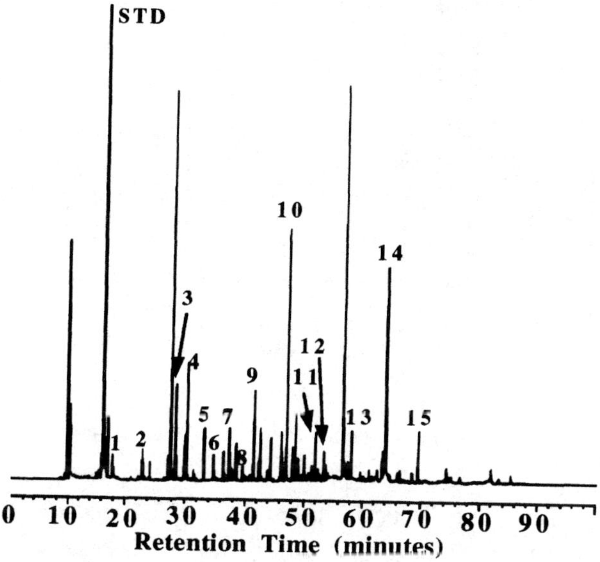

1.Methanal	6. Butanal	11. *Trans*-Hexanal
2. Ethanal	7. Isopentanal	12. Heptanal
3. Acetone	8. Crotanaldehyde	13. Octanal
4. Propanal	9. Pentanal	14. Nonanal
5. acrolein	10. Hexanal	15. Decanal

Courtesy of the Journal of Chromatography [ref. 6]

Figure 2 Chromatogram of Carbonyl Compounds from Sunflower Oil Monitored as Pentofluorophenylhydrazine Derivatives

It is seen that an excellent resolution was obtained between the individual aldehydes, and the detection limits reported ranged between

1×10^{-14} and 1×10^{-12} mol/ml per aldehyde using the electron capture detector. The results obtained from the analysis of urine are shown in figure 3.

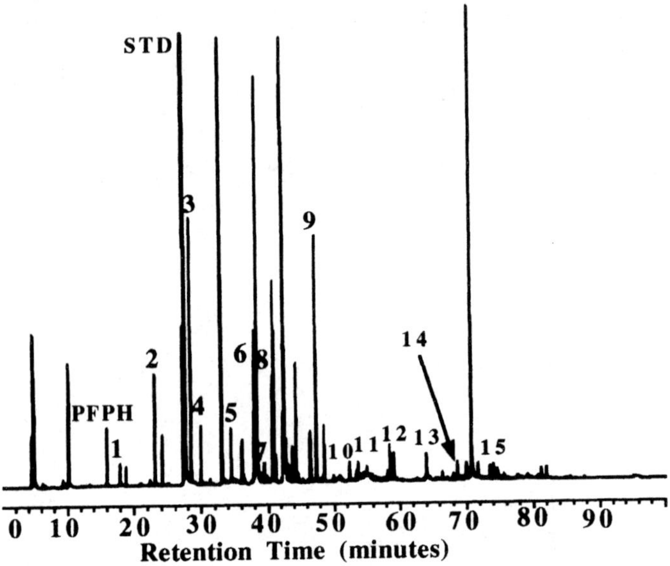

1. Methanal	6. Isopentanal	11. Heptanal
2. Ethanal	7. Crotanaldehyde	12. Octanal
3. Acetone	8. Pentanal	13. Nonanal
4. Propanal	9. Hexanal	14 Decanal
5. Butanal	10. *Trans*-2-Hexanal	15. Undecanal.

Courtesy of the Journal of Chromatography [ref. 6]

Figure 3 Chromatogram of Carbonyl Compounds Contained in Hexane Extract of Urine Monitored as Pentofluorophenylhydrazine Derivatives

Urine samples were taken every 2 hours from ten fasted men, 20-25 years old, after each had consumed 1.8 l of beer over an hour period. 5 ml of urine were derivatized at room temperature with 1 ml of pentofluorophenylhydrazine solution (0.22 M in 0.1 M HCl) for two hours. The pentofluorophenylhydrazine derivatives were isolated using SPE by passing the aqueous layer through a SPE cartridge containing an octadecyl bonded packing. The derivatives were recovered by eluting with 5 ml of hexane which was concentrated to 1 ml by evaporation at room temperature in a stream of nitrogen. The same column system and operating conditions were employed and it is seen that the same excellent resolution was obtained and in this case similar minimum detectable concentrations. It was also found that the recovery obtained using SPE ranged from 81-98% whereas the recovery employing solvent extraction only ranged from 51-79%

Determination of Insecticides in Drinking Water Using Dual Electron Capture and Nitrogen Phosphorus Detection [8]

A number of common insecticides are used to control a wide range of insects that attack cereals, fruit, vines, vegetables, coffee, cotton, rice, flax, beet sugar cane and other crops and, as a result, many of them find their way into drinking water [7]. The insecticides *carbaryl* (1-naphthyl methyl carbamate), *malathion* (diethyldimethoxythiophos-phoroylthiosuccinate), *parathion* (O,O-diethyl O-4-nitrophenyl phos-phorothioate), *fenitrothion* (O,O-dimethyl O-4-nitro-*m*-tolylphos-phorothioate) and *diazinon* (O,O-diethyl O-2-isopropyl-6-methyl-pyrimidin-4-yl phosphorothioate) are all regulated and their presence must be monitored in drinking water.

The general analytical procedure is to use an LC assay, as carbaryl tends to decompose at the temperatures required for development in GC. However, Oh-Shin *et al.* [8] developed a method using the pentafluoropropionic derivatives of the insecticides which allowed them to be eluted in stable form and each solute could be monitored either by an electron capture or nitrogen phosphorus detector. 200 ml of water was treated with 20 μl of the internal standard, parathion-methyl (100 μg/ml in acetone) and the mixture extracted with 13 ml of methylene chloride by shaking for three minutes. The organic phase was separated and evaporated to dryness under a stream of nitrogen. The dry residue was dissolved in 100 μl of ethyl acetate and 100 μl of pentafluoropropionic anhydride added. The mixture was then heated at 50°C for 60 min, evaporated to dryness under nitrogen and redissolved in 50 μl of ethyl acetate. 2 μl samples were injected onto the column. The separation obtained for both derivatized and underivatized samples are shown in figure 4.

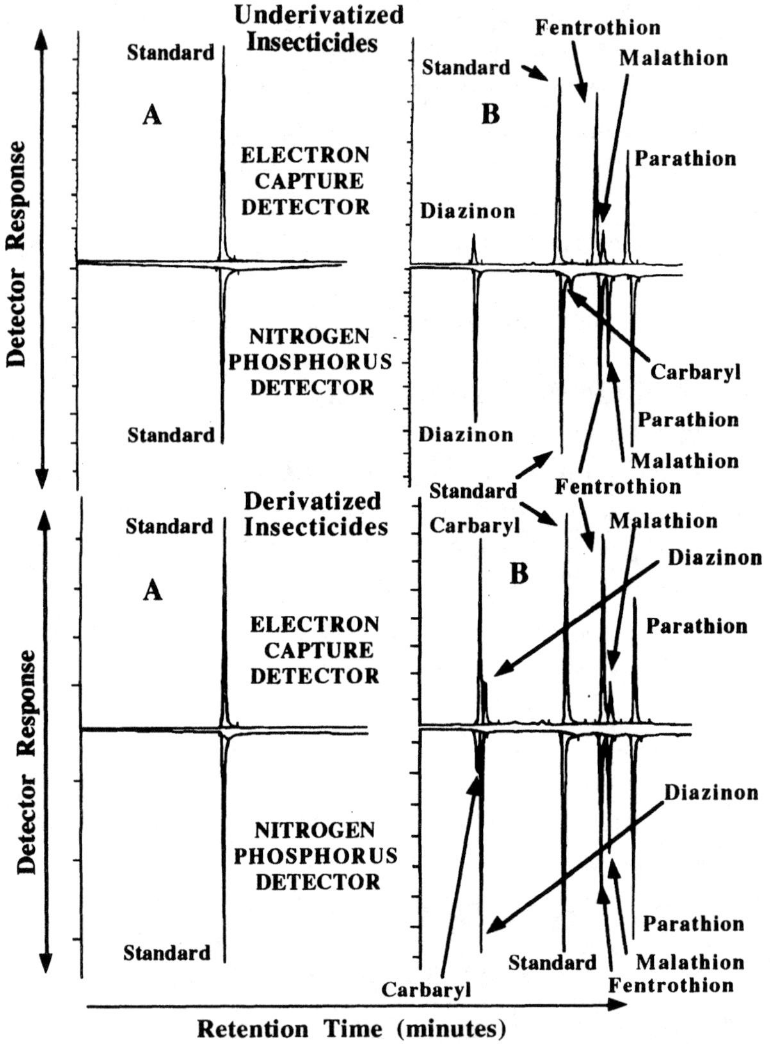

A Water Blank B Water Sample with 10 ng/ml of Each Insecticide

Courtesy of the Journal of Chromatography [ref. 8]

Figure 4 The Separation of Some Underivatized and Derivatized Insecticides Contained in Drinking Water

The HP-Ultra column that was used was 25 m long, 0.2 mm I.D. carrying a film of stationary phase 0.33 μm thick. The initial column temperature was 100°C which was programmed up to 280°C at 10°C per min. The carrier gas was nitrogen and the column flow rate 0.5 ml/min. It is seen that all the pesticides are well resolved and the derivatized carbaryl easily detected by the electron capture detector. The detection limit for all pesticides was below 0.1 ng/ml and pesticide recovery ranged from 79 to 87%. The particular value of this pesticide assay is that it allows gas chromatography to be used, as opposed to liquid chromatography, which permits the high resolution and high speed obtainable from open tubular columns to be realized. Above all, advantage can be taken of the high sensitivity of the electron capture detector and the nitrogen phosphorus detector to provide accurate trace analysis with relatively small samples.

Determination of Chlorobenzenes in Soil by Headspace Solid Phase Extraction and Ion-Trap Mass Spectrometry [11]

Chlorobenzenes find their way into the environment as solvents, dielectric fluids, deodorants and as intermediates from the production of pesticides, phenols and dyestuffs. They appear in water, soils, sediments, sewage sludges and in aquatic biota. The chlorobenzenes are listed as priority pollutants and some, for example, hexachlorobenzene are known to bc human carcinogcns [9,10]. It follows that their occurrence needs to be carefully monitored. The analysis of many of these matrices, soils, sediments, sludges, etc. usually requires several steps involving extraction preconcentration and complicated cleanup procedures. Santos *et al.* [11] developed a procedure using headspace solid phase micro-extraction followed by gas chromatographic analysis with ion trap mass spectrometric detection. The technique of headspace solid phase micro-extraction has been described in detail on page 33 in *Sample Preparation*, Chapter 2. A polydimethylsiloxane fiber, 100 μm thick, provided by Supelco was used. 0.1 g of soil was combined with 200 μl of water in a 40 ml sample vial. The water was shown to aid in extraction particularly when very small amounts of the chlorobenzenes were present. The fiber was conditioned at 250°C in the GC injection port, prior to use.

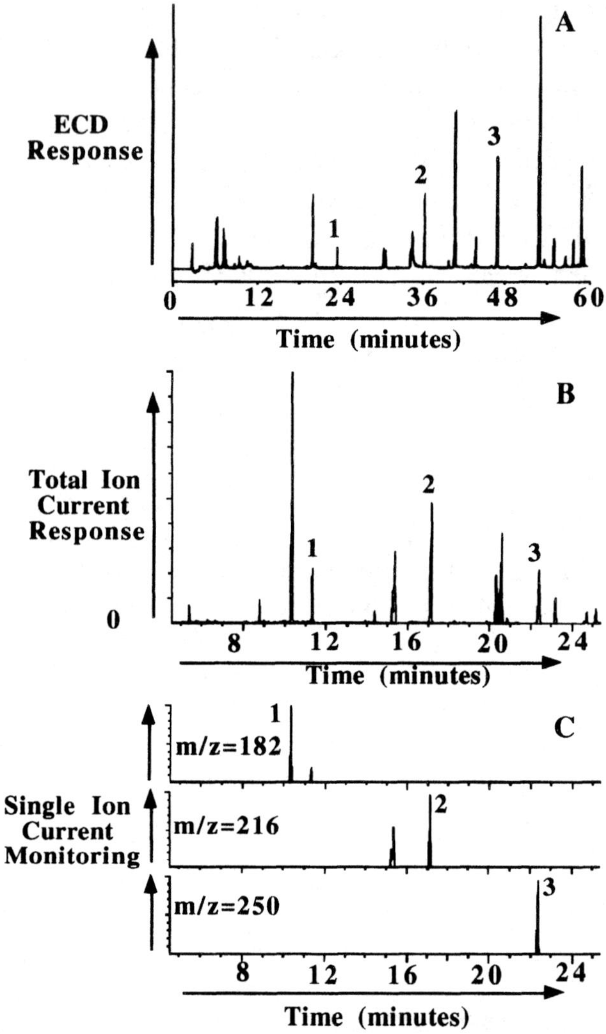

1. 1,2,3-trichlorobenzene 2. 1,2,3,4-tetrachlorobenzene 3. Pentachlorobenzene.
I.S. Internal Standard, 1,3,5-tribromobenzene.

Courtesy of the Journal of Chromatography [ref. 11]

Figure 5 The Separation of Some Chlorobenzenes on an Open Tubular Column

The vial was thermostatted at 30°C for 10 min and then the fiber was plunged into the headspace over the wet soil sample and held there for about 15 min. The fiber was then withdrawn, and the needle passed through the injector and the sample was desorbed from the fiber onto the column for 3 min at 250°C. The results obtained are shown in figure 5. The column employed was a Varian 3400 VX open tube, 30 m long, 0.25 mm I.D. coated with a film of 5% phenyl-95% methyl-polysiloxane 0.25 µm thick. Helium was used as the carrier gas at 31 cm/s and the column was held at 50°C for one minute, programmed to 90°C at 20°C/min and finally raised to 280°C at 25°C/min. An electron capture detector and a mass spectrometer were use to monitor the separation. Both methods of detection work well, the best resolution being obtained, as one would expect, from the mass spectrometer using single ion monitoring. Quantitative reproducibility was shown to be between 3 and 5 % and the minimum detectable concentration ranged from 0.03 to 0.1 ng per gram of soil. The advantageous use of headspace solid phase micro-extraction is clearly demonstrated. However, it must be remembered that the amount extracted will depend on many factors, such as the distribution coefficient of the solutes of interest between the wet soil and extraction fiber which, in turn, will depend on the geometry of the extraction chamber and its temperature. Consequently recovery will vary significantly from one solute to another.

Determination of Fentanyl in Whole Blood at the Subnanogram Level Using Nitrogen Sensitive and Mass Spectrometric Detection [12]

Fentanyl (a 4-anilinopiperidine derivative) is a rapid acting narcotic analgesic that has minimal effect on the cardiovascular system. It is 550 times more potent than meperidine, has a therapeutic index of 277 and as a result is the analgesic of choice for surgery, where it is important to produce anesthesia without cardiac depression. Analgesic and anesthetic effects are produced at serum levels of 0.5 to 20 ng/ml and the common techniques used for assaying fentanyl at these level are radio immunoassay, GC with nitrogen-selective detectors or GC/MS using single ion monitoring. The latter two methods have been compared by Watts and Caplan [12]. They employed a research compound as an internal standard that was structurally similar to alfentanil but contained a propyl instead of an ethyl bridge between the two heterocyclic rings. Peak area ratios were used to quantitatively compare peaks when using the nitrogen phosphorus detector and base peak ratios were used when employing a mass spectrometric method of detection. All glassware was rinsed with deionized water, methanol and then dried and finally silanized with hexamethyldisilazane. Splitless fused-silica injection port liners were soaked in nitric acid overnight and then rinsed with dionized water. The liners were then rinsed with methanol, hexane and then immersed in 10% dichlorodimethylsiloxane in toluene for 1 hr. The liners were then rinsed with toluene, ethanol, and hexane and then air dried. The liners were then packed with silanized glass wool as shown in figure 6.

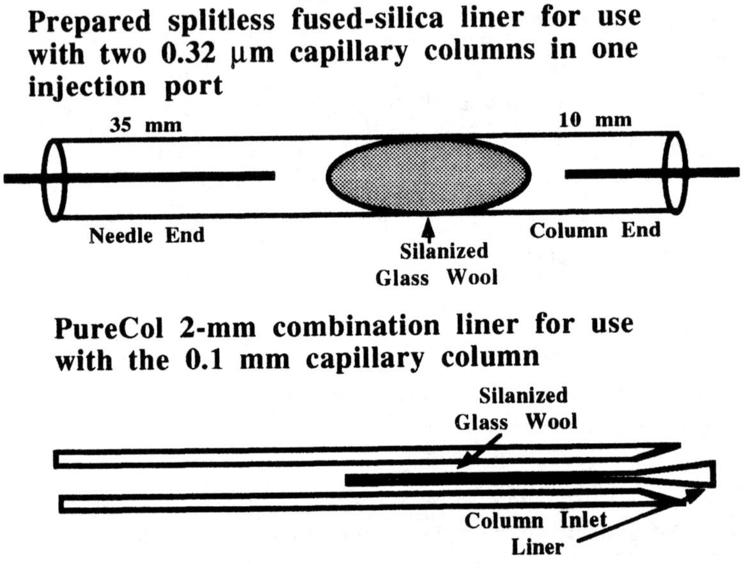

Figure 6 Design of Liners for Sample Injection

A 2 ml sample of blood was placed in a 17 ml screw-top culture tube and 10 ng of internal standard and 2 ml of saturated borate buffer (pH 9.0) added. The mixture was vortexed for 30 s and 10 ml of *n*-chlorobutane added and the tube rotated for a further 10 min. The tube was then centrifuged and the *n*-chlorobutane separated and placed in a second 17 ml culture tube containing 3 ml of 1 N sulfuric acid. The tube was rotated for 10 min and the *n*-chlorobutane discarded.

The acid phase was then washed with 3 ml of hexane and the hexane discarded and strong caustic soda solution added to raise the pH to 10.0. The contents were then re-extracted with 5 ml of *n*-chlorobutane for 10 min and after centrifugation the *n*-chlorobutane was transferred to a 5 ml conical tube and evaporated to dryness under nitrogen at 69°C. The residue was dissolved in 200 µl of a 50% ethanol solution in hexane and transferred to a 1 ml vial and evaporated to dryness in a heating block at 40°C. The residue was dissolved in 20 µl of a 50% ethanol solution in hexane and 1.5 µl was injected onto the chromatograph.

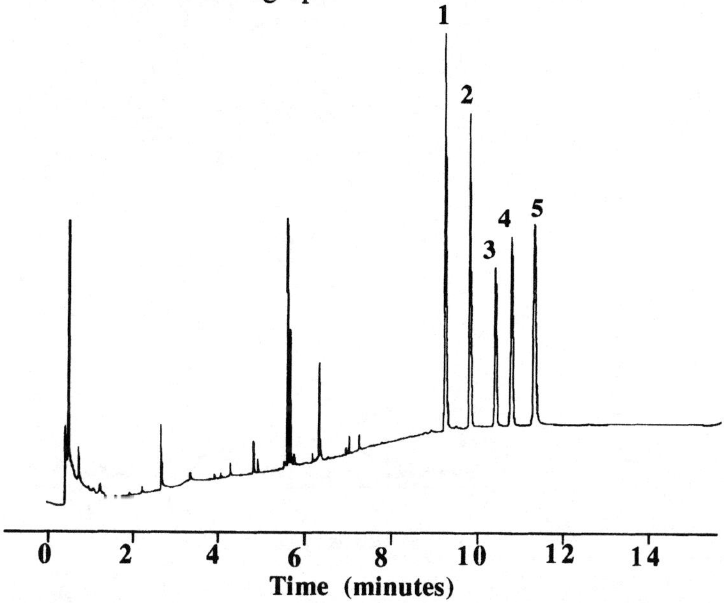

1. Fentanyl, 2. Sufentanil, 3. Caefebtanil, 4. Isofentanil and 5. Alfentanil.

Courtesy of the Journal of Analytical Toxicology [ref. 12]

Figure 7 The Separation of Some Fentanyl Compounds Monitored by the Nitrogen Phosphorus Detector

A chromatogram of fentanyl separated from four similar compounds is shown in figure 7. The two columns used were 12.5 m long, 0.32 mm I.D., the non-polar column (Ultra-2) carrying a film of 5 % phenyl methyl silicone 0.5 µm thick and the polar column (HP 17) carrying a film of 50% phenyl methyl silicone 0.26 µm thick. The linear velocities in each column at 250°C were 34 and 33 cm/s respectively. A two-holed-graphite ferrule sealed both columns into the same injection port. After injection the temperature of the columns was held at 90°C for one minute and then heated to 270°C at 30°C/min. and then held at 270 °C or 8 min. It is seen that an excellent separation was obtained. The overall recovery was found to be in excess of 75% over the concentration range of 0.25 to 2.5 ng/ml. Employing single ion monitoring the sensitivity obtained is shown by the chromatogram in figure 8.

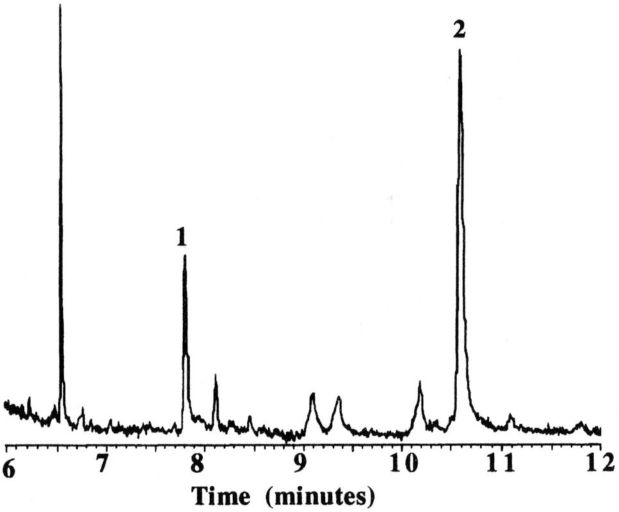

Courtesy of Journal of Analytical Toxicology [ref. 12]

Figure 8 Chromatogram from GC/MS Single Ion Monitoring of 2 ml Blood Sample Containing (1) 0.25 ng/ml of Fentanyl and (2) 5.0 ng/ml of Internal Standard

It is clear that a lower limit of detection of 0.25 ng/ml was readily obtainable, the resolution more than adequate and the analysis time less than 12 minutes. The *within* run precision determined at concentrations between 0.25 and 1 ng/ml ranged from 8 to 15% whereas the *between* run precision ranged from 3 to 12%.

The Measurement of Within-Tree Variation in Lignin Components by Pyrolysis Gas Chromatography [15]

Lignin is a three-dimensional cross-linked polymer consisting of various phenylpropane units. In hardwoods, such as eucalyptus and beech the lignin consists of guaiacylpropane units and syringylpropane units containing one or two methoxy groups respectively whereas in softwoods such as pine and spruce the lignin consists largely of guaiacylpropane units. The lignin content and its chemical structure have a significant influence on the pulping process of woods. In particular the ratio of syringyl and guaiacyl units (the S/G ratio) in hardwood lignin affects not only the pulping efficiency, but also the strength of the manufactured pulp and papers [13,14]. It follows that a simple and reliable method to determine the S/G ratio is required by the paper industry. Previously, various methods involving chemical degrading followed by chromatographic analysis had been used which was tedious, time consuming and required relatively large amounts of sample.

Yokoi *et al.* [15] developed a pyrolysis-gas chromatography procedure that allows the S/G ratio to be accurately determined using only 100 μg of powdered wood. A diagram of the vertical micropyrolyser is shown in figure 9. The sample is initially contained at about room temperature in a platinum sample holder (position 1, figure 9). The sample is then allowed to fall to position 2 where it is heated to, and maintained at, a preset temperature. The pyrolyser is connected directly to the column of the gas chromatograph which is fitted with a flame ionization detector. About 100 μg of cryo-milled wood is pyrolysed at one time using an inert gas such as helium as the carrier gas. The optimum pyrolysis temperature was found by experiment to be about 450°C. A metal capillary column was employed, 30 m long, 0.15 mm I.D. coated with 0.25 μm of polydimethylsiloxane immobilized by chemical cross-linking. The 50 ml/min of helium that flowed through the pyrolyser then passed though a splitter that allowed 1 ml/min to flow through the capillary column. The separation of the pyrolysis products from eucalyptus wood is shown in figure 10.

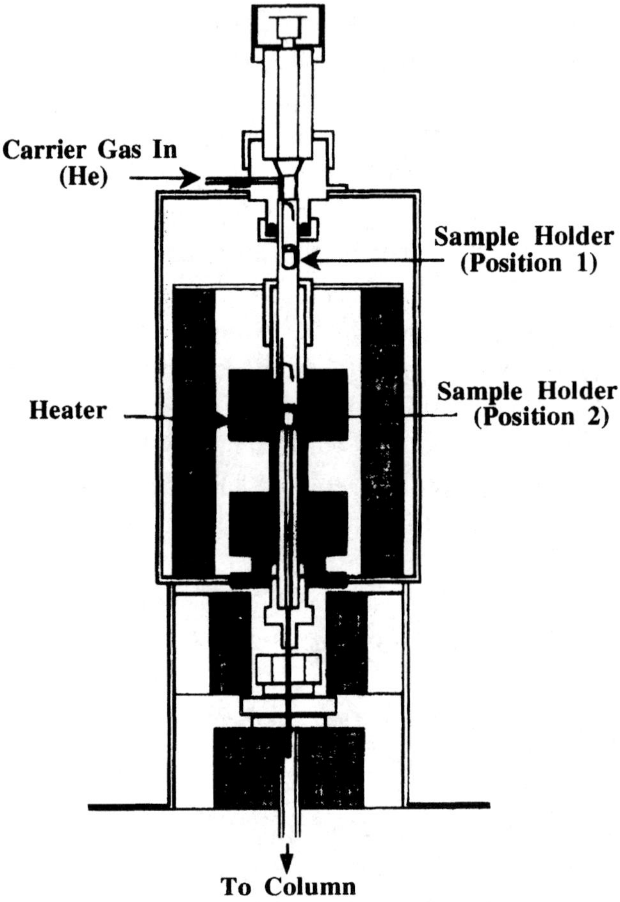

Courtesy of the Royal Society of Chemistry [ref. 15]

Figure 9 The Vertical Microfurnace Pyrolyser

The column temperature was programmed from 50°C to 280°C at a rate of 5°C per min. It is seen that an excellent resolution of the pyrolysis fragments was obtained and the S/G ratio could be determined from the fragments in about an hour and found to have a standard deviation of about 2%.

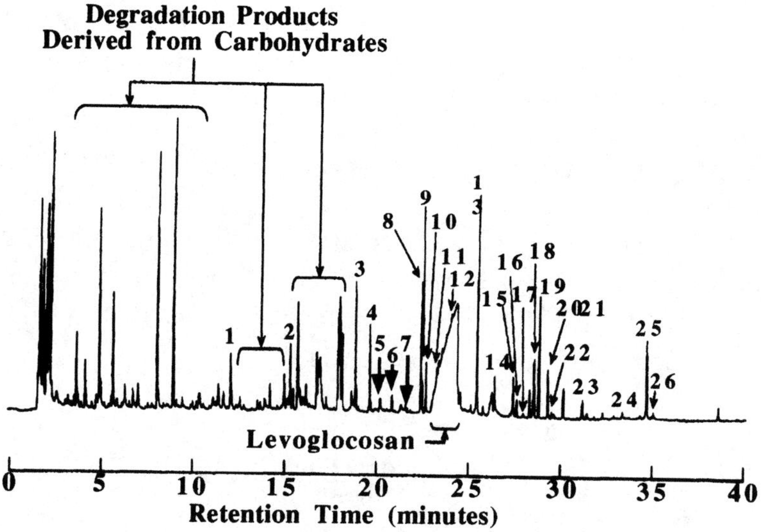

1. Guaiacol
2. 4-Methylguaiacol
3. Vinykguaiacol
4. Syringol
5. Eugenol
6. Vanillin
7. *cis*-Isoeugenol
8. Methylsyringol
9. Homovanillin
10. *trans*-Isoeugenol
11. Acetoguaiacone
12. Guaiacylacctonc
13. Vinylsyringol
14. Allylssyringol
15. Syringaldehyde
16. *cis*-Propenylsyringol
17. *cis*-Coniferyl Alcohol
18. Homosyringaldehyde
19. *trans*-Propenylsyringol
20. *trans*-Coniferaldehyde
21. Acetosyringone
22. *trans*-Coniferyl Alcohol
23. Syringylacetone
24. *cis*-Sinapyl Alcohol
25. *trans*-Sinapaldehyde
26. *trans*-Sinapyl Alcohol

Courtesy of the Royal Society of Chemistry [ref. 15]

Figure 10 A Typical Chromatogram from *Eucalyptus* Wood Obtained at 450°C

The peaks in figure 10 were identified by connecting the column exit directly to the inlet of a mass spectrometer. This application is an elegant example of the use of a pyrolyser as the sampling system of a gas chromatograph. There are a range of different types of pyrolysers available to the analyst from which to choose and the optimum for a particular application can only be ascertained by experiment. However, despite the random nature of the pyrolysis process, the technique can be made reproducible and, as in the example given above, allow a chromatographic technique be applied to substances that are neither volatile nor soluble.

Use of Capillary Columns for Sample Microextraction [16]

A coated open tube can be used for sample microextraction in much the same way as solid phase extraction cartridges. The use of the coated capillary in this way was examined by Tan *et al.* [16]. The extraction apparatus is shown in figure 11.

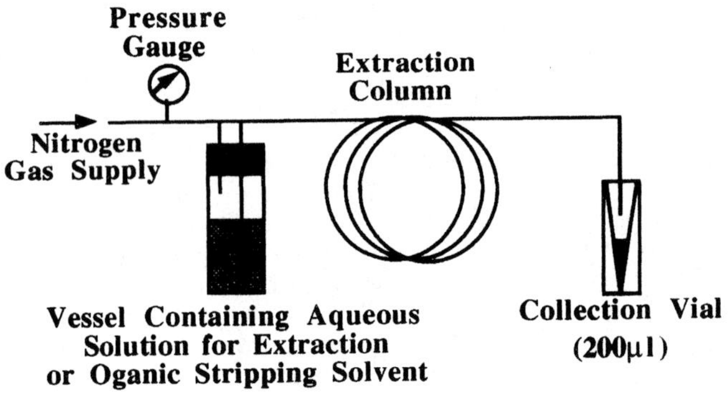

Courtesy of the Royal Society of Chemistry [ref. 16]

Figure 11 Solid Phase Extraction Apparatus Using a Coated Capillary

Basically, the apparatus consisted of a supply of nitrogen gas at elevated pressure, a sample container, the extraction column and a sample receiver. The extraction column, which was heavily loaded with stationary phase, was 1 m long, 0.25 mm I.D. carrying a film of polyethylene glycol 3.5 μm thick. In the determination of phenols in water, the water sample (about 100 μl) was forced through the column until all the water was expelled. It was found that the addition of salt (sodium chloride) and buffering the water sample to a pH of 2.0 significantly increased the efficiency of the extraction. The tube was then dried by the passage of nitrogen at 20°C (higher temperatures might cause the extracted material to elute). When dry the adsorbed material was back extracted into hexane by passing about 20 µl through the column, a known weight of standard added and an appropriate volume injected onto the capillary column. A chromatogram of some extracted phenols is shown in figure 12.

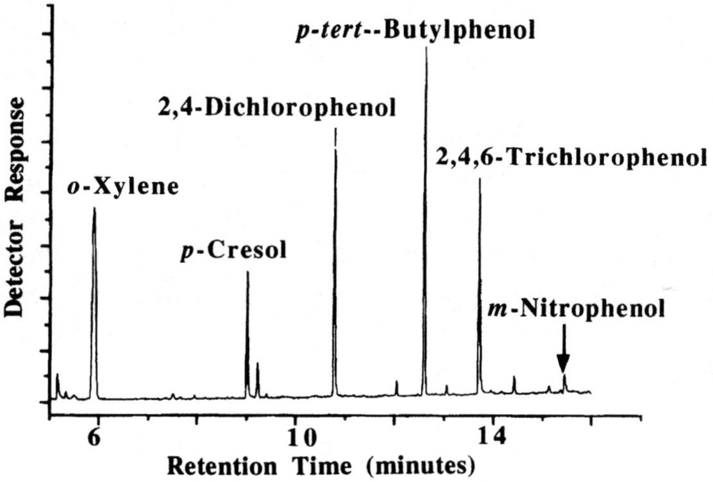

Courtesy of the Royal Society of Chemistry [ref. 16]

Figure 12 Chromatogram of Some Phenols Extracted from Water by Use of a Coated Capillary Tube

The column used was 30 m long, 0.25 mm I.D. coated with a film of BPX5 0.25 µm thick. The flow rate was 1.9 ml/min which is equivalent to a linear velocity of about 37 cm/s. The separation was carried out with an initial temperature at 93°C which was raised to 190°C at 10°C per min and held at 190°C for one minute. It is seen that an good separation was obtained in less than 30 min.

This procedure, using a short length of coated capillary, is an interesting alternative to the standard SPE cartridge but does not appear to offer any great advantages. There is significantly more physical manipulation required and, in addition, the apparatus is more complex. The authors also suggest that the method may be used for measuring the distribution coefficients of solutes between a solvent and the stationary phase and include examples in their paper.

Extraction Technique for the Analysis of Protective Clothing Breakthrough Indicator Pads [17]

Many of the problems that arise in quantitative GC analysis involve the extraction of the materials of interest from their natural matrix and isolating them in a suitable form (usually as a solution in an appropriate solvent) for injection onto the column. The matrices may be many and varied and obviously can not be discussed comprehensively. However, special examples have been included to illustrate how different matrices have been handled and perhaps give the analyst with a novel application some idea where to start. This procedure was developed by Stephen *et al.* [17] for the extraction of materials that have been adsorbed on carbon based test pads.

To provide a little background to this work, respiratory and dermal exposures are common in the workplace and a major focus has been on respiratory monitoring despite the dangers involved in other forms of contact such as dermal exposure. To monitor dermal exposure 'protective clothing indicator pads' are attached to the hands before gloving and subsequently the pads are examined for any solvents or other material that may have diffused through the glove walls. The pads usually contain the adsorbent carbon and in the example given here the solvents acetone, ethyl methyl ketone, trichloroethylene, toluene, *m*-xylene and D-limonene are considered the solvents of interest. To test the process the pads were spiked by injecting a few microliters of a dilute solution of the solvents onto the surface of the pads. The pads were then immersed in 2 ml of extraction solvent (hexan-2-ol) contained in an extraction vessel and the vessel capped. The extraction vessel was then placed in the microwave extractor and extracted for 5 min at 40°C with 20% power. The extract was then allowed to cool to room temperature, the extraction vessel opened and the extract filtered using syringe filters and the filtrate placed in a capped vial. An appropriate volume of sample of the extract (*ca.* 0.5

μl) was then placed on the GC column. The separation obtained is shown in figure 13.

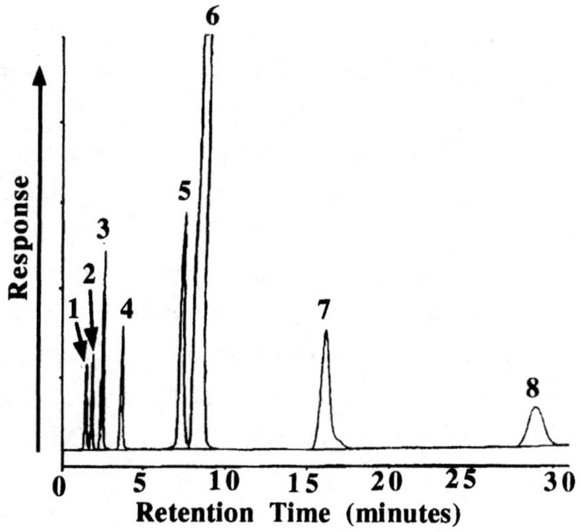

1. Acetone
2. Ethyl Methyl Ketone
3. Trichloroethylene
4. Toluene
5. *m*-Xylene
6. D-Limonene

Courtesy of the Royal Society of Chemistry [ref. 17]

Figure 13 Chromatogram of Pad Contaminant after Microwave Extraction and Recovery

A packed column was used, 3.05 m long, 3.18 mm I.D. containing 80/120 Carbopack B on which was coated 3% of SP 1500. The separation was developed isothermally at 225°C using helium as the carrier gas at a flow rate of 25 ml/min. It is seen the solvents chosen for test are adequately separated and the analysis time is about 30 min. The authors made an interesting observation that the recovery efficiency decreased as the boiling point of the solvent increased which is contrary to what one might expect. They also noted that the recovery could be increased for the higher boiling solvents if the extraction temperature was also increased. No explanation was given for this unexpected result. The need for microwave heating is not clear. It is likely that the same results would be obtained using other means of heating.

Analysis of Pethidine and Methadone in Human Urine by Solid Phase Microextraction and Gas Chromatography [18]

Pethidine (meperidine) and methadone are analgesic drugs widely used in the relief of pain. Pethidine can be used in any situation where an opioid analgesic is required. Methadone is also used in the treatment of opioid abstinence syndromes and heroin users. Athletes have been known to abuse these substances as a means of enhancing their competitive performance. The International Olympic Committee Medical Commission and other sports federations have defined these drugs as prohibited classes of pharmacological agents. Consequently, a rapid and highly sensitive method for their determination in biological fluids, *e.g.* blood and urine, is required.

Myung *at al.* [18] developed a method involving solid phase micro-extraction that simultaneously extracts and concentrates the drug from its blood or urine matrix. The technique of headspace solid phase micro-extraction has been described in detail on page 33 in *Sample Preparation*, chapter 2, so only specific details will be given here. The fibers that were used either carried a film of polyacrylate 85 μm thick or polymethyldisiloxane 100 μm thick. All fibers were conditioned in the hot injection port of the gas chromatograph prior to use. The optimum extraction conditions were found, by experiment, to be a 30 min immersion in the sample after the addition of sodium chloride to produce a concentration of 15% g/g and caustic potash to raise the pH to 11.0. The procedure was as follows: the internal standard (diphenylamine) was added to the urine sample as a methanol solution (15 μl of a solution containing 100 μg/ml) and the methanol was then removed in a stream of nitrogen; 5 M caustic potash was added to bring the pH to 11.0 and sodium chloride added to bring the salt concentration to 15% w/w; the fiber was then immersed in the prepared sample for 30 min at room temperature. After extraction the fiber was then transferred to the inlet of the gas chromatograph where the drugs were desorbed at 240°C. The nitrogen phosphorus detector was used to monitor the separation and the column was held initially for 30 s at 100°C and then programmed up to 300°C at 20°C per min and held there for 3 min. Examples of the chromatograms obtained are shown in figure 14.

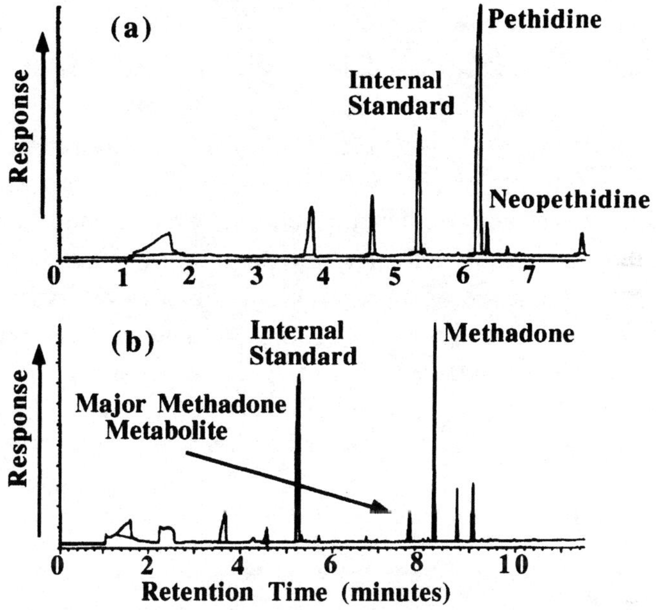

Courtesy of the Royal Society of Chemistry [ref. 18]

Figure 14 Chromatograms of Urine Sample from Addicts Showing (a) Pethidine and (b) Methadone

It is seen that more than adequate resolution is obtained indicating the analysis time might be shortened even further. In any event, the separation was fairly rapid, the analyses being completed in less than 10 min. The sensitivity of the method to both drugs was less than 1 ng/ml of urine which makes the technique very suitable for rapid screening in anti-doping tests in sport and for clinical and forensic use. The within day relative standard deviations (n=5) in the working range of 1-12 μg/ml for pethidine and 0.05 and 0.8 μg/ml for methadone was 1.8-5.9 and 4.0 and 8.7% respectively.

The Determination of Methyl Mercury and Inorganic Mercury in Whole Blood by Head Space Cryofocusing Gas Chromatography with Atomic Adsorption Detection [19]

Mercury is widely distributed in nature occurring in a range of different forms displaying various degrees of toxicity. Mercury can appear in both inorganic and organic forms, the inorganic form is often converted to organic form by natural biological processes. One of the more toxic organomercury compounds is methyl mercury which is the most dominant mercury compound found in blood. In monitoring the presence of mercury compounds in blood the amount of organic and inorganic mercury need to be determined separately as the two types differ significantly in toxicity. Brunerro *et al.* [219] developed a procedure for determining the two types of mercury using the apparatus shown in figure 15.

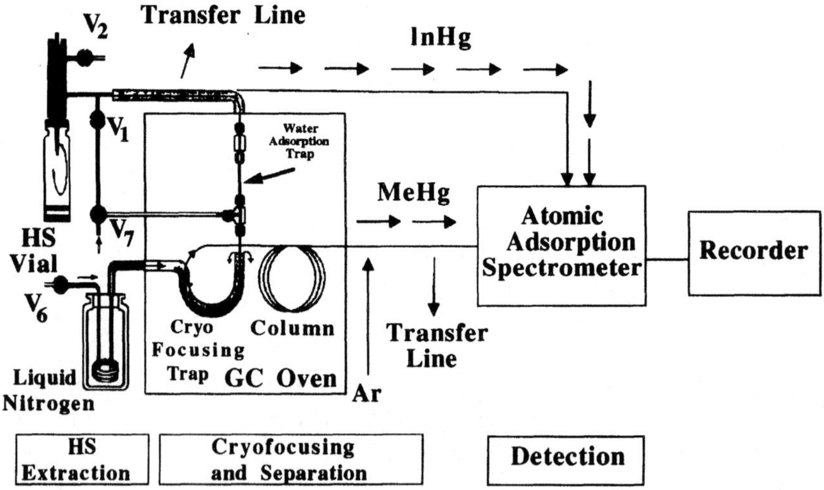

Courtesy of the Royal Society of Chemistry [ref.19]

Figure 15 Apparatus for Determining the Mercury Species in Whole Blood Samples

The methods used for measuring inorganic mercury and organic mercury were a follows.

Determination of Inorganic Mercury in Whole Blood

2 ml of whole blood were treated with 0.25 ml of concentrated sulfuric acid and 0.4 ml of stannous chloride (4 g of stannous chloride dissolved in 10 ml of concentrated hydrochloric acid) in a head space vial and mechanically stirred for 2 min. The contents were then heated to 65°C for 3 min. During this period the head space system is in standby position (figure 1) and a carrier stream of nitrogen at 32 p.s.i. flows through the transfer line at 150°C to the quartz cell. The head space vial is then placed in the apparatus and programmed to the pressurization position. The sample needle, preheated to 150°C is allowed to pierce the septum of the sample vial and the vial pressurized at 32 p.s.i. or 3 min. The flow of nitrogen is then stopped, the vapor from the vial allowed to flow for 24 s through the transfer line to the atomizer and the absorbance signal measured by the atomic absorption spectrometer.

Determination of Organic Mercury in Whole Blood

2 ml of whole blood were placed in a head space vial together with 10 mg of iodoacetic acid and 0.3 ml of concentrated sulfuric acid, vigorously shaken for 1 min and heated at 110°C for 6 min. After pressurizing for 2 min the over-pressured sample in the head space vial is applied to the gas chromatograph for 24 s, the transfer line being maintained at 150°C. During sampling the front of the column is cooled by liquid nitrogen which focuses all the material onto the front part of the column. The open tubular column was 30 m long, 0.53 mm I.D. carrying a film of polyethylene glycol ester as the stationary phase 1.2 μm thick. Helium was used as the carrier gas at a flow rate of 14 ml/min. During the development the column was held at 100°C for 1.5 min and then programmed up to 170°C at 15°C per min. and finally held at 170°C for 10 min. The results obtained for a standard water sample and blood sample spiked with 10 μg/l of methyl and ethyl mercury are shown in figure 16.

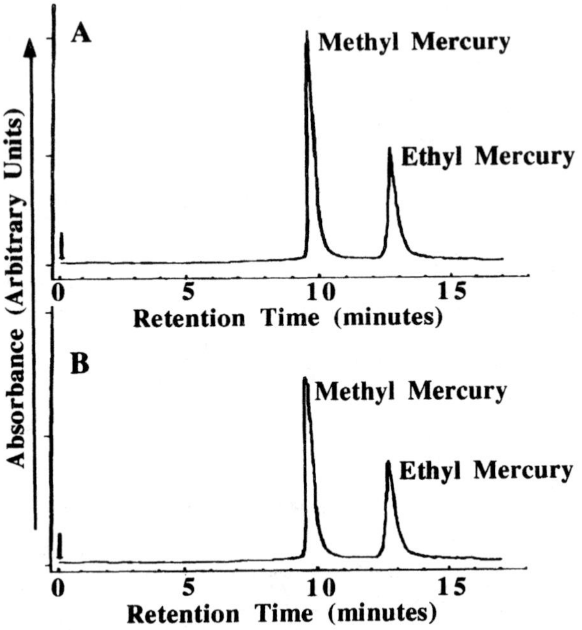

A/ Water Standard Containing 10 mg/l of Inorganic Mercury and Methyl Mercury.
B/ Whole Blood Containing 10 mg/l of Inorganic Mercury and Methyl Mercury.

Courtesy of the Royal Society of Chemistry [ref. 19]

Figure 16 Typical Chromatograms Showing the Separation of Methyl Mercury and Ethyl Mercury

The recovery of methyl and ethyl mercury ranged, within a few percent, around 100%. It was found that the minimum detectable concentration of inorganic and organic mercury, using the above technique, was 0.6 μg/ml and 0.2 μg/ml respectively.

The Determination of β-Alaninediacetic Acid in Waste Water and Aquatic Environments [20]

β-Alaninediacetic acid is a strong chelating agent and consequently a potential substitute for the more conventional chelating compounds such as ethylenediaminetetraacetic acid and diethylenediaminepenta-acetic acid. These chelating or complexing agents are used, among other products, in the paper industry to remove or inactivate metal ions which catalyze the decomposition of added bleach (ozone, peroxide or dithionate). These substances and, in particular, ethylenediaminetetraacetic acid degrade relatively slowly in waste water treatment plants and thus significant amounts can be released into the environment. It follows that an alternative chelating agent that would degrade more readily would be highly desirable and it has been thought that β-alaninediacetic acid might be an appropriate substitute. β-alaninediacetic acid has been used satisfactorily in certain photographic processes and it has been shown that it degrades much more readily (up to 99 %) in waste water plants. In order to study the behavior of β-alaninediacetic acid in pulp and paper processes, water waste treatment plants and in aquatic environments a sensitive method for its determination was required. A suitable analytical method was developed by Sillanpää *et al.* [20], using a gas chromatographic technique with a nitrogen phosphorus detector. The samples were prepared using the following procedure.

Water Samples

Twenty milliliters of water were evaporated to dryness at 110°C ± 5°C and then cooled to room temperature. 1 ml of esterifying agent is added [2.5 ml of concentrated sulfuric acid, 30 ml of ethyl alcohol (adding standard if required) and made up to 50 ml] and the mixture placed in a screw-capped file and heated for two hours on a water bath. After esterification the sample was extracted with 1.5 ml of toluene and neutralized with 10 ml of 1 M potassium hydrogen carbonate. The extract was then dried with anhydrous sodium sulfate.

Sediments

Sediments were dried in an oven at 90°C overnight and 1 g of the dry material was then extracted with 20 ml of 0.02M sodium dihydrogen phosphate for 10 min. The extract was then filtered free of the sediment and analyzed in the same manner as that described above for a water sample.

One microliter of the esterified sample in dry toluene was injected onto an HP-5 open tubular column, 30 m long, 0.25 mm I.D. carrying a film of stationary phase 0.25 μm thick. Initially the column was held at a temperature 100°C for 1 min and then programmed up to 200°C at 60°C/min and held there for a further 9.3 min. Nitrogen was used as the carrier gas at a flow rate of 3.5 ml/min. An example of an analysis is shown in figure 17.

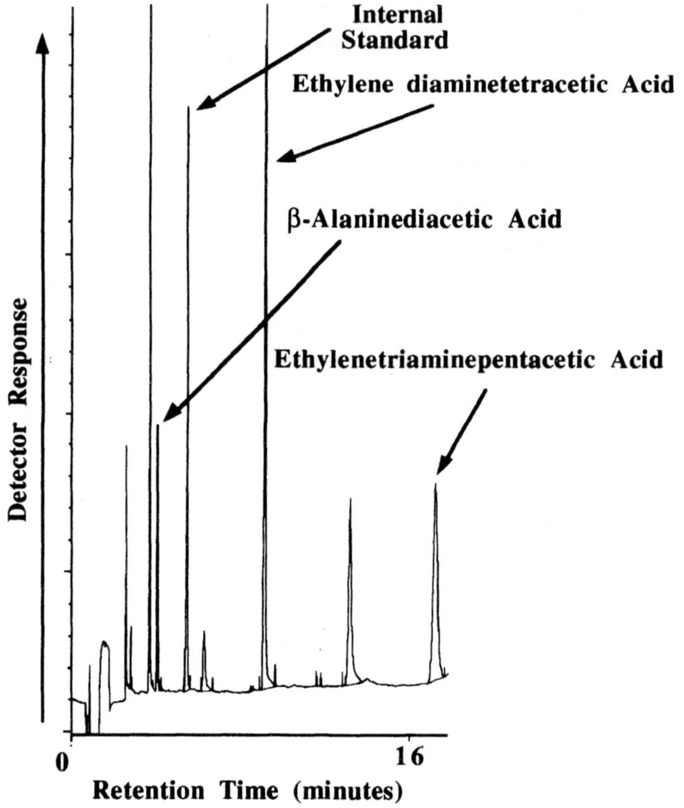

Courtesy of the Royal Society of Chemistry [ref. 20]

Figure 17 Typical Chromatograms Showing the Separation of Ethylenediaminetetraacetic Acid, Diethylenediamine-pentaacetic acid and β-Alaninediacetic Acid from a 20 ml Sample of Water

The detection limits were found to be about 2 μg/l for water samples and 0.21 μg/g for sediments. The run to run repeatability was quoted as about 7% except near the limits of detection where it could be somewhat greater.

Method for Analyzing Organochlorine Pesticides in Water Using Solid Phase Microextraction and Pulsed Discharge Electron Capture Detection [21]

The need for speed in environmental analysis is ever increasing and the relatively slow methods that existed at the time of this work significantly restricted the sample throughput of the monitoring laboratories. Jackson and Andrews [21], developed a fast screening method that involved solid phase microextraction, cryo-concentration and capillary column separations using a pulsed-discharge electron capture detector. Previous workers utilized salt addition and pH control of the sample to increase the efficiency of separation but Jackson and Andrews found this unnecessary. A diagram of their apparatus is shown in figure 18.

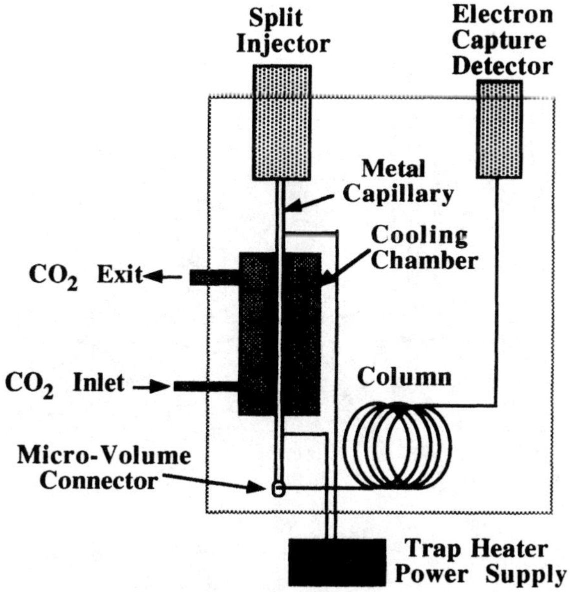

Courtesy of the Royal Society of Chemistry [ref. 21]

Figure 18 Diagram of Solid Phase Micro-Extraction and Solute Focusing Apparatus

Again solid phase micro-extraction has been described in detail on page 33 in *Sample Preparation*, Chapter 2, and so only specific details will be given here. Three different fibers were examined, having diameters 7, 30 and 100 μm respectively. The 7 μm fiber was conditioned for two hours at 320°C and the 30 and 100 μm fibers at 250°C for one hour. It was found that better precision and sensitivity were obtained with the 100 μm fiber. The apparatus shown in figure 18 was operated as follows. The sample was placed in a headspace vial and the fiber exposed to the headspace for 2 min. The fiber was then placed in the split injector (with no split flow) and heated up to 250°C for 2 min. During this time the cryotrap was kept cool and the oven held at 220°C. The solutes were thus focused on the cool part of the trap. After 2 min the split was opened, the coolant turned off and the oven ramped up to 266°C at 20°C per min.

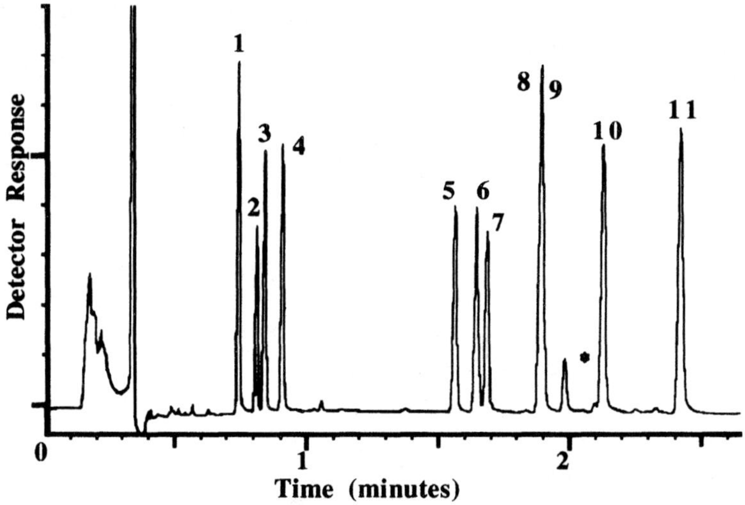

1. Tetrachloro-*m*-xylene
2. α-Benzenehexachloride
3. β-Benzenehexachloride
4. δ-Benzenehexachloride
5. *cis*-Chlordane
6. *trans*-Chlordane
7. *p,p'*-DDE
8. *p,p'*-DDD
9. Endrin
10. Endosufan Sulfate
11. Endrin Ketone

*. Endrin Aldehyde

Courtesy of the Royal Society of Chemistry [ref. 21]

Figure 19 Chromatogram of Pesticide Calibration Mixture

A chromatogram of a water sample spiked with a standard pesticide calibration mix is shown in figure 19. Each pesticide was present in the original sample at a concentration of 100 ng/ml. The column used was 5 m long, 0.1 mm I.D. carrying a film of methyl 5% phenyl silicone column 1.0 µm thick. It is seen that an excellent separation of all the pesticides of interest was obtained and all were eluted in less than 3 min. With an extraction time of 2 min, this allowed an analysis to be completed every 10 min. The limit of detection was reported to be about 10 ng/l. Unfortunately, Endrin decomposition in the desorption stage could not be kept less than 20% but, it would appear, could be kept constant and controllable.

Use of Supercritical Fluid Extraction for the Determination of Steroids in Animal Tissue [22]

In most countries, the use of anabolic steroids is prohibited in foodstuffs for food producing animals. Analytical methods using liquid/liquid extraction in conjunction with solid phase extraction and gas chromatography are available for monitoring steroids in animal foods and animal tissue. Most tissue samples, however, contain a relatively large amount of fat which, due to its comparatively high solubility in liquid carbon dioxide (the solvent most commonly used in supercritical liquid extractions), can carry through and, as a consequence, impair the final chromatographic separation. However, by employing a combined technique of supercritical liquid extraction and solid phase extraction the steroid residues can be selectively retained whereas the fat is not. A limitation of this technique is that the sample size may be restricted when tissue samples are used and, as a result, the desired detection limits not realized This problem has also been overcome, however, by freeze drying the tissue and using the freeze dried material for subsequent analysis. An interesting method has been develop by Stolker *et al.* [22], based on this approach. The scheme for processing samples of muscle, fat and skin tissue are outlined in figure 20. Only the procedure for tissue samples will be discussed here. A 5 g sample of the tissue is freeze dried in an Epsilon

2 freeze drier and the freeze dried sample is ground and blended with 3.5 g of Extrelut, 1 ml of water and appropriate internal standards.

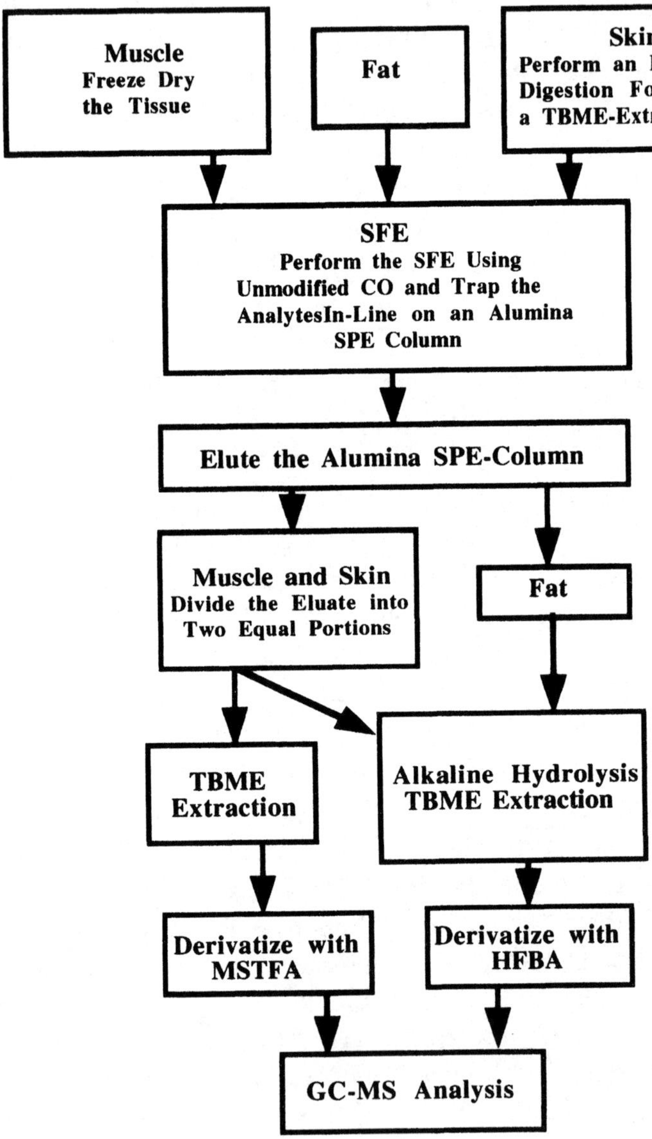

Courtesy of the Royal Society of Chemistry [ref. 22]

Figure 20 Scheme for Sample Pretreatment of Animal Tissues for Steroid Analysis.

The supercritical liquid extraction apparatus is connected directly to a solid phase extraction cartridge packed with aluminum oxide. The solid phase extraction cartridge is removed from the liquid extraction system and the steroids eluted with 6 ml of methanol/water mixture (63–35% v/v). The extract is divided into two equal portions and each treated as follows.

Portion 1

The portion is evaporated to dryness, and dissolved in 0.2 ml of an alkaline hydrolysis solution (5.6 g of potassium hydroxide in 100 ml of methanol) and then heated for 30 min at 37°C. The hydrolysis is arrested by the addition of 1 ml of acidic buffer (1.7 ml of concentrated hydrochloric acid in 98.3 ml of acetate buffer (2 mol/l)). The mixture is then extracted twice with 6 ml of tributyl-methylether and the two extracts bulked and evaporated to dryness under a stream of nitrogen. The residue is then transferred to a derivatizing vial with 0.5 ml of ethanol, evaporated to dryness, and 0.05 ml of heptafluorobutyric anhydride/acetone (1–4 v/v) is added. The vial is vortexed and heated for 1 hr at 60°C. The mixture is then evaporated to dryness under a stream of nitrogen at 5°C and the residue dissolved in 25 µl of isooctane and transferred to an injection vial of the chromatograph.

Portion 2

This portion is evaporated to dryness at 50°C under a stream of nitrogen. 2 ml of water is then added, the mixture vortexed for 30 s. The liquid is then extracted twice with 6 ml of tributylmethylether, the two extracts bulked and then evaporated to dryness under a stream of nitrogen. The residue is then transferred to a derivatizing vial with 0.5 ml of ethanol and again evaporated to dryness. 25 µl of *N*-methyl-*N*-(trimethyl-silyl)trifluoroacetamideammonium iodidedithioerythritol (1000+2+4, v/w/w) is added. The vial is mixed for 30 s, heated for 1 hr at 60°C and evaporated to dryness under a stream of nitrogen at 50°C. The derivatized residue is dissolved in 25 µl of isooctane and transferred to an injection vial of the chromatograph. The chromatograms obtained by single ion monitoring of a range of steroids and deutrated steroids is shown in figure 21.

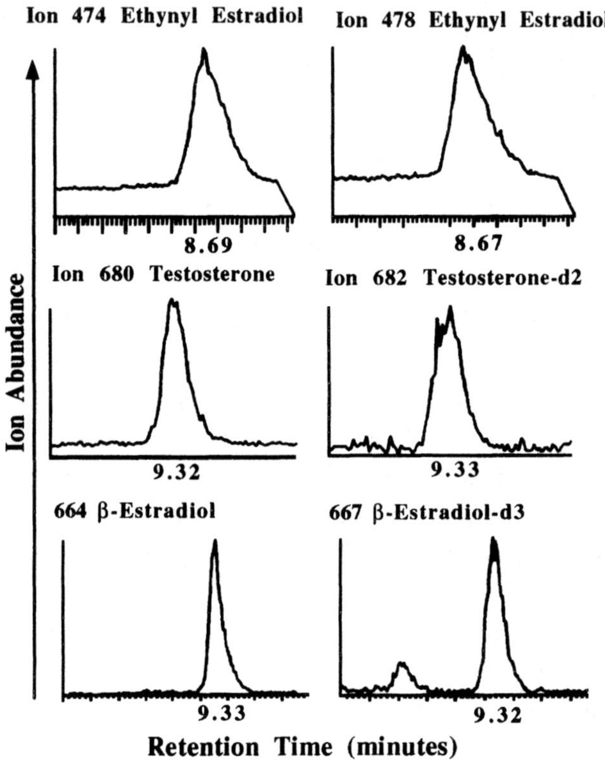

oxide.

Courtesy of the Royal Society of Chemistry [ref. 22]

Figure 21 Single Ion Monitoring Chromatograms of the Heptafluorobutyric Derivatives of Steroids and Deuterated Internal Standards from Spiked Samples of Bovine Skin (5 μg/kg)

It is seen that the different steroids and their deuterated isomers are easily discriminated from one another and the overall retention time is less than 10 min. The limit of detection was reported to be between 2 and 5 μg/kg and the repeatability to 42% depending on the specific steroid.

Synopsis

Due to its inherent high sensitivity and capacity for high efficiencies, gas chromatography is the technique of choice for many types of chemical assay. The need for all solutes to be volatile at the elution temperature, however, severely limits its field of application despite the popular use of derivatization to render solutes normally involatile sufficiently volatile for separation. The high efficiencies obtainable from open tubular columns have made gas chromatography essential for the separation of complex mixtures such as the ubiquitous polychlorinated biphenyls and the many complicated hydrocarbon mixtures produced by the petroleum industry. Gas chromatographic methods have been reported for nearly 50 years so the literature can usually provide help in the choice of apparatus and operating conditions for any new analytical challenge. A series of assays reported in the literature are described that highlight certain gas chromatographic techniques. An example of the use of derivatization and solid phase extraction in the determination of carbonyl compounds formed in the peroxidation of lipids and the metabolism of alcohol and other components of ale is described. The use of derivatization in conjunction with solvent extraction is exemplified in the determination of traces of insecticides in drinking water, using the dual electron capture and nitrogen phosphorus detectors to achieve the necessary high sensitivity. The determination of the chlorobenzenes in soil illustrates the advantage of combining headspace analysis with solid phase microextraction to achieve the necessary isolation and concentration of the substances of interest. The required high sensitivity was achieved by employing single ion monitoring using the ion trap mass spectrometer as the detector. The combination of selective solvent extraction and the use of the nitrogen phosphorus detector and mass spectrometer demonstrated that subnanogram quantities of the analgesic fentanyl could be determined in whole blood. The measurement of within-tree variations in lignin displayed the use of in-line pyrolysis coupled with an open tubular column and a flame ionization detector for the identification of chemical changes in involatile materials. The use of the coated capillary as an extraction device was demonstrated in the analysis of phenols in drinking water. The high sensitivity of the gas chromatographic detector was used to measure solvent contamination of carbon packed break-through pads. The pads were microwave extracted, syringe filtered and the extract analyzed on a packed column. A method for the analysis of pethidine and methadone in blood or urine established the use of solid phase microextraction in conjunction with the nitrogen phosphorus detector to provide separations with a sensitivity of 1 ng/ml in less than 10 minutes. Head-space analysis can often result in band dispersion on injection which can be reduced, in fact virtually eliminated, by cryo-focusing. This technique is demonstrated in the use of head-space analysis coupled with cryo-focusing in the determination of inorganic and organic mercury in whole blood. Involatile materials dissolved in a liquid can be easily concentrated by evaporation, then derivatized to render volatile and subsequently injected onto the gas chromatographic column. This procedure is exemplified by the determination of β-alaninediacetic acid in waste water and sediments. The impressive combination of solid phase microextraction, cryo-focusing, fast capillary columns and the pulsed discharge detector is elegantly demonstrated in the determination of a range of organochlorine pesticides in less than 10 min (sample preparation included) at levels of 100 ng/ml. Finally, the technique of supercritical fluid extraction is demonstrated in the determination of steroids in animal tissue

References

1. B. Larsen, *Int. J. Environ. Anal. Chem.*, **47**(1)(1992)47.
2. R. G. Webb and A. C. McCall, *J. Chromatogr. Sci.*, **11**(1973)366.
3. M. D. Eickson, *Analytical Chemistry of PCB's*, CRC Press, Boca Raton and New York (1997).
4. R. Fussle, S. Bhadki, A. Sziegoleit, J. Tranum-Jensen, T. Kranz and H. J. Wellensiek, *J. Cell Biol.*, **91**(1981)83.
5. H. Esterbauer and H. Zollner, *Free. Radic. Biol. Med.*, **7**(1989)197.
6. E. E. Stashenko, M. C. Ferreira, L. G. Sequeda, J. R. Martínez and J. W. Wong, *J. Chromatogr.*, **779**(1997)360.

7. The Pesticide Manual, Crop Protection Publications, (Ed., C. Tomlin) Cambridge, 10th edition (1994).
8. Yun-Suk Oh-Shin, Moonhee Ko and Ho-Sang Shin, J. *Chromatogr.*, **769**(1997)285.
9. U.S. Environmental Protection Agency, *Health Assessment Document for Chlorinated Benzenes;* EPA-600/8-84-015F, Washington D.C. (1986).
10. World Health Organization, IPCS, *Chlorobenzenes other than Hexachlorobenzene*, Geneva (1991).
11. F. J. Santos, M. N. Sarrión and M. T.Galeeran, *J. Chromatogr.*, **771**(1997)181.
12. V. Watts and Y. Caplan, *J. Analytical Toxicology*, **72**(1988)246.
13. D. Collins, C. Pilotti and A. Wallis, *Appita. J.*, **43**(1990)193.
14. T. Ona, T. Sonoda, K. Ito, M. Shibata, Y. Tamai and Y. Kojima, *Appita. J.*, **49**(1996)325.
15. H. Yokoi, Y. Ishidi, H. Ohtani, S. Tsuge, T. Sonoda and T. Ona, *Analyst*, **5**(1999)669.
16. B.C.D.Tan, P.J.Marriott, H. K. Lee and P. D. Morrison, *Analyst*, **6**(1999)941.
17. E. V. Stephen, P. Berardinella and R. C. Hall, *Analyst*, **5**(1999)651.
18. S. W. Myung, S. Kim, J. H. Park, M. Kim, J. C. Lee and T. J. Kim, *Analyst*, **9**(1999)1283.
19. M. R. Brunerro, J. R. Luna, A. Zambrano, M. Gallignani, J. L. Burguera and Y. Petit de Peña, *Analyst*, **10**(1999)1493.
20. M. Sillanpää, V. Vickackaite, Jaakko Rämö and L. Niinistö, *Analyst*, **10**(1998)2161.
21. G. P. Jackson and A. R. J. Andrews, *Analyst*, **5**(1998)1085.
22. A. A. M. Stolker, P. W. Zoontjes and L. A. van Ginkel, *Analyst*, **12**(1998)2671.

Part 3

Quantitative Liquid Chromatography Analysis

Chapter 8

Liquid Chromatographic Apparatus for Quantitative Analysis

The vast majority of quantitative LC analyses are carried out using the simple, *basic* liquid chromatograph. It follows that the first priority of this discussion will be to describe the essential parts and function of the basic LC instrument. Subsequently, a more sophisticated form of the liquid chromatograph will be considered and its relevance to quantitative LC analysis will be reviewed.

The Basic Liquid Chromatograph

The simple, basic chromatograph has been used very successfully for quantitative analysis, in both industry and university, for the last twenty years. It has been used very successfully for both product control and quality assurances.

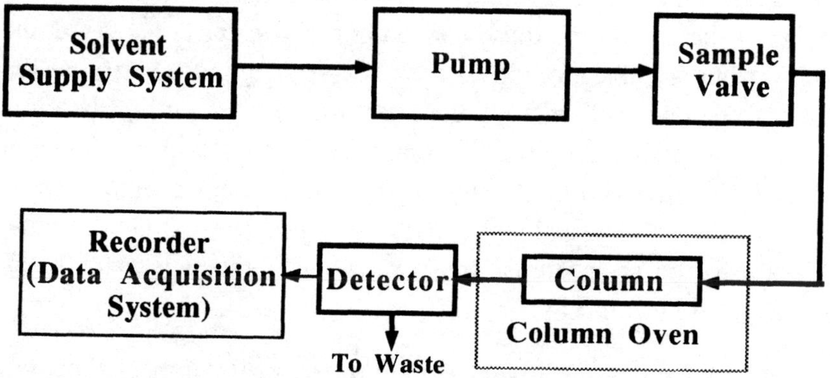

Figure 1 The Basic Liquid Chromatograph

In fact, providing the individual components are carefully selected, the simple form of the liquid chromatograph can provide the necessary accuracy and precision that is appropriate for the majority of LC analyses (0.5%-1.0%). A diagram of the basic liquid chromatograph is shown in figure 1 and consists, fundamentally, of six units, the gradient system, the pump, the sample valve, the column, the detector and the recorder or computer data acquisition and processing network.

The Solvent Supply System

The solvent supply can range from a simple glass reservoir for isocratic development to a two or three solvent-programmer. All reservoirs are usually fitted with a device to allow helium to be bubbled through the solvent to remove dissolved oxygen and nitrogen. The presence of dissolved air can cause bubbles to form in the mobile phase during passage through the column which can reduce the column efficiency. The most serious effect of bubbles, however, is on the detector, where they can produce serious, and in many instances completely unacceptable, noise. Sparging with helium virtually eliminates this problem. In many analyses, where the polarity range of the components of interest is not large, the separation can be achieved isocratically at the expense of analysis time. It should be pointed out that for simple mixtures, it will often be more time consuming to search for an optimum solvent program to reduce the analysis time than to tolerate the extra time involved and employ isocratic development. This is particularly true where there are only a few analyses of the same type to be carried out. If, on the other hand, the mixture contains substances that cover a wide range of polarity, or the procedure is to be used on a regular basis for quality control, then it is well worth while spending the necessary effort in method development to reduce the analysis time to a minimum. If a gradient programmer is employed, then a three solvent capability is usually sufficient for the vast majority of separations. In fact, two solvents will usually be adequate for 90% of all LC applications.

There are two types of solvent programmer, programmers that mix the solutes at high pressure and those that mix the solvents at low pressure. The two types are depicted diagramatically in figure 2. The *low pressure* programmer draws each of the three solvents from their respective reservoirs through three electrically actuated valves by means of a single pump. The flow through each valve is determined by its opening time and thus by the frequency and wave form of the driving power supply. The advantage of this programmer arises from the use of a single pump although the multi-valve system is almost as complex as a pump, and nearly as expensive.

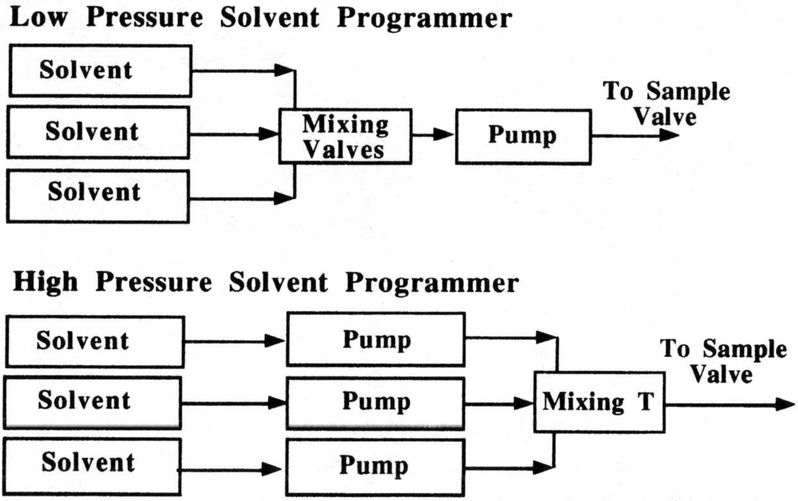

Figure 2 The Two Types of Solvent Programmer

The *high pressure* programmer has a pump for each solvent and the solvent program is formulated by exclusively adjusting the pumping rate of each. The pumps are usually operated by stepping motors and the flow rate from each pump is controlled by the frequency of its respective power supply generated by the computer. As there is a pump for each solvent, this type of programmer is more expensive and cumbersome than the former, but in general, it is thought to give more accurate and reproducible solvent programs.

Gradient programs are usually employed in conjunction with columns packed with bonded phases and in particular the reversed phases. The binary solvent mixtures most commonly used with reversed phases are methanol/water and acetonitrile/water. As methanol is a *less* dispersive solvent than acetonitrile, methanol/water mixtures are used for eluting the *more polar solutes.* Conversely, acetonitrile/water mixtures are used for eluting *more dispersive solutes.* Mixtures of solutes that have very similar interactive properties, albeit predominantly polar or dispersive, may need a third solvent to facilitate a finer adjustment to the interactive character of the solvent. The third solvent chosen is often tetrahydrofuran (THF) which is more dispersive and less polar than acetonitrile. There are a number of software packages available that suggest an experimental routine of separations using different solvent mixtures, and from the results, the optimum solvent mixture that will achieve the separation in the minimum time is identified. These programs can be useful for particularly recalcitrant mixtures, but in most applications a satisfactory program can usually be identified from a few simple solvent programs based on rational choice.

Liquid Chromatography Pumps

Early liquid chromatography columns were packed with particles of relatively large diameter and, consequently, the mobile phase could be passed through them by gravity flow. However, when the advice of Martin and Synge was eventually heeded, and particle diameters were reduced to a few microns, high pressure pumps were required to accommodate the increased column impedance and provide adequate column flow. LC pump design has evolved over the ages and there are now a number of different types available. The first two high pressure pumps to be utilized for LC purposes were the pneumatic pump and the syringe pump. The syringe pump is now more or less extinct in general LC, although it is still used to supply carefully controlled reagent flows in certain automatic derivatization procedures. The pneumatic pump, however, has survived, but largely for column packing and, to some extent, to provide the high flow rates necessary for preparative LC. It is simple in design, can be made to provide extremely high pressures and is relatively inexpensive to manufacture. However, it is rarely used for quantitative LC analyses.

All modern pumps depend for their efficient performance on the use of non-return valves. The non-return valves ensure that the liquid only passes in one direction through the pump. A diagram of a non-return valve is shown in figure 3.

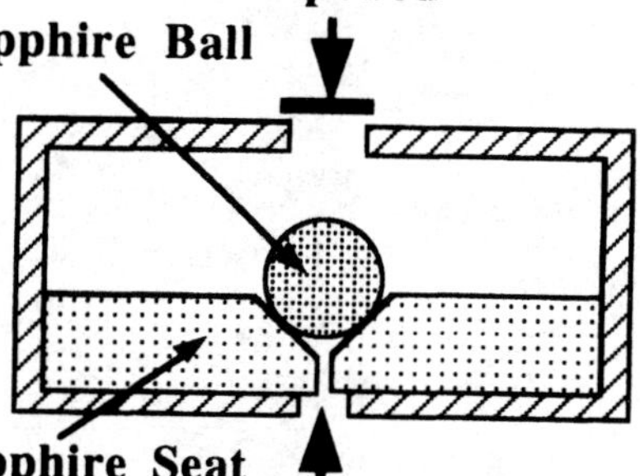

Figure 3 A Non-Return Valve

The non-return valve consists of a sapphire ball resting in a flat conical seat. The solvent enters through the valve seat and displaces the ball from the seat, allowing the liquid to pass by it. When flow ceases, or the inlet pressure falls, the ball is driven back onto the seat by the back-pressure on the valve. As a consequence, the valve is sealed preventing any back flow. On the next stroke of the pump, when the pressure from the pump exceeds the back-pressure on the valve, the ball is again displaced, and flow continues through the valve. It is clear that with this type of valve, the higher the back-pressure, the better the seal. In the majority of valves, the seat and the ball are both made of sapphire but in some valves the ball is of sapphire and the seats are made of stainless steel. To improve the efficiency, two valves are sometimes used in series which are fabricated as one dual integrated unit that can be directly fitted to the pump.

The simplest, most inexpensive, and consequently one of the more popular LC pumps is the single piston pump. This type of pump was one of the first continuous high pressure pumps to be used with columns packed with small particles. A diagram of a single piston pump is shown in figure 4.

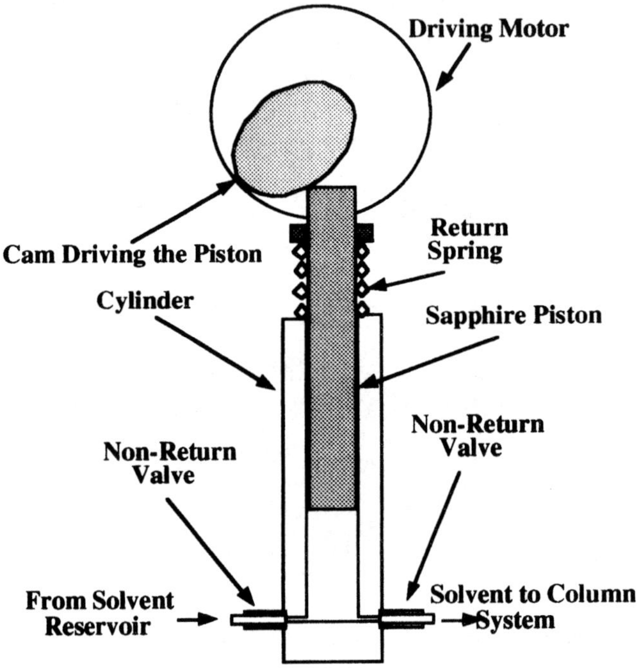

Figure 4 The Single Piston High Pressure LC Pump

The single piston pump consists of a sapphire piston usually 3-4 mm in diameter and about 4 cm long fitted into a stainless steel cylinder. The piston is actuated by a rotating cam which has a specially cut profile that provides a steady linear forward thrust and a rapid non-linear return. The return stroke of the pump that refills the cylinder is actuated by a return spring. The displaced volume of the pump is usually about 250 µl, but can be reduced to 2-5 µl if required for use with small bore columns. As with all LC pumps, the inlet and outlet are connected directly to two non-return valves similar in design to those described above. Single piston pumps can be designed to provide pressures of 6,000 p.s.i. or more and flow rates that range from 0.02 to 2 ml/min for use with small bore columns and from 0.2 to 10 ml/min for columns of wider bore.

The single piston pump is adequate for most LC applications and, unless more stringent specifications are called for, such pumps are recommended for general use. Due to the dynamics of the refill cycle, the flow is halted for a short period while the cylinder is replenished and before flow is started again and this can cause slight pulsations in flow. Pulsed flow is not detrimental to most separations, providing the columns contain a reasonable volume of mobile phase to dampen the pulses or if a pulse dampener is used. Flow pulses can, however, cause noise on some detectors and may prevent the maximum detector sensitivity from being used. This is not usually a serious problem in quantitative analyses as the use of the maximum detector sensitivity is not often necessary. Problems associated with flow pulses tend to be exaggerated and consequently the need for a more expensive, pulse free pump should be well established before the single piston model is rejected.

The problems induced by pulsed flow in LC columns (real or imaginary) provoked the development of two special types of pumps, both of which were designed to achieve a virtually pulse-free flow but function in entirely different ways. The first approach to pulse free operation was the design of the original engineers of Waters Chromatography who coupled two single stroke pumps, operating out-of-phase, so that one pump was delivering solvent while the other was refilling. By the clever design of the cams that drove the pistons, a virtually pulse free delivery was achieved. This twin pump system rapidly became extremely popular and was probably the major factor in establishing Waters Chromatography as the prime LC instrument manufacturer of the period. The original pump, with little change in functional design, is still manufactured by Waters Chromatography and remains one of the more popular and reliable LC pumps available.

The second design of pulse free pump also utilized two pistons and cylinders but was, perhaps, more innovative in style. Only one pump was employed for the mobile phase supply; the other was used to refill the first pump extremely rapidly. The pump was appropriately called the *rapid refill pump*. A diagram of a rapid refill pump is shown in figure 5. The pump consists of two cylinders and a single common piston. The expression of the solvent to the column is depicted in the upper part of figure 5. As the piston progresses to the left, solvent is pumped to the column system but at the same time fresh solvent is being withdrawn into the right-hand chamber from the solvent supply system.

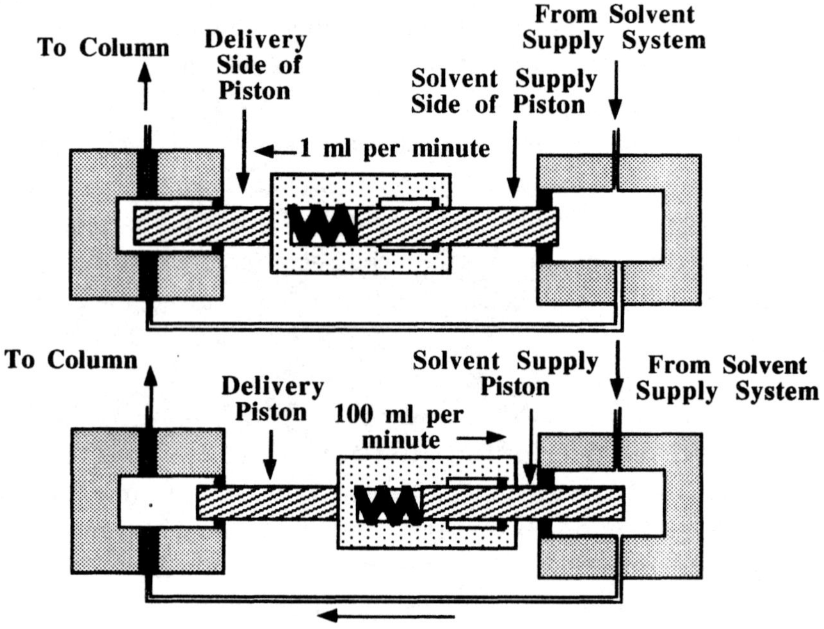

Courtesy of Perkin Elmer Inc.

Figure 5 The Rapid Refill Pump

At the point where the piston arrives at the extent of its travel, a step in the driving cam is reached and the piston is rapidly reversed. As a result the contents of the chamber on the right-hand side are conveyed to the left-hand chamber. This situation is depicted in the lower part of figure 5. The transfer rate of the solvent to the left-hand chamber is 100 times as fast as the delivery rate to the column and consequently reduces the pulse on the refill cycle very significantly. Furthermore, if a solvent gradient is being used and the right-hand chamber is being filled with a solvent mixture, excellent mixing is achieved during the refill of the left-hand chamber. This is another popular contemporary

LC pump, which is less costly than the double pump, but significantly more expensive than the single piston pump.

Sample Valves

Similar to those devices used for injecting samples onto a GC column for quantitative analysis, the sample valve in LC is also one of the more critical parts of the chromatograph and must be very precisely constructed. The *accuracy* of the device is not particularly important but, for the effective use of *external standards*, the sample volume must be delivered to the column with *very high precision.* Fortunately, it is easier to construct a sample valve that is very precise than it is to construct one that is highly accurate.

The early methods of sample injection employed a septum injection system in much the same way as GC. However, as the particle size of the packing was reduced, and the inlet pressures increased, the septum injection techniques were abandoned in favor of the sample valve. The two types of sample valve, the *internal loop valve* and the *external loop valve* are used in LC in the same way as GC, but are constructed in a more robust form to accommodate the high pressures. In order to improve the seating and eliminate leaks, the valve faces are sometimes made from appropriate ceramics. The *internal loop valves* are largely used with small bore columns, that is to say columns having internal diameters of less than 1.5 mm. The *external loop valves* are used for larger diameter columns up to semi-preparative columns.

The sample volume in the *internal loop valve* that is contained in the connecting slot of the valve rotor can range from 0.1 μl to about 0.5 μl. The injection procedure is exactly the same as that used for internal loop valves in GC. The *external loop valve* is the more common sampling system and, again, operates on exactly the same principle as the GC valve, but offers a wider choice of sample sizes. A modified form of the external loop sample valve has become very popular for quantitative LC analysis, a diagram of which is shown in figure 6.

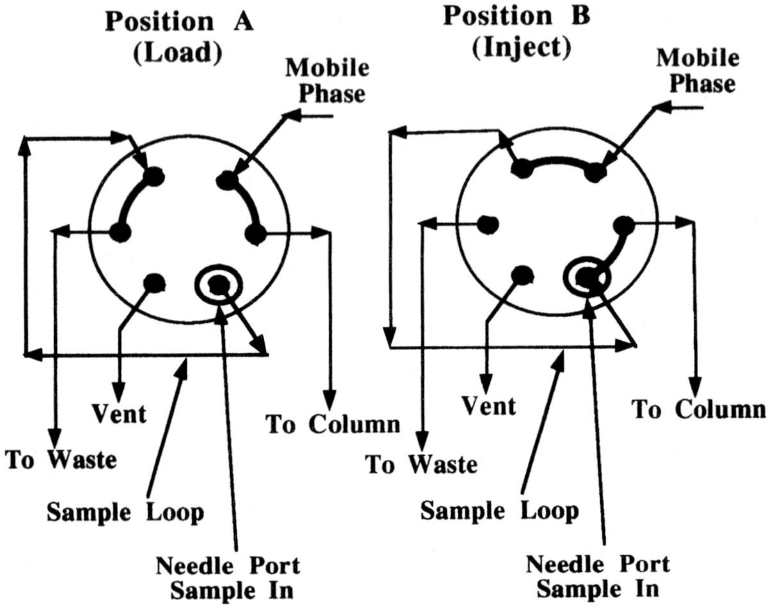

Courtesy of Rheodyne Instruments Inc.

Figure 6 The Modified External Loop Sample Valve

The basic difference between this type of valve and the normal external loop sample valve is the incorporation of an extra port at the front of the valve. This port allows the injection of a sample by a syringe directly into the front of the sample loop. Position (A) shows the load position. Injection in the front port causes the sample to flow into the sample loop. The tip of the needle passes through the rotor seal and, on injection, is in direct contact with the ceramic stator face. Note the needle is chosen so that its diameter is too great to enter the hole. After injection, the valve is rotated to position (B) and the mobile phase flushed the sample directly onto the column. The sample is actually forced out of the beginning of the loop so it does not have to flow through the entire length of the loop. This type of injection system is ideally suited for quantitative LC, and is probably by far the most popular injection system in use. Sample valves based on this design are available from a number of manufacturers.

Column Ovens

The effect of temperature on LC separations is often not nearly so profound as its effect in GC separations, but can be *critical* when closely similar substances are being separated. In LC a change in temperature will change the free energy of the solute in both phases (generally in a commensurate manner) and so the *net* change in the free energy difference with temperature, which controls the magnitude of the *absolute* retention, can be relatively small. Its effect on *relative retention*, however, can be very significant and, in fact, be the determining factor in achieving a satisfactory resolution [1-3]. The effect of temperature on diffusivity will be similar in both GC and LC. An increase in temperature will increase the diffusivity of the solute in both phases and thus *increase* the dispersion due to longitudinal diffusion and *decrease* dispersion due to resistance to mass transfer. As a result, at the *optimum velocity*, the efficiency of both the LC and GC column will be largely independent of temperature, however, the optimum velocity will be higher and provide the potential for faster analyses. Due to the lesser effect of temperature on *solute retention* in LC (compared to that in GC), temperature is not nearly so critical in governing absolute *retention time* but is often essential in achieving adequate *resolution*, particularly between closely eluting solutes such as optical and other types of isomers. In contrast to the GC column, the thermal capacity of an LC column is much higher as the specific heats of liquids are much greater than those of a gas. As a consequence, a high heat capacity thermostatting fluid is necessary and if retention measurements need to be precise, air ovens would not be ideal for thermostatting LC columns. On the other hand, liquid thermostatting media are rather messy to use and tend to make column changing difficult and lengthy. However, if accurate data are required, good temperature control may be essential. If precise retention measurements are not required, an air thermostatting oven might be a reasonable compromise.

General Comments on Detectors

The detector used in quantitative LC analysis will be discussed in detail later in this chapter, however, some general comments should be made at this time. The essential specifications for an LC detector are the same as those for a GC detector and have already been discussed. However, LC detectors have intrinsically far lower sensitivities than their GC counterparts and the sensitivity difference between an LC detector and a GC detector may extend over several orders of magnitude.

In general, LC bulk *property* detectors also have much lower sensitivities than the *solute property* detectors among which are found those with the highest sensitivity. The fluorescence detector, probably one of the most sensitive LC detectors, has a sensitivity that is probably 3 to 4 orders of magnitude less than that of the electron capture GC detector. This great difference in sensitivity between LC and GC detectors accounts for the extensive use of GC in quantitative environmental testing and forensic chemistry. From a practical point of view, the sensitivity of LC detectors needs to be enhanced. Enlisting the use of modern electronic and ionic systems in support of some of the earlier detecting principles long since forgotten (such as the transport detectors) might evoke some interesting new possibilities in LC detection.

Data Acquisition and Processing

Data acquisition and processing in LC is virtually identical to that in GC. The same basic system is employed the only difference being that the software used is slightly different for LC in that the temperature programming routines used in GC must also include simultaneous gradient elution routines. The same quantitative techniques are used and peak heights or peak areas can either be taken from the computer measurements or directly from the recorder chart if one is used.

The Modern Versatile Liquid Chromatograph

The fully equipped liquid chromatograph, at today's prices, will cost in excess of $20,000 depending on the number of alternative detectors and ancillary apparatus that is purchased with it. These types of instruments are not recommended for routine quantitative analyses where the requirements can be clearly defined and the specific equipment can be purchased that is necessary solely for the intended analyses. The versatile chromatograph is necessary, however, for method development where all helpful and accessible ancillary equipment should be available in order to identify the specific system that is optimum for a particular sample.

When purchasing equipment, price is not a reliable guide to the performance of any particular part of the equipment. Instruments should be bought on the basis of their specifications and if these are unavailable this should give cause for concern. Any instrument intended for purchase should be taken "on loan for assessment" before purchase and this is also important for gas chromatographs and all other analytical instruments.

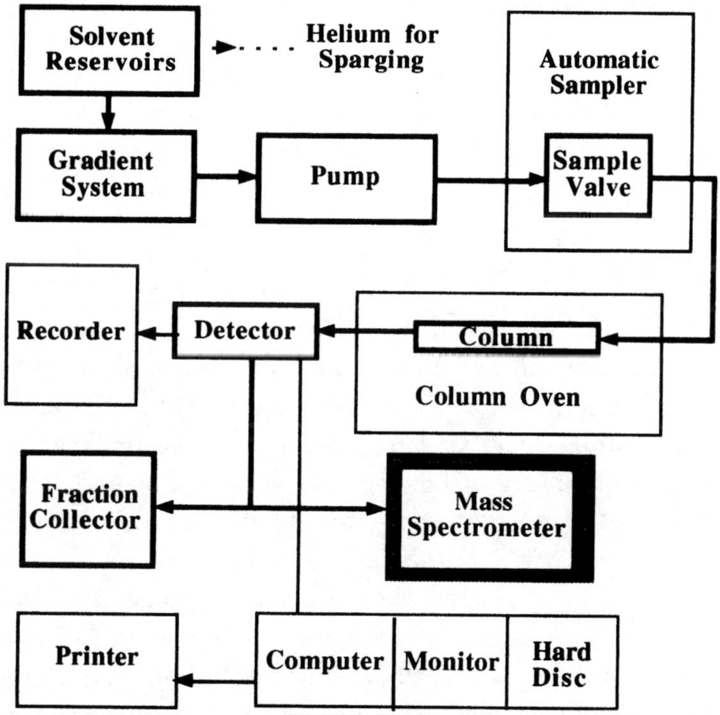

Figure 7 The Modern Versatile Liquid Chromatograph

Most companies have instruments available for assessment loan and if not, the instruments should be tested and assessed at the manufacturer's establishment. A diagram of a modern versatile liquid chromatograph is shown in figure 7. The solvent reservoirs, usually made of glass, but sometimes made of stainless steel, are situated in a flameproof enclosure and arranged to be sparged with helium. The flameproof enclosure may be necessary if employing very inflammable solvents such as *n*-hexane which could be used with silica gel columns. For solvents carrying high concentrations of water and buffers for ion-exchange chromatography, the reservoirs must be made of glass and connecting tubes usually of Teflon. This is only necessary when corrosion is likely to take place from ionic materials. The gradient system can be high pressure or low pressure, both are very effective, but, as already noted, the general impression is that the high pressure programmers, with a pump for each solvent, give more precise gradient programs.

The pump should have a wide delivery range from perhaps 2-3 μl/ min to 20 ml/min and provide inlet pressures up to at least 6,000 p.s.i. and preferably up to 10,000 p.s.i. This range will accommodate on the one hand the small bore column having diameters less than a millimeter and on the other preparative columns having diameters of 2 or more centimeters.

It has been claimed that when certain proteins come in contact with stainless steel, they can become denatured and thus all parts of the apparatus that are in contact with the sample should be fabricated from bio-compatible material. This problem may easily have become exaggerated, particularly when one recalls the number of stainless steel screws, bolts and braces that are, at this time, holding many patients bones together. Nevertheless, protocol may require that the sample valve should be constructed of titanium or some other material that is considered biochemically acceptable if it is to be used for biochemical purposes. The versatile chromatograph should also be fitted with an automatic sampler. This does not imply it will be used for routine analysis, but the automatic sampler can accelerate the process of method development considerably. An automatic sampler is extremely valuable when carrying out sequential optimization experiments that can identify the best solvent mixture and operating conditions to separate specific materials in the minimum time.

If accurate retention data are required, the column should be thermostatted with a high heat capacity fluid as already discussed and the mobile phase should be passed through an efficient heat exchanger (*e.g.* low dispersion tubing [4,5]) before entering the sample valve. It is very important to bring the temperature of the mobile phase to the oven temperature before it enters the sample valve, as due to the high heat capacity of liquids, the temperature at the beginning of the column can easily become significantly lower than the mean temperature. This stringent control of temperature is only necessary when highly precise retention data are required which is not always necessary in quantitative analysis. One final and important design feature of the oven should be mentioned. The oven should be constructed in such a way that there is an absolute minimum length of connecting tube between the sample valve and the column and between the column and the *detector sensor cell* (not just the detector as a whole). Dispersion in a short length of connecting tube can be commensurate with the total dispersion that takes place in the column and must be reduced to a minimum if the full resolution of the column is to be realized [6]. The connecting tube between the sample valve and the detector should be no greater than 0.010 in. I.D. and preferably less and the length should kept to a few cm, preferably 1-2 cm. In a similar manner the connection between the column and the detector should be about 1-2 cm, if possible, and with the same diameter restrictions. Unfortunately, even today, the importance of extra-column dispersion is not fully appreciated and is not taken into account in either detector or chromatograph design in many modern instruments. A general value for the dispersion that takes place in a connecting tube can be calculated from theory [7] and is shown in table 1 for a range of cylindrical tubes of different radii and different lengths. It should be pointed out, however, that connecting tubes less than 0.002 in I.D. (0.00508 cm) can easily become blocked and thus the smaller diameter tubes can only be used in an extremely clean chromatographic system.

Table 1 Standard Deviation of Connecting Tubes of Different Sizes

	Standard Deviation of Tube Dispersion				
Tube Diameter	l=1 cm	l=2 cm	l=5 cm	l=10 cm	l=15 cm
0.001 in 0.00254 cm	22.3 nl	31.5 nl	49.9 nl	70.5 nl	86.4 nl
0.002 in 0.00508 cm	47.6 nl	67.3 nl	106.4 nl	150 nl	184.4 nl
0.003 in 0.00762 cm	107 nl	151.3 nl	239.2 nl	0.34 µl	0.41 µl
0.005 in 0.01270 cm	298 nl	421 nl	0.67 µl	0.94 µl	1.15 µl
0.010 in 0.02540 cm	1.19 µl	1.68 µl	2.66 µl	3.76 µl	4.61 µl

The subject of LC detectors will be discussed in a separate section at the end of this chapter, but it should be said at this time that the versatile liquid chromatograph should be able to accommodate the diode array UV detector, the refractive index detector, the fluorescence detector and the electrical conductivity detector. These detectors will accommodate the vast majority of sample types presented for method development and for general quantitative analysis.

A fraction collector is always useful in method development to allow specific solutes to be collected and their identity confirmed by off-line spectrometric examination. An associated mass spectrometer with an LC interface is, of course, an extremely expensive ancillary instrument that is likely to cost several times that of the chromatograph itself. Furthermore, it is not generally used in quantitative LC. To some extent it is a luxury, unless the instrument is to be used in regular analyses, the results of which are required for litigation purposes and the identity of the individual solutes must be confirmed by alternative methods. Even then, solute identity can usually be confirmed off-line, more economically, by employing normal MS services.

A computer data acquisition system, similar to that used in GC, is quite adequate even for general quantitative use. The software may be more involved, in that the computer will control the entire chromatograph, including the gradient system, the pumps, the automatic sampler, the detector and also be on line with the laboratory management computer if there is one. The latter service can be important for clinical forensic services. Unfortunately, the computer can sometimes divorce the analyst from the reality of the physical chemical technique itself. The instrument becomes a "black box" and if this happens the technique can never be properly used. The computer is literally only a robot, it adjusts the flow rates, sample size, detector sensitivity, etc., in exactly the same way as it would be done manually using knobs, switches or valves. It integrates curves in a very laborious manner and produces a result in the same way as manually measuring a peak. It carries out its action very clumsily and in a far more involved way, but because it does its calculation at several MHz, it is very fast and thus conceals its basic inefficiency. The keyboard should be thought of as an extension of the hands; instead of leaning over and switching the pump on, instructions are typed in on the keyboard; instead of turning the switch to adjust the flow rate, it is actuated via the keyboard. In most instances it takes much longer to type the instruction in than it would to make the adjustment manually, but of course its is not nearly so technically impressive and certainly not 'state-of-the-art'. On the other hand this derisive attitude towards computers should not be carried too far; computers are essential to modern life. They store data extremely efficiently, carry out complex sequences of operations accurately, rapidly provide smart easily readable reports, organize analytical service and generally make science and industry more efficient and effective. In scientific measurement, however, the computer can very easily obscure the logic and technology that is essential to the success of the process and, in doing so, can easily hide sources of error and inaccuracy and even provide misleading results. It is important that the computer should not be allowed to obscure the science it is supporting and its mode of interaction should not be allowed to isolate the analyst from the significance of the chemistry that is being controlled.

Liquid Chromatography Detectors

As already stated the vast majority of quantitative LC analyses are monitored by one of four detectors, the UV detector, the fluorescence detector, the electrical conductivity detector and the refractive index detector. Other detectors have found certain areas of application, for example the high sensitivity of the electrochemical detector to those substances that readily oxidize or are reduced have been popular for certain types of trace analysis. The so called "mass (molecular weight) detector" that functions on the basis of scattered light measurement has also found use in the detection of large molecular weight biopolymers. However, the four detectors mentioned above will be those selected for some detailed discussion as they have the greatest pertinence to quantitative LC.

The UV Detectors

The UV detectors are predominantly 'solute property detectors' but if the solvent used as the mobile phase contains a weak UV chromaphore, then there will also be a background signal from the mobile phase. Consequently, the detector will also respond to both the bulk property of the mobile phase and the UV adsorption of the solutes. In practice, the mobile phase is chosen to be completely transparent to the wavelength of the UV light selected for detection. This is essential if gradient elution is to be employed, as any adsorption due to one of the solvents will result in a steady and continuous baseline drift throughout the program. There are basically three types of UV detector, the *fixed wavelength detector*, the *variable wavelength detector* and the *diode array detector*. Each of these detectors, although measuring the amount of UV light adsorbed by the column eluent are quite different in design and therefore will be discussed separately.

The Fixed Wavelength UV Absorption Detector

Ideally, the fixed wavelength detector is supposed to operate largely with light of a single wavelength. However, in practice, this is not possible. Although the majority of the light may be emitted at one wavelength by most practical light sources, most single wavelength UV lamps also emit light at other wavelengths, albeit at a much lower intensity.

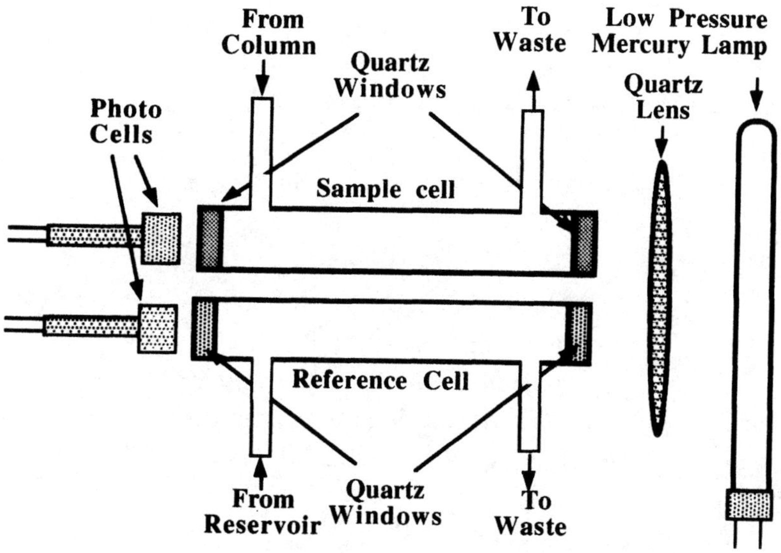

Figure 8 The Fixed Wavelength UV Detector

The important aspect of the "single wavelength UV lamp" is that the light emitted at the specific wavelength has a very high intensity compared with that emitted at the same wavelength by broad spectrum emission lamps. This provides the fixed wavelength UV detector with the propensity for a much higher sensitivity but at the expense of limited detector versatility. A diagram of the optical system of a fixed wavelength UV detector is shown in figure 8.

The light source is usually a discharge lamp, such as a low pressure mercury vapor lamp, and the light is focused through two absorption cells onto two photoelectric sensors. The output from the sensors is electronically modified and then passed to a recorder or a computer data acquisition system. One absorption cell carries the column eluent and the other is a reference cell that carries the pure mobile phase. The volume of the cell must be kept as small as possible to reduce band dispersion and maintain the integrity of the separation. Modern high efficiency columns can restrain the peak volume to a few microliters and thus, if the cell volume is too large, more than one peak can enter the cell at one time and the resolution would be completely lost. The sensing volume of a modern LC detector cell should not be greater than 5 μl and preferably less than 2 μl. The intensity of the UV light adsorbed in the cell is a function of the concentration of the solute in the cell, which by Beer's Law, is given by,

$$I_T = I_o e^{-\varepsilon lc} \quad \text{or} \quad \ln I_T = \ln I_o - \varepsilon lc \qquad (1)$$

where (I_o) is the intensity of light entering the cell,
(I_T) is the intensity of light transmitted through the cell,
(l) is the path length of the cell,
(c) is the concentration of solute in the cell,
and (ε) is the molar absorption coefficient of the solute at the specific wavelength of the UV light.

If equation (1) is put in the form $$I_T = I_o 10^{-\varepsilon' lc} \qquad (2)$$

then (ε') is termed the molar extinction coefficient. Differentiating equation (1)

$$\frac{\partial\left(\frac{I_T}{I_o}\right)}{\partial c} = -\varepsilon l$$

It is seen that the sensitivity of the detector, as measured by the transmitted light, will be directly proportional to the extinction coefficient (ε) and the path length of the cell (l). It follows, that to increase the sensitivity of the system, (l) must be increased. However, there is a limit to which (l) can be enlarged, as the total volume of the cell must be kept to a minimum, to ensure there is no loss of resolution as previously discussed. It follows that to increase (l), and thus the sensitivity, the radius of the cell must also be concurrently reduced to maintain a small cell volume. Reduction in cell radius also reduces the amount of light falling on the photo cell and increases the noise and thus reduces the signal to noise ratio and consequently, the detector sensitivity. The two effects oppose one another and thus the process of increasing detector sensitivity by lengthening the cell is limited. Most modern UV cells have path lengths lying between 1 and 10 mm and internal diameters of 1 mm or less.

From equation (2),

$$\log\frac{I_T}{I_0} = \varepsilon' l c = A$$

where (A) is the *absorbance.*

(ΔA) is commonly employed to define the detector sensitivity where the value of (ΔA) is the change in absorbance that produces a signal to noise ratio of 2.

Thus, $$\Delta A = \varepsilon' l \Delta c$$

where (Δc) is the detector concentration sensitivity or the "minimum detectable concentration", which is the parameter of importance to chromatographers.

It follows that, $$\Delta c = \frac{\Delta A}{\varepsilon' l}$$

It is clear that two detectors having the same sensitivity defined as the minimum detectable change in absorbance will not necessarily have the same sensitivity with respect to solute concentration. Only if the path lengths of the two cells are identical will they also exhibit the same concentration sensitivity. This can cause some confusion to the chromatographer who might expect that instruments that are defined as having the same spectroscopic sensitivity to have equivalent chromatographic sensitivity. To compare the sensitivity of two detectors given in units of absorbance, the path lengths of the cells in each instrument must be taken into account.

There are two other UV lamps that can be used in fixed wavelength detectors. The zinc lamp which has major emission lines at 213.9 nm which would be ideal for the detection of proteins and polypeptides. This lamp also emits light of significant intensity at 307.6 nm. The presence of the higher wavelength light would not affect the sensitivity of the detector at 213.9 nm (in fact it would add to the lamp's detection versatility) and thus a fixed wavelength detector operating with a zinc lamp could be useful in the separation of some biopolymers. The other light source, the cadmium lamp, has a major emission wavelength at 228.8 which could also be useful for the detection of protein-like substances. The cadmium lamp does, however, emit light at a large number of other wavelengths and is, in fact, almost a wide spectrum emission lamp in the sense that the discrete emissions cover a wide wavelength range. The cadmium lamp, if viewed from the point of view of detection versatility would be far more effective than the low pressure mercury lamp and would respond to a much wider range of solute types. The reason for its preclusion from use in fixed wavelength detectors is not clear, but is possibly due to cost. Nevertheless, in principle, the essential design of the lamp does not seem expensive, and it could provide the basis of a fixed wavelength detector which would respond to a far greater range of solutes than detectors utilizing the low pressure mercury lamp. Ignoring the specific uses of a diode array detector to obtain UV spectra (which will be discussed below), a fixed wavelength UV detector employing a cadmium lamp could provide an effective low cost alternative to the detection versatility of the diode array detector. Typical specifications for the fixed wavelength detector employing a low pressure mercury lamp as the light source are given in the following table.

Typical Specifications for a Fixed Wavelength Detector Employing a Low Pressure Mercury Lamp as the Light Source

Sensitivity (Toluene)	**5 x 10^{-8}g/ml**
Linear Dynamic Range	**5 x 10^{-8} to 5 x 10^{-4} g/ml**
Response Index	**0.98–1.02**

The Variable Wavelength Detector

The variable wavelength detector employs a lamp that emits light over a wide range of wavelengths and by using a monochromator, light of a particular wavelength can be selected for detection purposes. This can improve the sensitivity by selecting a wavelength at which the solute(s) of interest has particularly strong absorption. Alternatively, for identification purposes the flow rate can be arrested, the solute trapped in the detection cell and scanned through a range of wavelengths to produce an absorption spectrum. A diagram of a variable wavelength detector is shown in figure 9.

Light from a deuterium lamp is focused by means of two curved mirrors onto a diffraction grating and the selected wavelength is reflected back onto the second curved mirror. The light then passes to a third curved mirror and is then focused via a plane mirror and a quartz lens through the detector absorption cell. After leaving the absorption cell, the transmitted light is focused by a second quartz lens onto a photo cell. Rotation of the grating allows the wavelength of the light passing through the cell to be selected to provide the maximum sensitivity. There are three basic advantages to a variable wavelength detector.

1. A UV absorption spectrum can be obtained for any, or all, the eluted solutes. Such spectra can sometimes help to confirm solute identification. Unfortunately, UV spectra are of limited assistance in solute identification because the majority of substances (excluding, perhaps, those containing aromatic nuclei) exhibit very similar absorption curves and are in most cases insufficiently different to arrive at confident conclusions on solute identity.

2. Confirm peak purity. The spectra of a solute can be taken at various points across a peak as it is slowly moved through the cell and if the spectra are consistently the same, the peak is pure. It will be seen later that this can be achieved more elegantly using a diode array detector.

3. The variable wavelength detector can be simply used as a wavelength selector to increases sensitivity by selecting the optimum wavelengths for maximum absorption. In fact, the optimum wavelength can be chosen for each solute and a program used to

change the wavelength to that most appropriate for the peak being eluted during the development of the chromatogram. This is probably the most popular use of a variable wavelength detector. Typical specifications for a variable wavelength detector are given in the following table.

Typical Specifications of a Variable Wavelength UV Detector.

Sensitivity (Toluene)	**1×10^{-7}g/ml**
Linear Dynamic Range	**1×10^{-7} to 5×10^{-4} g/ml**
Response Index	**0.98 - 1.02**

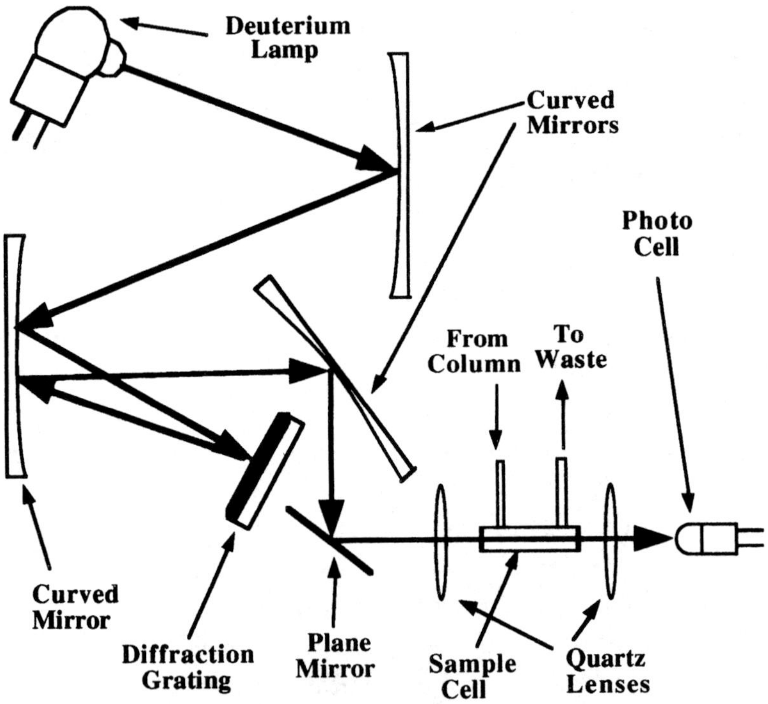

Figure 9 The Variable Wavelength Detector

The Diode Array Detector

The diode array detector also utilizes a deuterium or xenon lamp that emits light over the UV spectrum range as a light source. Light from the lamp is focused by means of an achromatic lens through the sample cell and onto a holographic grating. The dispersed light from the grating is arranged to fall on a linear diode array. A diagram of a UV diode array detector is shown in figure 10. The resolution of the detector ($\Delta\lambda$) will depend on the number of diodes (n) in the array (which is determined by the semiconductor manufacturer), and on the range of wavelengths covered ($\lambda_2 - \lambda_1$) (defined by the optical arrangement, the grating that is used and the width of the solid state array).

Thus

$$\Delta\lambda = \frac{\lambda_2 - \lambda_1}{n}$$

The diode array detector takes a UV spectrum of the eluent continuously throughout the complete development of the chromatogram which can be used to a great advantage.

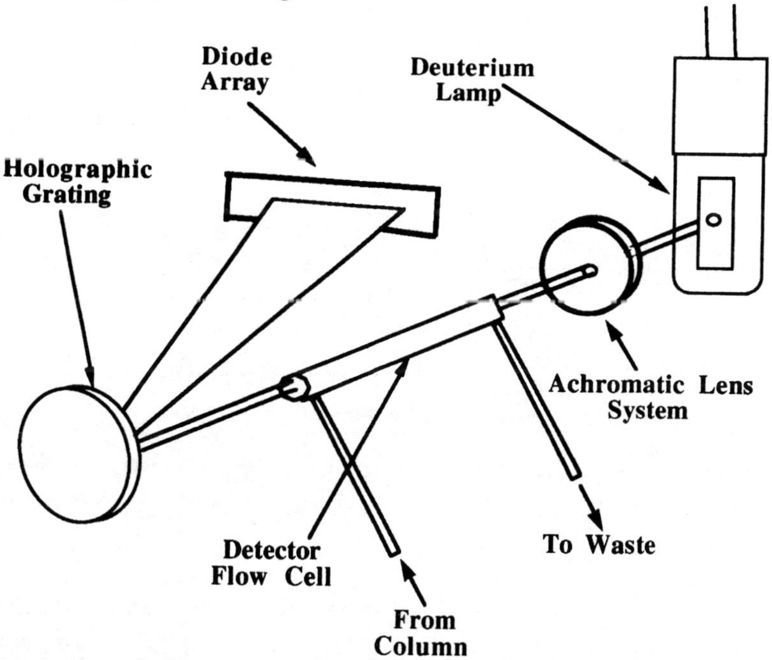

Figure 10 The UV Photo Diode Array Detector

It follows that a chromatogram can be reconstructed by monitoring at a specific wavelength [strictly ($\Delta\lambda$), a narrow group of wavelengths defined by the solid state diode array chip] and thus depicts only those substances that adsorb UV light at the chosen wavelength. Consequently, the chromatogram can be arranged to display only those substances that have unique absorbance characteristics. The diode array detector can be used in a number of other unique ways and an example of its use to verify the purity of a given solute is shown in figure 11. The chlorthalidone was isolated from a sample of tablets and separated by a reverse phase (C18) on a column 4.6 mm I.D., 3.3 cm long, using a solvent mixture consisting of 35% methanol and 65% aqueous acetic acid solution (water containing 1% of acetic acid).

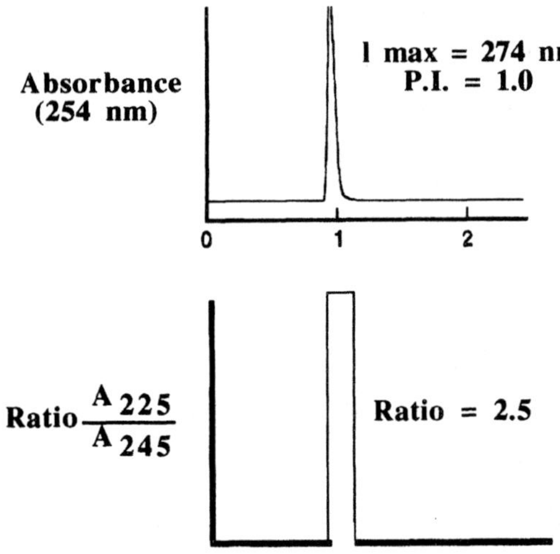

Courtesy of the Perkin Elmer Corporation

Figure 11 Dual Channel Curves from a Diode Array Detector for Chlorthalidone Showing the Separation and Confirming Peak Purity

The flow rate was 2 ml/min and the separation monitored at 274 nm is shown in the upper part of figure 11. As a diode array detector was employed, it was possible to ratio the output from the detector at different wavelengths and plot the ratio simultaneously with the chromatogram monitored at 274 nm. Now, if the peak was pure and homogeneous, the ratio of the absorption at the two wavelengths (those selected being 225 and 245 nm) would remain constant throughout the elution of the entire peak. The lower diagram in figure 11 shows this ratio plotted on the same time scale and it is seen that a clean rectangular peak is obtained which unambiguously confirms the purity of the peak for chlorthalidone. It should be pointed out that the wavelength chosen to provide the confirming ratio will depend on the UV absorption characteristics of the substance concerned, relative to those of the most likely impurities to be present. Another example of the use of the diode array detector to confirm the integrity of an eluted peak is afforded by another application published by the Perkin Elmer Corporation showing the separation of a mixture of aromatic hydrocarbons. The separation they examined is shown in figure 12.

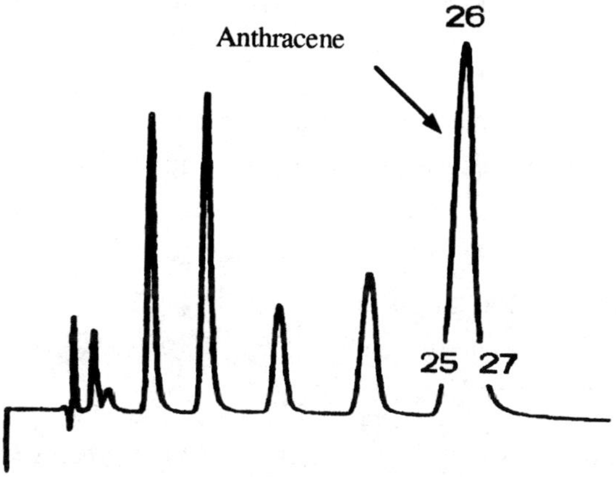

Courtesy of the Perkin Elmer Corporation

Figure 12 The Separation of Some Aromatic Hydrocarbons

This separation was carried out on a column 3 cm long, 4.6 mm in diameter and packed with a C18 reversed phase on particles 3 μ in diameter. It is seen that the separation appears to be good and all the peaks represent individual solutes.

However, by plotting the absorption ratio, $\frac{250\,\text{nm}}{255\,\text{nm}}$, for the anthracene peak, it became clear that the peak tail contained an impurity.

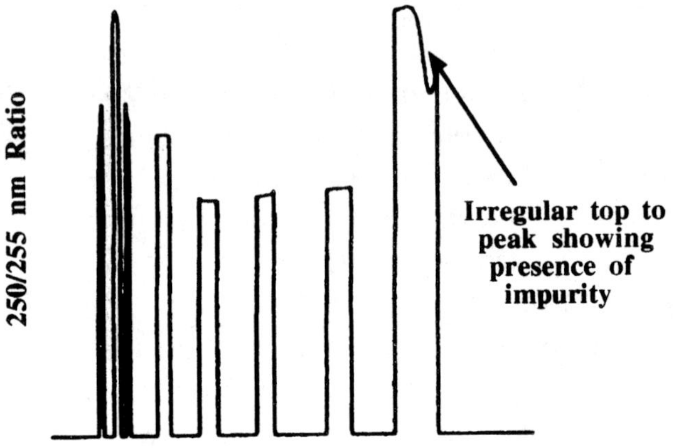

Courtesy of the Perkin Elmer Corporation

Figure 13 Curves Relating Absorption Ratio, 250nm/255nm to Time

The absorption ratio peaks are shown in figure 13. The ratio peaks in figure 13 clearly show the presence of an impurity by the sloping top of the anthracene peak and further work identified the impurity as t-butyl benzene at a level of about 5%. The characteristics of both the variable wavelength and the diode array detector are very similar and consequently have very comparable specifications. Their more important properties are listed below.

Typical Specifications of a Diode Array Detector

Sensitivity	**1.5×10^{-7}g/ml**
Linear Dynamic Range	**1.5×10^{-7} to 5×10^{-4} g/ml**
Response Index	**0.97-1.03**

The Fluorescence Detector

The fluorescence detector is probably the most sensitive detector generally available to LC, but is so at the sacrifice of versatility. By definition, the fluorescence detector will only detect those materials that will fluoresce or, by appropriate derivatization, can be made to fluoresce. Nevertheless, due to its sensitivity it is one of the most important detectors for use in trace analysis both in environmental and forensic analysis. A diagram of a simple form of fluorescence detector is shown in figure 14. The UV lamp that provides the excitation radiation is often a low pressure mercury lamp as it is comparatively inexpensive and provides relatively high intensity UV light at 253.7 nm. Many substances that fluoresce will, to a lesser or greater extent, be excited by light at this wavelength. The light is focused by a quartz lens through the cell and another lens situated normal to the incident light focuses the fluorescent light onto a photo cell.

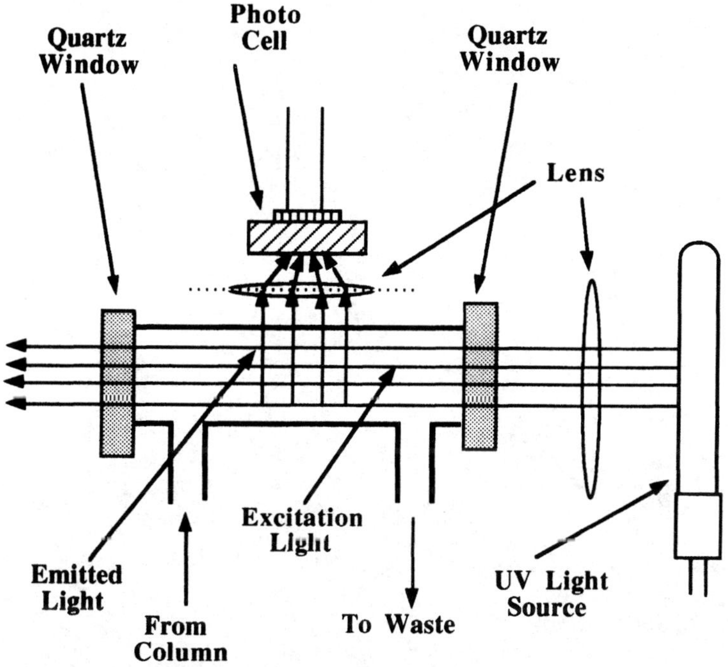

Figure 14 The Simple Fluorescence Detector

Typical specifications for a fluorescence detector are as follows:

Typical Specifications for a Fluorescence Detector

Sensitivity (Anthracene)	1×10^{-9} g/ml
Linear Dynamic Range	1×10^{-9} to 5×10^{-6} g/ml
Response Index	0.96-1.04

A far more sophisticated fluorescence detector is shown in figure 15.

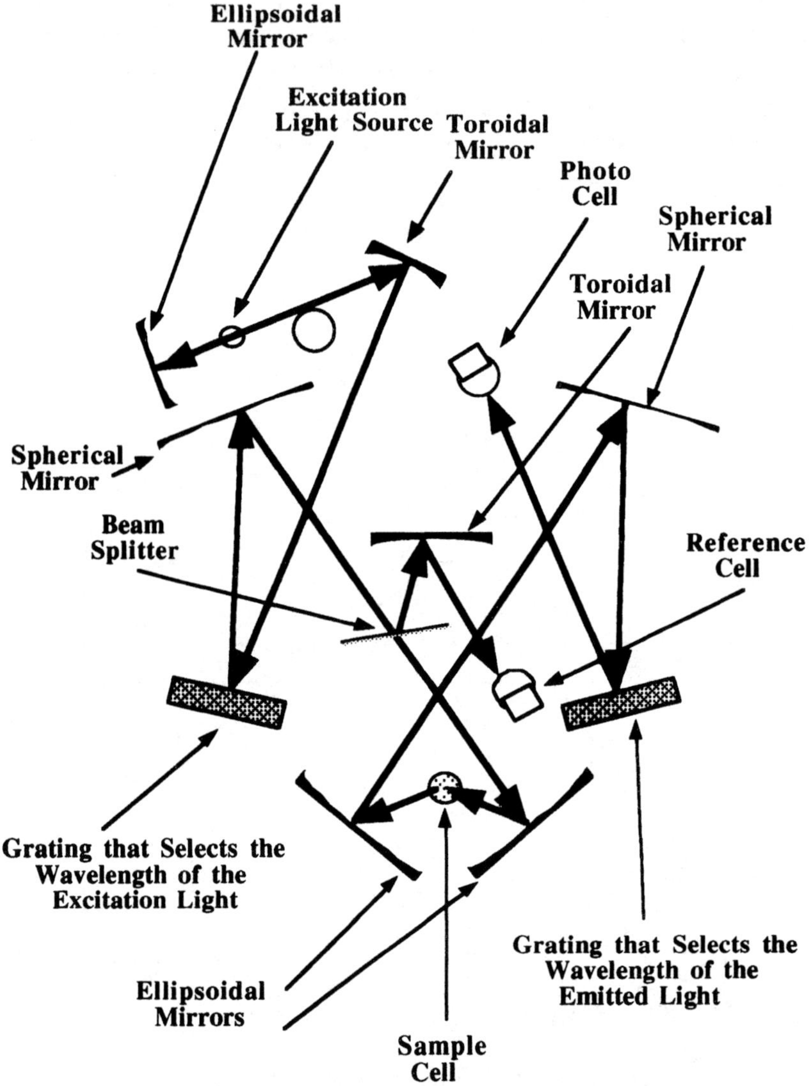

Figure 15 The Fluorescence Spectrometer Detector

The diagram depicts the basic fluorescent spectrometer but because it incorporates two distinctly different light paths, the optical system appears quite complicated. If the different light paths are considered separately, that is the path of the excitation light and the path of the fluorescent light, the diagram is easier to understand. The excitation source (usually a deuterium lamp which emits UV light over a wide range of wavelengths) is situated at the focal point of an ellipsoidal mirror shown at the top left-hand corner of the diagram. The parallel beam of light is arranged to fall on a toroidal mirror that focuses it onto a grating on the left-hand side of the diagram.

This grating allows the frequency of the excitation light to be selected or the whole spectrum scanned providing a complete range of excitation wavelengths. The selected wavelength then passes to a spherical mirror and then to a ellipsoidal mirror at the base of the diagram which focuses it onto the sample. The excitation light path is largely situated on the left-hand side of the diagram. Between the spherical mirror and the ellipsoidal mirror, in the center of the diagram, is a beam splitter that reflects a portion of the incident light onto another toroidal mirror. This toroidal mirror focuses the portion of incident light onto the reference photo cell providing an output that is proportional to the strength of the incident light. The path of the fluorescent light is largely on the right-hand side of the diagram. Fluorescent light from the cell is focused by an ellipsoidal mirror onto a spherical mirror at the top right-hand side of the diagram. This mirror focuses the light onto a grating situated at about center right of the figure. This grating can select a specific wavelength of the fluorescent light to monitor or scan the fluorescent light and provide a fluorescent spectrum. Light from the grating passes to a photoelectric cell which monitors its intensity. The instrument is extremely complex and as a result rather expensive, however, from the point of view of measuring fluorescence it is extremely versatile. An excellent example of its use is given in figure 16 which depicts the separation of a group of priority pollutants using programmed fluorescence detection.

The separation was carried out on a column 25 cm long, 4.6 mm in diameter and packed with a C18 reversed phase. The mobile phase was programmed from 93% acetonitrile/7% water to 99% acetonitrile/1% water over a period of 30 minutes. The gradient was linear and the flow rate was 1.3 ml/min. All the solutes are separated and the compounds, numbering from the left are given in the table. The separation illustrates the clever use of wavelength programming to obtain the maximum sensitivity.

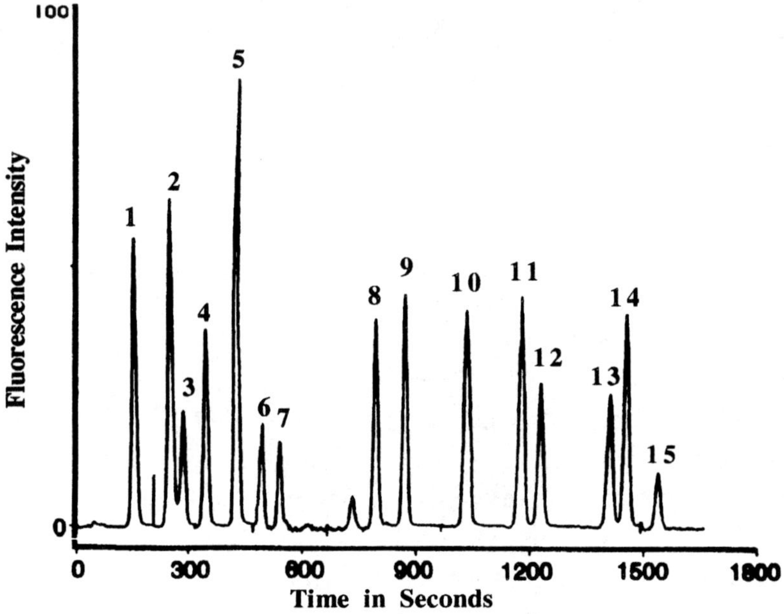

Courtesy of the Perkin Elmer Corporation

Fifteen Priority Pollutants

1 Naphthalene	**9 Chrysene**
2 Acenaphthene	**10 Benzo(b)fluoranthene**
3 Fluorene	**11 Benzo(k)fluoranthene**
4 Phenanthrene	**12 Benzo(a)pyrene**
5 Anthracene	**13 Dibenz (a,h)anthracene**
6 Fluoranthene	**14 Benzo(ghi)perylene**
7 Pyrene	**15 Indeno(123-cd)pyrene**
8 Benz(a)anthracene	

Figure 16 Separation of a Series of Priority Pollutants with Programmed Fluorescence Detection

The program used for this particular mixture was rather complex. During the development of the separation, both the wavelength of the excitation light and the wavelength of the emission light that was monitored were changed. This ensured that each solute, as it was eluted, was excited at the most effective wavelength and then, consequently, monitored at the strongest fluorescence wavelength. The actual program that was used is shown below. It is seen that the analysis is an elaborate procedure carried out with a very complex and expensive instrument. Nevertheless, if the analysis is sufficiently important it may be necessary to resort to this type of solution. The system can also provide a fluorescence spectra, should it be required, by arresting the flow of mobile phase when the solute resides in the detecting cell and scanning the fluorescent light as with the variable wavelength detector. It should also be remembered that, although very sensitive, the fluorescence detector, however sophisticated, has a limited linear dynamic range and thus is not ideal for quantitative accuracy and precision and great care must be exercised when using this detector.

Fluorescence Detector Program

Time (seconds)	Wavelength of Excitation Light	Wavelength of Emitted Light
0	280 nm	340 nm
220	290 nm	320 nm
340	250 nm	385 nm
510	260 nm	420 nm
720	265 nm	380 nm
1050	290 nm	430 nm
1620	300 nm	500 nm

The Electrical Conductivity Detector

The electrical conductivity detector was the first in-line detector to be developed (Martin and Randall 1951 [8]) and is still used extensively in ion exchange chromatography today. The basic sensor, being so simple, has changed little in the intervening period. The detecting volume has become much smaller but it still consists of two electrodes situated in the column eluent, the resistance (or more strictly the impedance) of which is measured by a suitable electronic circuit. A diagram of a conductivity sensor is shown in figure 17.

It is seen that the basic sensor is, indeed, very simple and can be designed to have an effective sensing volume of a few nanoliters. For this reason it is often used as a detector in capillary electrophoresis and capillary electro-chromatography. The sensor is usually placed in one arm of a Wheatstone bridge, and the out of balance signal electronically modified to provide an appropriate output.

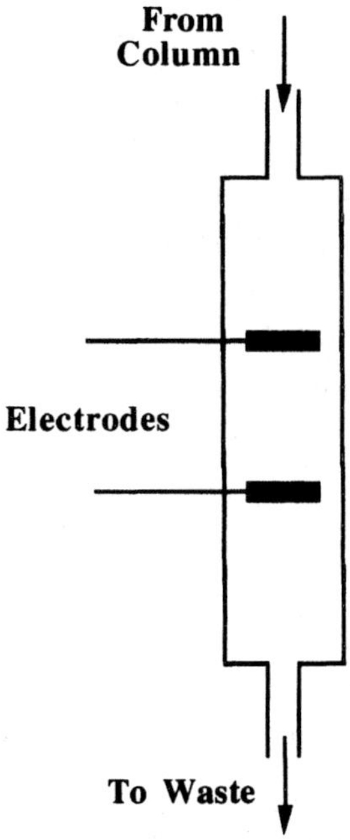

Figure 17 The Electrical Conductivity Detector

The out of balance signal that occurs when an ionic solute is present between the electrodes will be proportional to a change in electrical resistance between the electrodes and this will *not* be linearly related to the ion concentration. As a consequence an appropriate amplifier must be used to modify the signal and provide an output directly that is proportional to solute concentration. The resistance (R) of any conductor varies directly as its length (ℓ) and inversely as its cross-sectional area (a).

Thus,
$$R = \frac{\rho\ell}{a} \tag{3}$$

where (ρ) is the *specific resistance* of the conductor,

The *specific resistance* of a conductor is the resistance across two opposite faces of a 1 cm cube of the conductor material. The *specific conductance* (κ) of a solute is defined as the reciprocal of the specific resistance, thus

$$\kappa = \frac{1}{\rho} \tag{4}$$

Thus the conductance of a given solution (C) is given by

$$C = \frac{1}{R} = \frac{\kappa a}{\ell} \text{ ohms}^{-1} \tag{5}$$

Now the conduction of all the ions produced by 1 g equivalent of an electrolyte at any particular concentration can be evaluated by imagining two large parallel electrodes, 1 cm apart, and the whole of the solution placed between them. The conductivity of the system is called the equivalent conductance (Λ). Now suppose that 1 g equivalent of the electrolyte is dissolved in (v) ml of solution, then this will cover (v) sq. cm of electrode area. It follows that in the above equation (a) becomes (v) and (l) is unity and,

$$\Lambda = \frac{1000\kappa}{c} \text{ ohms}^{-1}\text{cm}^2$$

where (c) is the concentration of the electrolyte in gram equivalents /liter.

or,
$$\kappa = \frac{\Lambda c}{1000} \tag{6}$$

Substituting for (k) from (6) in (5) and rearranging,

$$C = \frac{1}{R} = \frac{\Lambda ca}{1000\ell} \text{ ohms}^{-1} \tag{7}$$

It is seen that the resistance of the cell is inversely proportional to the electrolyte concentration (c), the equivalent conductance of the electrolyte and the geometry of the cell {(a) and (l)}. It follows that the output of the cell, which is in fact proportional to the resistance change of the cell must be modified to provide an output that is linearly related to electrolyte concentration. The conductivity detector is a bulk property detector and, as such, responds to any electrolytes present in the mobile phase (*e.g.* buffers, etc.) as well as the solutes. It follows that the mobile phase must be arranged to be either nonconducting, which in many cases is difficult if not impossible to achieve, or the mobile phase buffer electrolytes must be removed prior to the detector. This technique of buffer ion removal is commonly called *ion suppression.* The first type of ion suppression involved the use of a second ion exchange column subsequent to the analytical column of the appropriate form for the removal of buffer cations, replacing them with hydrogen ions. Conversely, if cations were being separated a suitable anion exchanger would replace the buffer anions with hydroxyl ions. This approach works well, providing the ion suppressor (ion remover) does not cause too much peak dispersion and consequent loss of resolution.

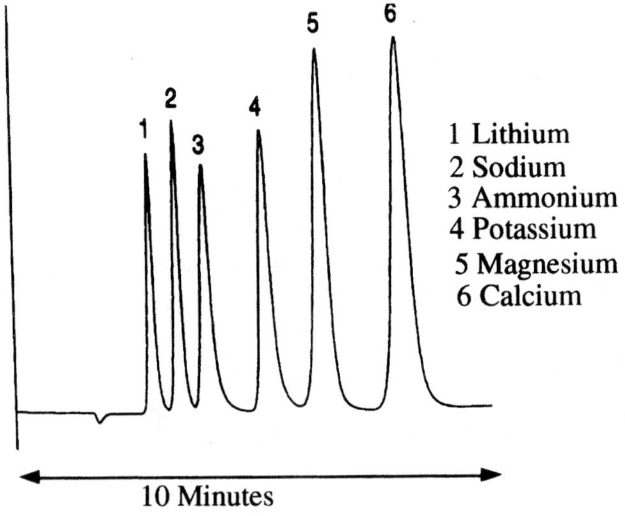

Courtesy of Dionex Inc.

Figure 18 Determination of Alkali and Alkaline Earth Cations

Another alternative, and far more efficient procedure, is to employ an organic buffer, such as methane sulfonic acid and a short length of reversed phase column subsequent to the analytical column. After the separation has been achieved, and the individual ions are eluted from the column, the methyl sulfonate that is eluted with them is absorbed almost irreversibly onto the reversed phaseleaving the ions of interest to enter the detector in the absence of any buffer ions. It is obvious that this type of suppressor column will have a limited lifetime but can be regenerated by passing acetonitrile through the suppressor volume followed by water. An example of the use of this type of suppression system is given in figure 18. A proprietary ion exchange column, IonPacCS12, was used and the mobile phase consisted of a 20 nM methanesulphonic acid solution in water. A flow rate of 1 ml/min was employed and the sample volume was 25 μl.

Although this technique of ion suppression is frequently used in ion exchange chromatography, depending on the nature of the separation, a specific type of ion suppression column or exchange membrane will be required that will be appropriate for the phase system that is chosen. A wide variety of different types of ion suppression columns are available for this purpose. It should be pointed out that any suppressor system introduced between the column and the detector that has a finite volume will cause band dispersion. Consequently, the connecting tubes and suppression column must be very carefully designed to eliminate or reduce this dispersion to an absolute minimum. Modern ion supression columns and exchange membranes are designed to reduce the extra-column dispersion to a minimum.

The Refractive Index Detector

The refractive index detector is probably the least sensitive of all the commercially available and generally useful detectors. It has the disadvantages of being sensitive to changes in ambient temperature, pressure and flow rate and, as it is a bulk property detector, cannot be used for gradient elution. Despite these disadvantages, the refractive index detector is extremely useful for monitoring those compounds that are nonionic, do not adsorb in the UV, and do not fluoresce.

There are a number of different optical systems used in refractive index detectors [9] but the simplest and probably the most common is the differential refractive index detector which is shown diagrammatically in figure 19. The differential refractometer responds to the deflection of a light beam that is caused by the differing refractive index between the contents of the sample cell and that of the reference cell. A beam of light from an incandescent lamp (usually a tungsten filament lamp) passes through an optical mask that confines the beam to the region of the cell.

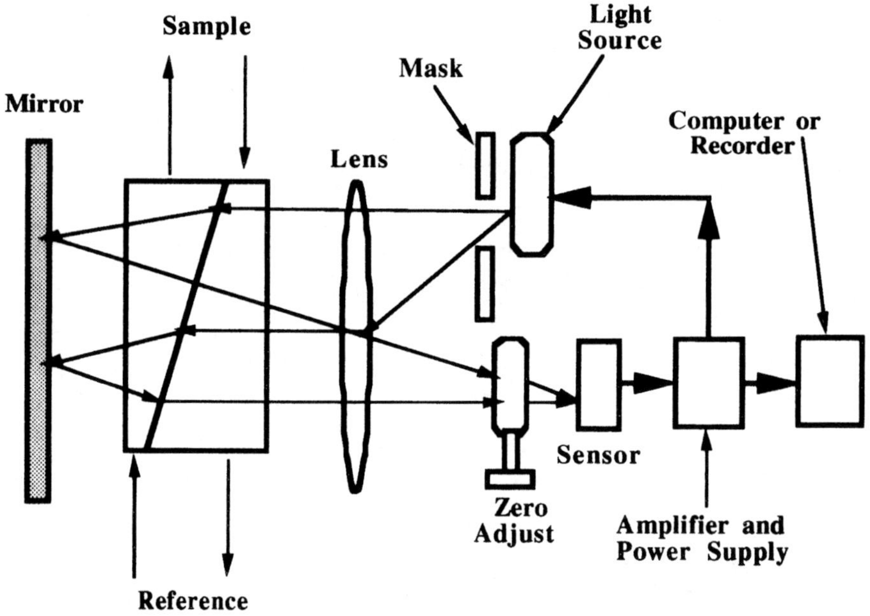

Courtesy of the Millipore Corporation

Figure 19 The Differential Refractometer

The lens collimates the light beam which then passes through both the sample and reference cells to a plane mirror. The mirror reflects the beam back through the sample and reference cells and is focused by another lens onto a photo cell. The location of the beam, rather than its intensity, is determined by the angular deflection of the beam caused by the refractive index difference between the contents of the two cells. As the beam changes its position of focus on the photoelectric cell, the output changes and the resulting difference signal is electronically modified to provide a signal proportional to the concentration of solute in the sample cell.

In general, the refractive index detector tends to be the 'choice of last resort' and is used where, for one reason or another, all other detectors are inappropriate or impractical. Nevertheless, the detector has certain areas of application where it is especially useful, one being for the separation and analysis of polymers. It has been shown that if a polymer contains more than six monomer units, the refractive index becomes directly proportional to the concentration of the polymer and is nearly independent of the molecular weight. Consequently, as there is no need for individual solute response factors, a polymer mixture can be analyzed by peak area normalization, a method that has been previously discussed. Some typical specifications for the refractive index detector are as follows:

Typical Specifications for a Refractive Index Detector

Sensitivity (benzene)	**1x 10^{-6} g/ml**
Linear Dynamic Range	**1 x $10^{-6}$to 1 x 10^{-4} g/ml**
Response Index	**0.97 - 1.03**

A typical example of the use of the refractive index detector is shown in figure 20. Carbohydrates do not have UV chromaphores (or at least only absorb at very low wavelengths), fluoresce or contain ions, and thus unless derivatives are made, one of the few detectors that will sense them is the refractive index detector. The column used was Supercosil LC-NH_2, an aminopropyl bonded phase that exhibits fairly strong polar interactions in addition to the dispersive interactions from the propyl chains. The support used was a spherical silica having a mean pore size of 100 Å. The column was 25 cm long, 4.6 mm I.D. and the particle size of the packing was 5 μm. The mobile phase consisted of a mixture of 75%v/v acetonitrile and 25%v/v water and the flow rate was 1 ml/min. The sample consisted of 20 ml of a solution containing 10 mg/ml of each analyte in deionized water. It is seen that the carbohydrates are well resolved in little over 16 minutes.

The refractive index detector is often used in the separation of carbohydrates, aliphatic alcohols, the hydrolysis products of starch and those natural products that do not have chemical groups that permit other methods of detection. The refractive index detector is, nevertheless, a catholic detector and would be used far more extensively if it could tolerate gradient elution.

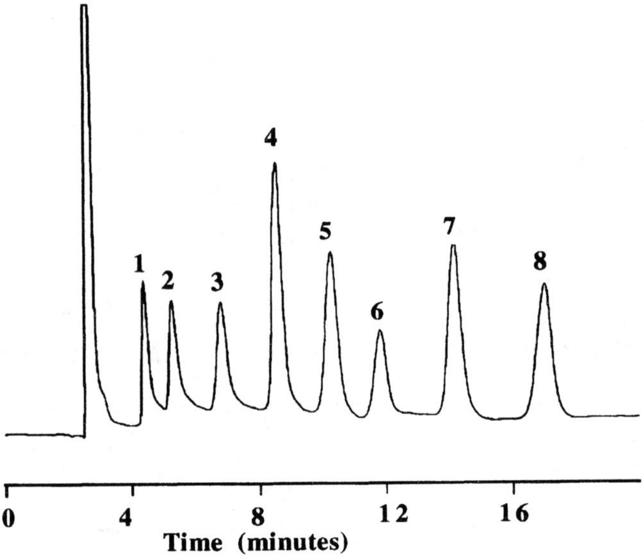

Courtesy of Supelco Inc.

1. Ribose	2. Arabinose	3. Galactose	4. Sucrose
5. Maltose	6. Isomaltose	7. Melezitose	8. Raffinose

Figure 20 The Separation of Some Carbohydrates

Synopsis

The majority of quantitative LC analyses can be successfully carried out with relatively simple and basic equipment. All that is necessary is a solvent supply system, a pump, a sample valve, a column and column oven, a suitable detector and a recorder or data acquisition system. The solvent supply system should include either a low pressure or high pressure solvent programmer, preferably having a three solvent capacity. Single piston pumps are usually satisfactory or, if detector noise from pump pulsations are a problem, then twin pumps or a rapid refill pump may be preferable. Sample valves with internal or external loops are appropriate, but the external loop valve fitted with a direct syringe injection port is particularly easy to operate. The more versatile chromatograph fitted with an automatic sampler, a fraction collector, a range of detectors, including a mass spectrometer and appropriate interface, will cope with almost any LC analytical challenge. Such equipment is particularly suitable for method development, but is rarely necessary for standard quantitative analyses. The most useful detectors for quantitative LC are probably the fixed wavelength UV absorption detector or the UV diode array detector, the former being slightly more sensitive, the latter being more versatile. The sensitivity of the UV detector is often defined in terms of minimum detectable adsorbance, however detectors having the same sensitivity defined in these terms may not have the same concentration sensitivity. Direct comparison can only be made on the basis of minimum adsorbance if the path lengths of the absorption cells are the same. The diode array detector, although somewhat less sensitive, has the advantage of being able to provide spectra and to confirm peak purity and thus increase the confidence level of the analysis. The fluorescence detector is probably the most sensitive LC detector, but it is specific to those substances that fluoresce, or by suitable derivatization can be made to fluoresce. The fluorescence detector has a very limited linear dynamic range (little more than two orders of magnitude) and even with a sensitivity of 1×10^{-9} g/ml is three or four orders of magnitude less sensitive than the equivalent GC detector. Employing a fluorescence spectrometer as a detector provides excitation and/or fluorescence spectra of each eluted solute and thus can confidently confirm solute identity. The electrical conductivity detector is almost exclusively used in ion chromatography, it is fairly sensitive but specific resistance is not linearly related to concentration. It follows that for accurate quantitative analysis, the ion concentration may warrant the use of a nonlinear amplifier to provide an output that is linearly related to ion concentration. The refractive index detector is a relatively insensitive detector and is sensitive to temperature, flow and pressure changes, but has survived as an effective LC detector due to its catholic response. For this reason it is often used in quantitative analysis where the solutes are not easily detectable by any other means. Many biopolymers give the same response irrespective of the number of monomers present and thus quantitative analysis can be accomplished using a normalization procedure eliminating the need for internal or external standards. The refractive index detector has sensitivity of about 1 x 10^{-6} g/ml and a linear dynamic range of little over two orders of magnitude.

References

1. T. E. Beesley and R. P. W. Scott, *Chiral Chromatography*, John Wiley and Sons, Chichester-New York (1998)39.
2. R. P. W. Scott and T. E. Beesley, *Analyst*, **124**(1999)713.
3. R. P. W. Scott, *Analyst*, **124**(1999)1745.
4. R. P. W. Scott and E. D. Katz, *US Patent* 5,0322,283 (1991).
5. E. D. Katz and R. P. W. Scott, *J. Chromatogr.*, **268**(1983)160.
6. R. P. W. Scott, *Liquid Chromatography Column Theory*, John Wiley and Sons, Chichester, New York (1992)153.
7. R. P. W. Scott, *Chromatographic Detectors*, Marcel Dekker, New York-Basel-Hong Kong (1996)47
8. A. J. P. Martin and S. S. Randall, *Biochem. J.*, **49**(1951)293.
9. R. P. W. Scott. *Chromatography Detectors*, Marcel Dekker, New York-Basel-Hong Kong (1996)247.

Chapter 9

Liquid Chromatography Applications

A very large number of liquid chromatography applications have been described in the technical literature over the last two decades. As the technique has developed, the applications reported have covered a continually increasing range of sample types and, in addition, the technique has been applied to wider and wider areas of analytical interest. Parallel to this growth has been a corresponding advancement in sampling techniques and sample preparation procedures. It follows that, in keeping with the GC applications that have already been described, and in order to keep the reader up to date with contemporary applications of the technique, all the examples that have been chosen for discussion will be taken from recently reported assays. Where appropriate, and where necessary, details of sampling routines and sample preparation procedures have been included in the discourse on each application.

It should be pointed out that it would be impossible to give application examples that comprehensively covered all types of LC assays, as they are so diverse. However, eleven examples have been chosen that include a range of sample preparation techniques as well as a spectrum of different LC analytical systems. Four journals are recommended for reference, and the reports published in them cover the subject of quantitative LC analysis very thoroughly. They include the two journals recommended for GC applications, namely, the *Analyst,* published by the Royal Society of Chemistry and *Analytical*

Chemistry, published by the American Chemical Society. In addition, the *Journal of Liquid Chromatography,* published by Marcel Dekker, Inc., is recommended, as it contains a wealth of information on LC analyses and general LC applications. The fourth journal, the *Journal of Chromatographic Science,* published by Preston Publications Inc., can also be extremely useful for reference purposes as it also includes applications and reports on many of the more novel LC applications.

The Determination of Bixin and Norbixin in Human Blood Plasma [1]

Bixin, a naturally occurring carotenoid, has a structure that is unique among carotenoids in that it contains two carboxylic groups, one of which is a methyl ester. Bixin occurs mainly in the seeds of the Anatto plant *Bixa Orellana* and norbixin, which is the water soluble, hydrolyzed derivative of bixin, is used commercially as Annato Food Colors. Annatto has been a normal ingredient of the diet of Latin America for many centuries and thus its uptake into the bloodstream and its rate of elimination is of physiological interest. The normal intake of bixin or norbixin is relatively low (probably less than 4 mg/day) but, nevertheless, its long term effect on human biochemistry needs to be known. This assay is an example of the use of liquid-liquid extraction techniques to isolate the material of interest for subsequent LC separation and assay by liquid chromatography.

The sample preparation involves three stages: (1) the addition of an internal standard (Sudan 1) to the sample of plasma, (2) the proteins contained in the serum are denatured by the addition of ethanol, (3) the bixin, norbixin and the internal standard are extracted into chloroform. The extracts are then pooled, evaporated to dryness and the solvent-free residue dissolved in methanol. The methanol solution is then injected onto the column and the amount of bixin and norbixin estimated by comparing the areas of the respective peaks to that of the Sudan 1 standard. The basic procedure was as follows, 10 ml of blood was drawn by venipuncture and added to a tube containing ethylenediaminetetraacetate and the plasma isolated in the usual way by centrifugation. 3 ml of the plasma was treated with 6 ml of ethanol, shaken, and allowed to stand for 30 min to precipitate the proteins. 3 ml of chloroform was then added, shaken in a separating funnel, and the lower layer run off. The residue was then treated with a further 3 ml of ethanol and 6 ml of chloroform, shaken, and the lower layer again separated and added to the previous extract. The residue was extracted a third time in the same manner and the bulked chloroform extract evaporated to dryness. The dry residue was then mixed with 1 ml of methanol, sonnicated and the supernatant liquid used for analysis.

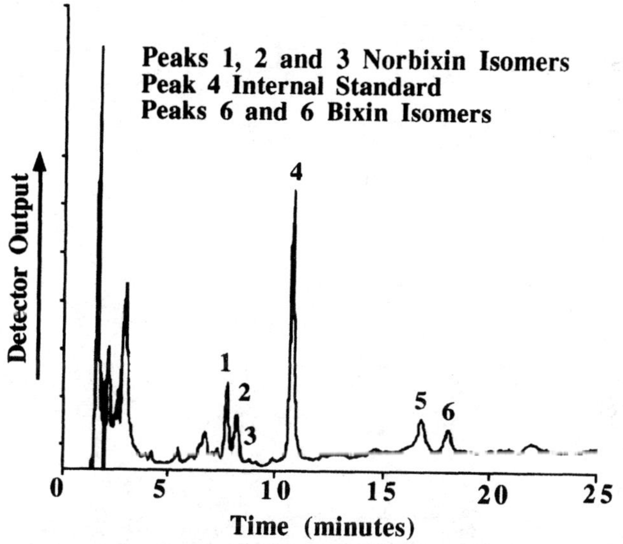

Courtesy of the Royal Society of Chemistry [ref. 1]

Figure 1 Chromatogram of Bixin and Norbixin Contained in Plasma After Ingestion of Annato Food Colors

The column used was a Spherisorb S50DS1 (reversed phase) in conjunction with a mobile phase consisting of a mixture 70%v/v acetonitrile 30%v/v 2% aqueous acetic acid at a flow rate of 1.5 ml/min. The sample volume was 10 μl and the solution of the standard (0.8 μg /ml) was obtained by dissolving 4 mg of Sudan 1 in 100 ml of methanol and diluting 1 ml of the solution to 50 ml with methanol.

The separation was monitored by the light absorption at 460 nm. A typical example of the separation of the materials of interest is shown by the chromatogram in figure 1. The quantity of bixin and norbixin isomers was obtained by comparing the peak areas of the respective peaks to the peak area of the standard Sudan 1 using similar equations to those already described in Part 1. The actual peak identities are shown in figure 2.

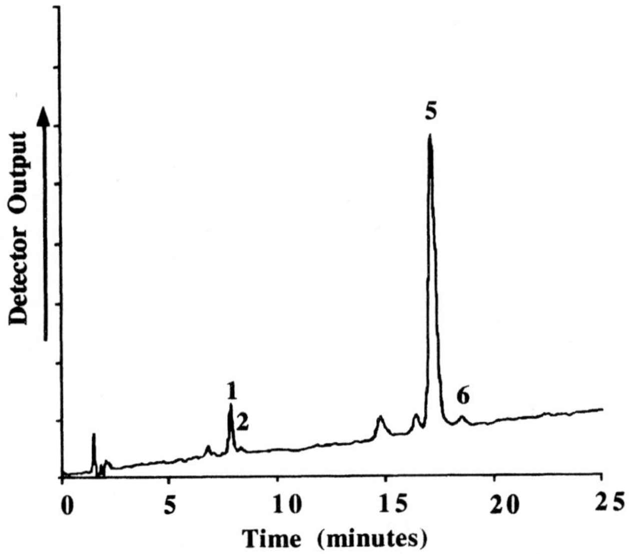

Courtesy of the Royal Society of Chemistry [*ref. 1*]

Peak 1	cis-norbixin
Peak 2	trans-norbixin
Peak 5	cis-bixin
Peak 6	trans-bixin

Figure 2 Chromatogram of Annatto Food Color Used for Ingestion

The assay is very similar to many used for determining a range of drugs, drug metabolites and food materials present in blood serum. This particular procedure depends largely on the substances of interest being efficiently extracted by chloroform from aqueous media. It also depends on the precipitated protein not being soluble or miscible with chloroform so that they do not need to be separated prior to extraction. The technique can be used to examine the metabolism of the bixin compounds over different ingestion periods. The chromatograms from the analysis of serum 2 hr and 4 hr after ingestion are shown in figure 3.

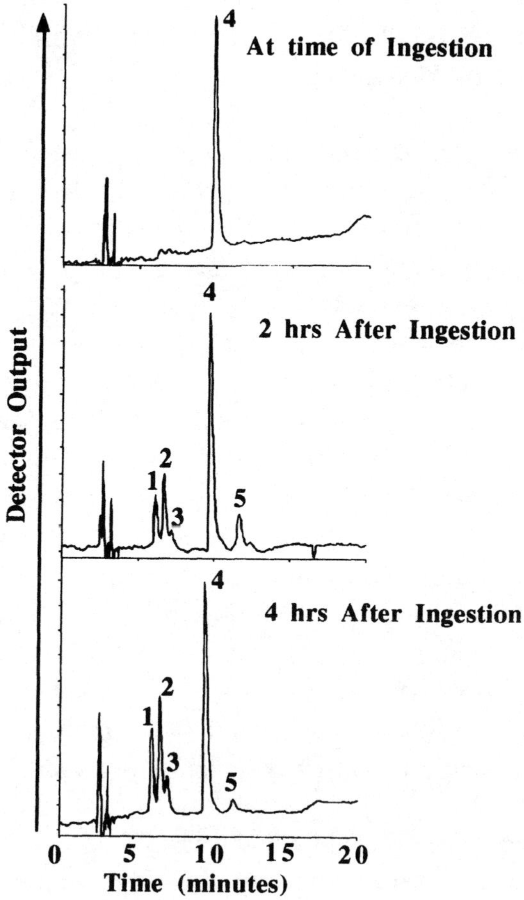

Courtesy of the Royal Society of Chemistry [ref. 1]

Figure 3 Chromatograms of Blood Serum Showing Bixin and Norbixin Contents, 2 hr and 4 hr After Ingestion

It is clear that the adsorption and elimination of the bixin compounds can be easily monitored using the technique to analyze blood serum. It is interesting to note that norbixin isomers gradually and continually build up in the serum over the 4 hour period. The bixin isomers, however, are rapidly adsorbed into the serum and then fairly rapidly eliminated, probably forming norbixin. After 4 hours one bixin isomer has been completely eliminated while the other has been reduced to a mere trace compared with its level 2 hr after ingestion.

Determination of the Enantiomers of Methamphetamine and Its Metabolites in Urine [2]

This application is an example of a relatively fast forensic test for illicit drugs using fluorescence as the method of detection which was developed by Al-Dirbashi *et al.* [2]. The enantiomeric discrimination of drugs is not only important for assessing biological activity, but is also vital for forensic toxicologists to determine whether a drug has been administered as part of a valid medical regime or has been taken illicitly. (S)-(+)-methamphetamine [(S)-(+)-MP], for example, is a controlled substance whereas its antipode, which can have only one-fifth the activity, is often an ingredient in many 'over-the-counter' products. Consequently, the presence of (S)-(+)-methamphetamine and its metabolites indicate the illegal use of the drug. Conversely, the presence of their antipodes could arise from the legitimate use of the drug, for example in the case of selegiline [(R)-(-)-N-(1-phenyl-2-propyl)-2-propinylamine, Diprenyl], a selective inhibitor of monoamine oxidase type B used in the treatment of Parkinson's disease. The fluorescent reagent that was employed was 4-(4,5-diphenyl-1H-imidazole-2-yl)benzoyl chloride (DIB-Cl).

The sample preparation was simple and fast. 10 µl of urine (or diluted urine) was placed in a brown screw-capped reaction vial and mixed with 10 µl of acetic acid. The mixture was evaporated to dryness under a gentle stream of nitrogen at room temperature. The mixture was then treated with 10 µl of carbonate buffer and 190 µl of a solution containing 100 µmol of DIB-Cl per liter of acetone, mixed and vortexed and maintained at 24°C-27°C for ten minutes. 5 µl portion of the solution were used for chromatographic analysis. An example of the chromatograms obtained are shown in figure 4.

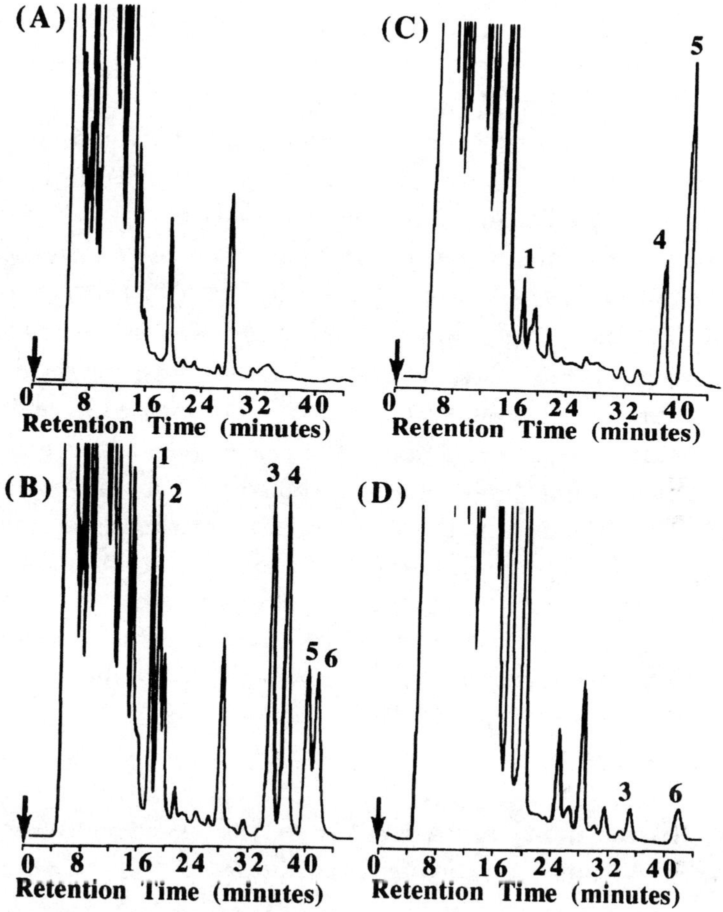

Courtesy of the Royal Society of Chemistry [ref. 2]

1. (S)-(+)-*p*-hydroxymethamphetamine, 2. (R)-(-)-*p*-hydroxymethamphetamine, 3. (R)-(-)-amphetamin, 4. (S)-(+)-amphetamin, 5. (S)-(+)-methamphetamine, 6. (R)-(-)-methamphetamine. (A) 10-fold diluted human urine (control), (B) 10-fold diluted human urine spiked with 10 μmol/l, (C) 10-fold diluted abuser's urine, (D) 10-fold diluted Parkinson's patients urine after ingestion of legal drugs.

Figure 4 Typical Chromatograms of DIB-Cl Derivatives of Methamphetamines

The most suitable chiral column was found to be Chiracel OD-R, 25 cm long, 4.6 mm I.D., particle size 10 μm. The separation was carried out isocratically at 35°C employing propan-2-ol-acetonitrile-phosphate-citrate buffer (pH 4.5) containing 50 mmol/liter of sodium hexafluorophosphate (1+59+40 v/v) as the mobile phase which was degassed prior to use. The sample volume was 5 μl and the eluted peaks were detected employing an excitation wavelength of 330 nm and a monitored emission wavelength of 440 nm. It is seen from figure 4 that the resolution between the isomers is more than adequate and the presence of legal forms of the drug in a patient's urine can be easily differentiated from those that have ingested illegal drugs. For accurate quantitative work calibration curves were obtained using spiked samples. Peak heights were employed as opposed to peak areas for quantitative assessment. Within- and between-day reproducibilities were obtained using drug-free urine spiked with methamphetamine isomers. The detection limits of the isomers, assuming a signal-to-noise ratio of 3, appeared to range between 10 and 40 nmol/liter of urine, which assuming a molecular weight of 149 (methamphetamine) is equivalent to about 1.5 to 6 μg per liter of urine. Within- and between-day reproducibilities were also reported as about 10 nmol/liter of urine or, in terms of methamphetamine, would be about 1.5 μg per liter of urine.

Trace Enrichment of Alkylthio-*s*-triazine Herbicides by Supported Liquid Membrane Techniques [3]

The triazine herbicides are one of the most common pre- and post-emergence weed control agents employed today and are used for a wide variety of crops including green vegetables. In fact, probably 30% of all herbicides that are produced are triazines. The triazines are relatively persistent and easily transported and, as a result, they are the most frequently found contaminants in soil, surface and ground water and consequently are a prime cause of lowering the quality of drinking water. Regulations now in force in Europe limit the total pesticide content to 0.5 μg/l, and for a single pesticide to 0.1 μg/l. It follows that to assay pesticides at this level, some form of concentrating procedure is generally necessary. Concentration has been successfully achieved by liquid-liquid extraction, supercritical fluid extraction and solid phase extraction in conjunction with various analytical separation procedures. Mergersa and Jönsson [3] developed a supported liquid membrane extraction technique for alkylthio-*s*-triazine herbicides with a wide range of polarity. The supported liquid membrane technique was originally developed by Addunsson [4] and consists of a porous poly(tetrafluoroethylene) (PTFE) membrane on which a water immiscible solvent or mixture of solvents is immobilized which forms a permeable barrier between two aqueous flowing phase systems. Certain solutes can traverse the solvent barrier and pass from one aqueous flow system to another. The extraction apparatus is shown in figure 5.

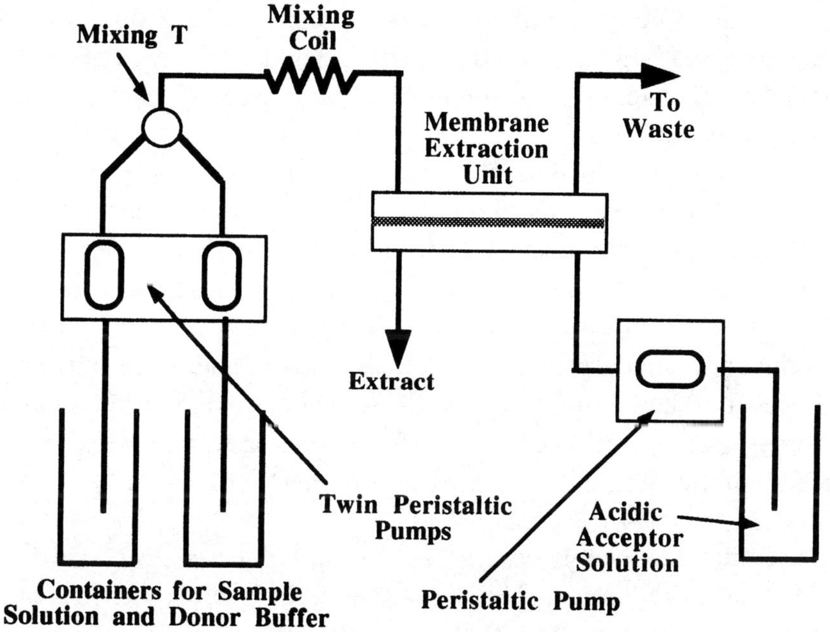

Courtesy of the Royal Society of Chemistry (ref. 3)

Figure 5 Liquid Membrane Extraction Apparatus

The flow rates of the donor (liquid to be extracted) and acceptor (extracting solvent) phases were controlled by two separate peristaltic pumps and the various parts of the extraction system were connected by PTFE tubing. The sample and buffer liquids were merged in PTFE

'T' and mixed in a short coil before entering the donor section of the extractor. The organic solvent absorbed on the membrane was undecane and dihexyl ether. The sample and buffer were pumped through the extractor at 1 ml/min for 20 min. This was repeated while the acceptor channel in the extractor was maintained stagnant. The system was then allowed to stand for another 10 min and the contents displaced by 2 ml 0.1 mol/l sulfuric acid solution at 0.4 ml/min. The pH of the extract was adjusted to 7.0 with 1 mol/-1 sodium hydroxide and 20 μl of the enriched sample placed on the LC column.

A reversed phase C18 column (Techsphere 5ODS) 25 cm long, 4.6 mm I.D. was employed in conjunction with a UV adsorption detector operating at 235 nm. The mobile phase consisted of a mixture of 56% acetonitrile and 44% 0.05 mol/l sodium acetate adjusted to a pH of 7.0 with 0.5 mol/l sulphuric acid solution and was used at a flow rate of 1 ml/min. The choice of the membrane solvent is critical and depends on the distribution coefficient of the herbicides between the solvent and the donor liquid. The optimum solvent was found to be undecane which gave a good extraction efficiency (ranging from 50% to 60% for all the herbicides examined) and provided a very stable membrane that could be used over a relatively long period of time. The pH of both the donor and acceptor solutions were also found to be important and needed to be optimized for particular solutes or groups of solutes. Chromatograms obtained from blank samples and samples of water spiked with the herbicides are shown in figure 6.

It is seen that the overall system works well and that all the herbicides of interest were extracted and separated. It is clear that, from a quantitative point of view, the extraction efficiency could strongly affect the accuracy and precision of the analysis. The use of an internal standard would be highly advisable as both the solutes of interest and the internal standard are likely to be extracted with the same efficiency and thus the effect of any changes in extraction efficiency would be minimized.

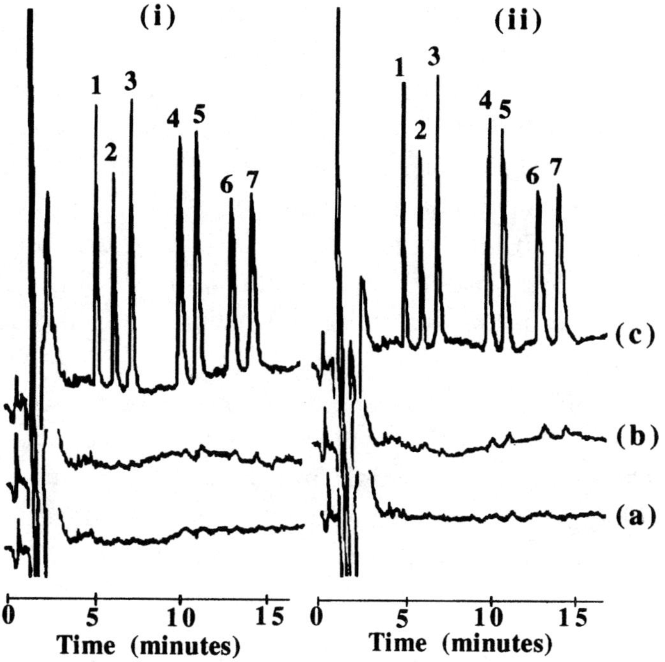

Courtesy of the Royal Society of Chemistry (ref. 3)

(i) Reagent in water, (ii) Spiked in Kävlinge River Water. 1. Desmetryn, 2. Metoprotryn, 3. Ametryn, 4. Prometryn, 5. Terbutryn, 6. Dimethametryn, 7. Dipropetrym. (a) Blank extraction, (b) Extraction of buffer followed by sample extraction and (c) Sample extraction.

Figure 6 Chromatograms of Seven Triazines Extracted from Water

Nevertheless, the efficiency of the extractions and the accuracy and precision of the analysis should be checked fairly frequently with calibration samples to ensure the system does not change and give false results. This is not a problem unique to this method of extraction but is likely to be common to all extraction and concentration techniques.

The Determination of Trace Amounts of the Transition Metals in Parenteral Solutions [5]

Certain pharmaceutical solutions, such as those used in intravenous fluid therapy are often infused in relatively high volumes. It follows that even if there are very small amounts of contaminants in the infusion, such procedures can ultimately lead to the introduction of relatively large amounts of undesirable materials into the bloodstream. Consequently, solutions employed in parenteral administration must meet very high standards of purity that should be far higher than those specified for other pharmaceutical preparations that are used for different purposes. It follows that the purity of any water that is used in such preparations must be very careful controlled, particularly with regard to the presence of trace metals. Almeida *et al.* [5] developed a method for determining traces of Cu, Pb, Zn, Co, Cd and Mn. The procedure involved a preconcentration step on a cation chelating guard column (a Nova-Pak C_{18} guard column coated with Chrome Azurol S) followed by separation on a Nova-Pak C_{18} permanently coated with sodium hexadecane sulphonate. A diagram of the chromatographic apparatus that was used is shown in figure 7.

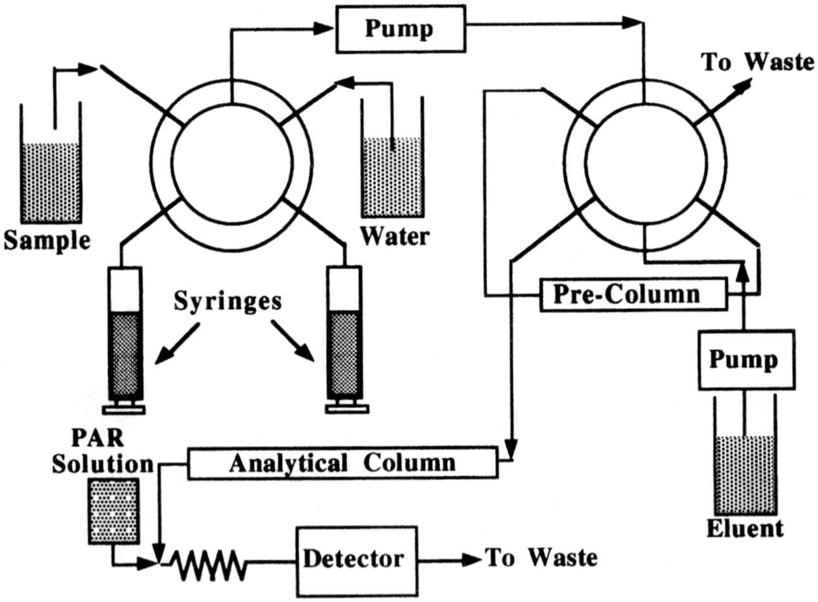

Figure 7 The Chromatographic System for Measuring Heavy Metals in Aqueous Solutions

The procedure is fairly obvious from the layout shown in the diagram. The sample of water was diluted (1+1) with 0.05 mol/l acetic acid-acetate buffer (pH 6.0) and was first passed through a Nylon filter (0.45 µm). The preconcentrator column (15 cm long, 3.9 cm I. D., packed with C_{18} reversed phase, particle diameter 4 µm coated with Chrome Azurol S) was preconditioned with purified water for about 10 min. Then 4 ml of the filtered sample was passed through the column at 1 ml/min and washed with purified water to remove the sample matrix. The preconditioning column was then switched in line with the analytical column (15 cm long, 3.9 cm I.D., packed with C_{18} reversed phase, particle diameter 4 µm coated with sodium hexadecane sulphonate). The eluent from the analytical column passed to a PTFE 'T' where the flow was joined by a stream of color reagent, (4-(2-pyridylazo)resorcinol (PAR) 0.2 mmol/l in 1 mol/l acetic acid and 3 mol/l aqueous ammonia), driven by an appropriate peristaltic pump. The mixture passes through a reaction coil and then to the cell (8 µl volume) of an absorption spectrometer that measured the light adsorbed at 520 nm.

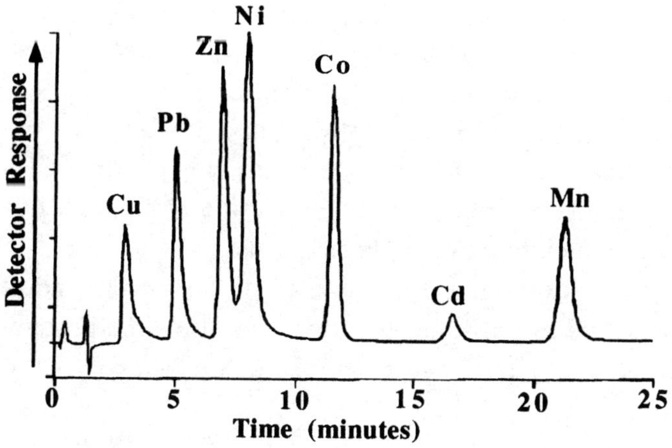

Courtesy of the Royal Society of Chemistry (ref. 5)

Figure 8 A Typical Chromatogram of a Sample Spiked with the Transitional Elements

The concentration of the ions on the preconcentrator column and their subsequent release and separation on the analytical column was achieved by appropriate pH control. After preconcentration, the ions were removed from the column and separated on the analytical column using a mobile phase consisting of 100 mol/l tartaric acid solution with the pH adjusted to 3.1 using sodium hydroxide. An example of the separations obtained is shown in figure. 8. It is seen that more than adequate resolution was obtained, particularly if peak height measurements were used for quantitative assay. The repeatability of the results when measuring concentrations of about 50 µg/l was found to range from 3.5% for Cd to 5.5% for Ni. The limits of quantitative detection employing a sample volume of 4 ml for preconcentration was found to be 0.4 µg/l (Cu), 2.1 µg/l (Pb), 0.5 µg/l (Zn), 0.33 µg/l (Ni), 0.23 µg/l (Co), 2.5 µg/l (Cd) and 2.6 µg/l (Mn).

Determination of Theophylline and Caffeine in Blood Serum by Direct Injection [6]

In general, most quantitative analytical methods utilize standard columns and well-established stationary phases. In this way, fully characterized columns will be readily available from reliable manufacturers and good *reproducibility* as well as good *repeatability* can be achieved and maintained. However, on occasion, certain special and unique stationary phases (usually adsorbed or attached to standard stationary phases or standard columns) are employed to meet special sample requirements. One requirement for the on-column injection of blood serum (which has already been discussed) is the clean separation of the serum proteins from the solutes of interest. In addition this must be achieved in a manner that does not allow the protein to build up on the column and cause blockage or changes in retention characteristics.

Umemura *et al.* [6] solved this problem by coating a standard reversed phase with a zwitterionic bile acid derivative, 3-[(3-cholamidopropyl)-dimethylammonio]-1-propane-sulphonate. The polar and ionic functional groups such as sulphonate, ammonium and the three hydroxyl groups produce a strong polar layer over the surface that interact exclusively with the aqueous environment, whereas the dispersive components of the modifier are directed towards the dispersive chains of the reversed phase. This results in the high molecular weight materials such as the proteins being excluded from the dispersive pores of the reversed phase. As a consequence the bulk of the proteins are unretained, or nearly unretained, when 0.2 mM sodium hydrogenphosphate (pH 7.4) is employed as the mobile phase. In contrast, small molecules such as some inorganic ions and aromatic compounds are retained and well resolved. An example of the selectivity of the system with regard to the early elution of large molecular weight proteins is depicted in figure 9.

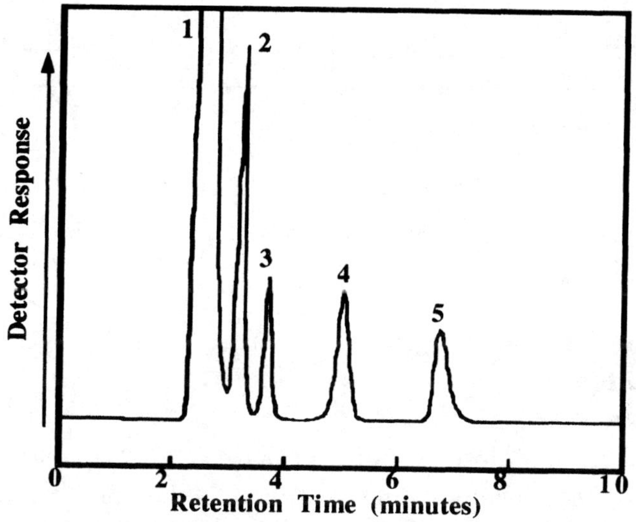

Courtesy of the Royal Society of Chemistry [ref. 6]

1. albumin, 2. NO_3^- ion, 3. I, 4. phenylalanine, 5 benzoic acid

Figure 9 Chromatogram of a Model Sample Mixture Showing the Early Elution of the Large Protein-Like Solutes

Practical example of the separation of serum samples containing the drugs of interest is shown in figure 10. The column used was 25 cm long, 4.6 mm I.D., packed with 5 μm particles of C18 bonded reversed phase. The mobile phase was 0.2 mM sodium hydrogenphosphate (pH 7.4) at a flow rate of 0.7 ml/min. It is seen that an excellent separation is obtained and the substances of interest can be easily assayed quantitatively.

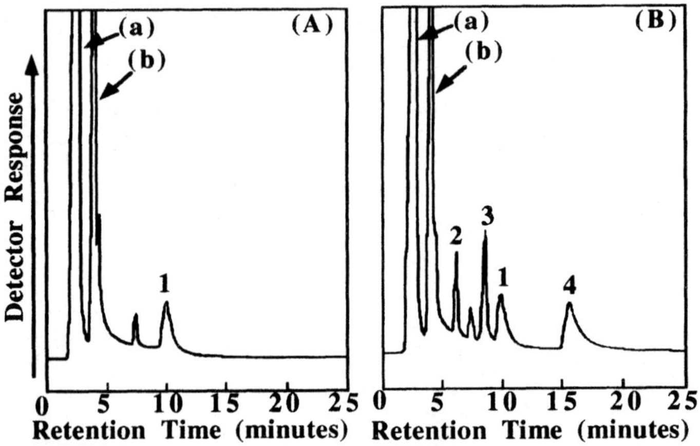

Courtesy of the Royal Society of Chemistry [ref. 6]

(A) Neat human serum, (B) human serum spiked with three types of drugs. (a). human serum protein, (b) inorganic ions, 1. tryptophan, 2. theobromine, 3. theophylline, 4. caffeine.

Figure 10 Chromatograms of Neat and Spiked Human Serum Samples

A UV absorption detector operating at 254 nm was used for monitoring. The detection limits for theophylline and caffeine were found to be 0.2 mg/l and 0.5 mg/l respectively. The standard deviations for the two solutes were found to be < 1.4% and < 2.2% respectively for 10 replicate measurements on serum spiked with 5 mg/l of each drug.

The Determination of Spectinomycin Residues in Various Tissue Types from Husbandry Animals [7]

Spectinomycin is a broad-spectrum aminocyclitol antibiotic that inhibits the synthesis of Gram-negative bacteria. The drug is administered to husbandry animals largely *via* oral and intramuscular acute toxicity and pharmacokinetic patterns. When husbandry animals are treated with this drug, residues of the antibiotic may be expected to turn up in edible animal parts including eggs, fat and organs. Residues of the drug are allowed in products from pigs, cattle and poultry up to a certain concentration called the *maximum residue level.* It follows that suitable assays are needed to monitor the drug level in animal products before they are brought to market.

A method for extracting spectinomycin residues from complex animal matrixes and determining their content by liquid chromatography has been developed by Bergwerff *et al.* [7]. The drug residues were extracted from homogenized tissue with 25 nM citrate at pH 4.0, trichloroacetic acid and dichloromethane. The extract was purified and concentrated using a carboxylic acid-bonded solid phase extraction column. The solid phase extraction eluate was separated by cation exchange using two columns and a column switching system in conjunction with post-column derivatization and fluorescence detection.

Details of the procedure are as follows. The tissues containing the drug residues are homogenized and 5 g portions contained in a 50 ml screw-capped polypropylene tube are treated with 25 ml of 25 mM citric acid at pH 4.0, 1.0 ml of trichloroacetic acid and 10 ml of dichlormethane. The combination is mixed on a platform shaker for 20 min and the mixture centrifuged for 10 min at 2300 g (tissue material) or at 4000 g (egg material). The supernatant liquid is filtered through a paper filter which is subsequently washed with 10 ml of water. The pH of the pooled filtrate is adjusted to 6.7 with 1 M caustic soda and then passed through a conditioned carboxylic acid solid-state extraction tube at 1 ml/min. Prior to use the tube is conditioned with 3 ml of methanol, followed by 5 ml of 25 mM potassium phosphate solution (pH 6.7). The extraction tube is then washed with 10 ml of 25 mM potassium phosphate solution (pH 6.7), run dry under pressure and the residue of liquid removed by centrifugation at 2500 g for 10 min. The extracted solutes are then desorbed with 3.5 ml of 50 mM citric acid at a pH 2.6 and the solution centrifugation at 2500 g for 10 min. The extracted solutes are then desorbed with 3.5 ml of 50 mM citric acid at a pH 2.6 and the solution collected in a 5 ml preweighed tube. Prior to taking a sample for analysis the tube is vortexed for about 10 s and centrifuged at 2500 g for 10 min. The actual analysis was carried out in the LC system shown in figure 11 which incorporates a post-column reactor and a fluorescence detector.

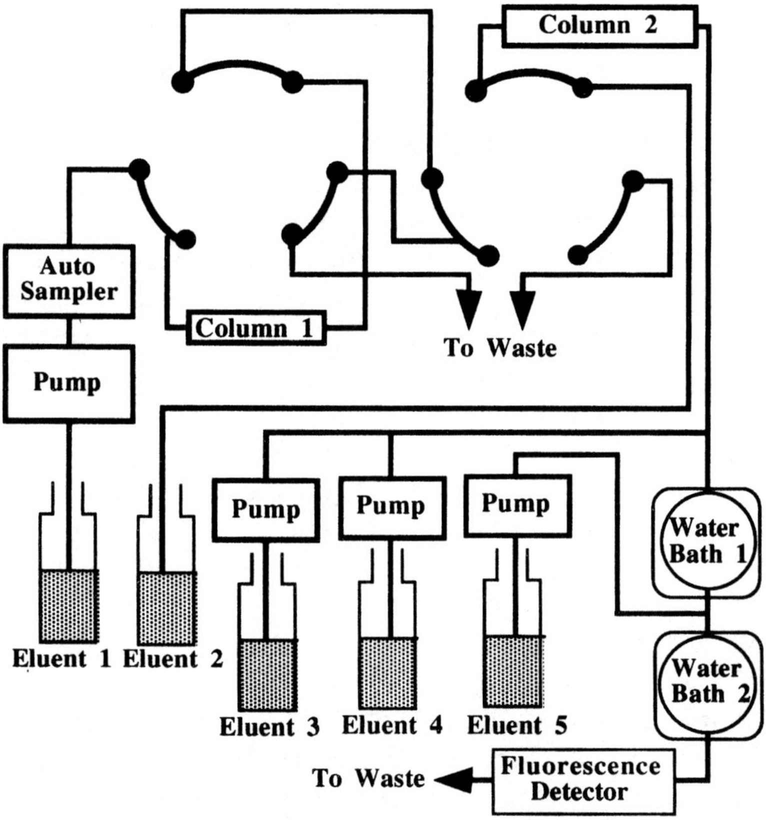

Figure 11 Sampling System Including LC Set–Up and Post Derivatization Apparatus

The liquid chromatograph includes four pumps, an autosampler, and a valve switching unit comprising two six port switching valves to which columns 1 and 2 are connected. The mobile phase flow rate through columns 1 and 2 are set at 0.8 ml/min and 1.50 ml/min respectively. For the analysis of egg material, column 1 is a Spherisorb SCX column 5 cm long, 4.6 mm I.D. packed with 5 μm particles fitted with an integral guard column 1 cm long, 4.6 mm I.D. For the analysis of tissues, a column 10 cm long, 3.0 mm I.D. is used (Chromopack 5 μm Ionsphere C glass cartridge) which is also fitted with a dedicated guard column 1 cm long, 2.1 mm I.D. Column 2 comprises a Sperisorb SCX, 25 cm long, 4.6 mm I.D. also packed with 5 μm particles. Eluent 1 is a mixture of 0.1 M aqueous solution of sodium sulfate and acetonitrile (8+2 v/v), and eluent 2 is a mixture of 0.15 M potassium phosphate at pH 3.5 and acetonitrile (8+2 v/v). For egg derived material eluent 1 is used with column 2. For all other tissues eluent 1 is used with column 1 and eluent 2 with column 2.

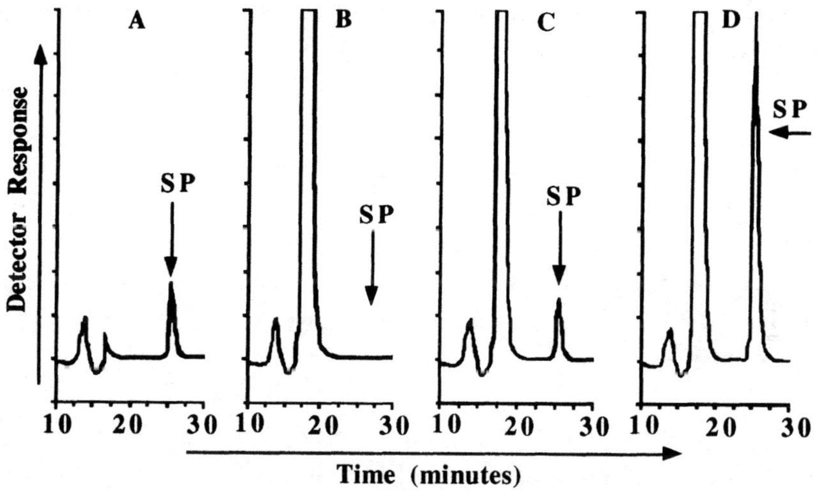

Courtesy of the Royal Society of Chemistry [ref. 7]

(A) a standard solution of spectinomycin (SP) (0.8 μg/ml), (B) blank porcine muscle tissue, (C) blank porcine muscle tissue, spiked with spectinomycin at 500 μg/kg, (D) spectinomycin-incurred porcine muscle tissue at the injection site, 48 hr after administration.

Figure 12 Typical Chromatograms Obtained for Spectinomycin Residues in Various Tissue types

For post-column derivatization, 1 mM potassium phosphate (pH=7) containing 10 μl/ml of sodium hypochlorite (eluent 3) at a flow rate of

0.25 ml/min joins a flow of 0.5 M sodium borate solution (pH=10.2) (eluent 4) at 0.75 ml/min and mixes with the column eluent. Oxidation takes place in a reaction coil 2.2 m long and 0.5 mm I.D. immersed in a water bath at 70°C. A flow of 0.3 M sodium borate solution (pH=10.2) containing 3.0 mM of *o*-phthaldialdehyde and 14 mM 2-mercaptoethanol (eluent 5) at a flow rate of 0.6 ml/min. joins the combined column eluent and oxidation flow and passes through a reaction coil 1.7 m long and 0.3 mm I.D. situated in a bath at 25°C. The fluorescent derivatives formed then pass through a fluorescence detector employing an excitation light of 340 nm and measuring the emitted light at 460 nm. An example of some results obtained by the technique is shown in figure 12.

It was found that using the technique as described the minimum detectable concentration of spectinomycin was about 0.05 mg/kg and recoveries were about 74%. The precision of the signal response was about 2% and interference effects from lincomycin were less than 2%. It is clear from this type of analytical challenge that where quantitative results are of prime interest and the matrix is complex and not amenable to simple and conventional sample preparation techniques, very elaborate analytical systems may need to be developed to provide analytical data having the required accuracy and precision in a reasonable time.

The Determination of Alternariol in Tomato Paste by Solid Phase Extraction and Liquid Chromatography Using Fluorescence Detection [8]

Alternaria spp. produce a wide variety of toxic metabolites with different chemical structures [9]. Dibenzo-α-pyrones, including alternariol (AOH), alternariol monomethyl ether (AME) and altenuene (ALT) have been some of the major *Alternaria* mycotoxins that have been identified and reported. These secondary metabolites have been isolated from a large number of food and feed materials that have been contaminated with various species of *Alternaria* and have been considered possible agents that affect human health. As

Alternaria is a common spoilable mold of tomatoes and, in addition, tomatoes inoculated with strains of *Alternaria* have been shown to produce alternariol (AOH) and alternariol monomethyl ether (AME), it was clear that a method was required to assay these mycotoxins in vegetables and fruits.

Fente *et al.* [8] developed a procedure for measuring *Alternaria* mycotoxins in tomatoes using solid phase extraction to isolate the mycotoxins, and subsequent separation on a reversed phase column using isoctratic development and fluorescence detection. The procedure used by Fente *et al.* was as follows. 10 g of tomato paste was centrifuged at 3500 rpm for 20 minutes and the upper layer was used for extraction purposes. The Oasis HLB 3 cc (60 mg) solid phase extraction cartridges were mounted in a vacuum manifold with the pressure set at 5 in. mercury.

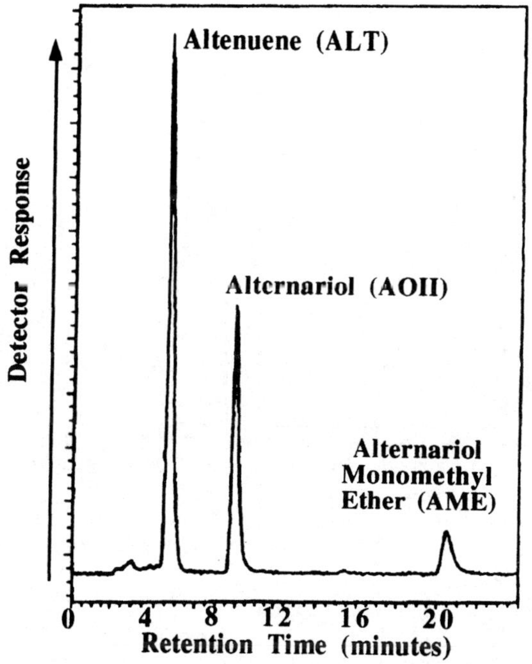

Courtesy of the Royal Society of Chemistry [ref. 8]

Figure 13 The Separation of the Standards Altenuene, Alternariol and Alternariol Monomethyl Ether

The column was conditioned with 2 ml of methanol, and then equilibrated with 2 ml of water. The supernatant liquid from the centrifuge tube was drawn through the cartridge which was subsequently washed with 2 ml of water and 2 ml of methanol. The materials of interest were displaced from the adsorbent with 5 ml of 1% acetic acid in methanol, using a syringe to control the flow at 1 drop per second. The eluate was evaporated to dryness at 45°C and the residue dissolved in 200 µl of mobile phase. A separation of the standards, altenuene, alternariol and alternariol monomethyl ether directly is shown in figure 13.The column used was the Spherisorb ODS 2 which was 25 cm long, 4.6 mm I.D., packed with particles 5 µm in diameter. The mobile phase was a methanol water mixture (32-68) acidified to a pH of 3.0 with orthophosphoric acid. The mobile phase, filtered and degassed by sparging with helium, was used at a flow rate of 0.8 ml/min.

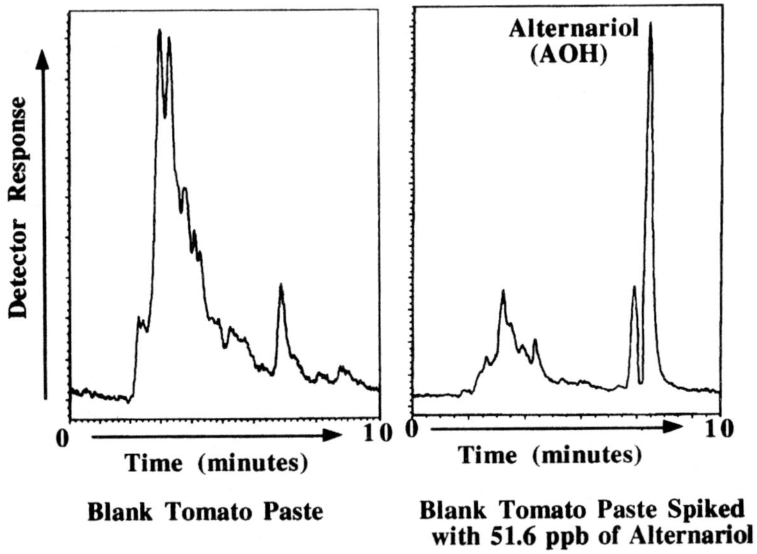

Courtesy of the Royal Society of Chemistry [ref. 8]

Figure 14 Chromatograms of Tomato Paste, Blank and Spiked with Alternariol

The solutes were detected by UV absorption at 254 nm and by fluoresence (excitation wavelength 300 nm and emission wavelength

430 nm). It is seen that the standards are eluted well resolved in about 20 min. The extraction and separation of alternariol from spiked tomato paste is shown in figure 14. It is seen that the solute of interest is eluted well clear of other contaminating materials that are co-extracted with the alternariol. The authors reported that the extraction efficiency at all levels was better than 78% and the minimum detectable concentration was about 2 ppb. The repeatability as reported was about 3% and the reproducibility about 9%.

Automated Determination of Amphetamine Enantiomers Employing Two-Dimensional Column Switching [10]

The psychoactivity of the amphetamines has already been discussed and they have been abused both in society in general and in sports in particular. However, a unique feature of the amphetamines is the wide difference in psychoactivity of the different enantiomers of the same amphetamine. It follows that the development of chiral analytical methods is an area of major interest in amphetamine analysis.

One method that involved the use of two-dimensional column switching was developed by Pastor-Navarro *et al.* [10] to determine the enantiomers of amphetamine in urine. They employed a column switching technique to first derivatize and separate the enantiomeric pair from other materials present in the urine, and a second to resolve and measure the individual enantiomers on a separate chiral column. In effect, this represents a concentration procedure using the first column not only for separation but also to isolate the materials of interest in a relatively small volume. The procedure that was employed is as follows.

A 50 μl sample of urine is injected onto the first column. The actual apparatus used has been described in detail by Herráez-Herrández *et al.* [11]. The first column acts as a trapping column and is connected to the analytical column by means of a switching arrangement that allows a back flushing procedure. The extraction column (Hypersil ODS-C_{18}) was 2 cm long, 2.1 mm I.D. packed with particles 30 μm in diameter. Initially the matrix constituents are flushed out using a sodium acetate buffer. Then a borate buffer containing *o*-phthalaldihyde and *N*-acetyl-L-cysteine is passed through the column and the flow arrested to allow reaction to take place.

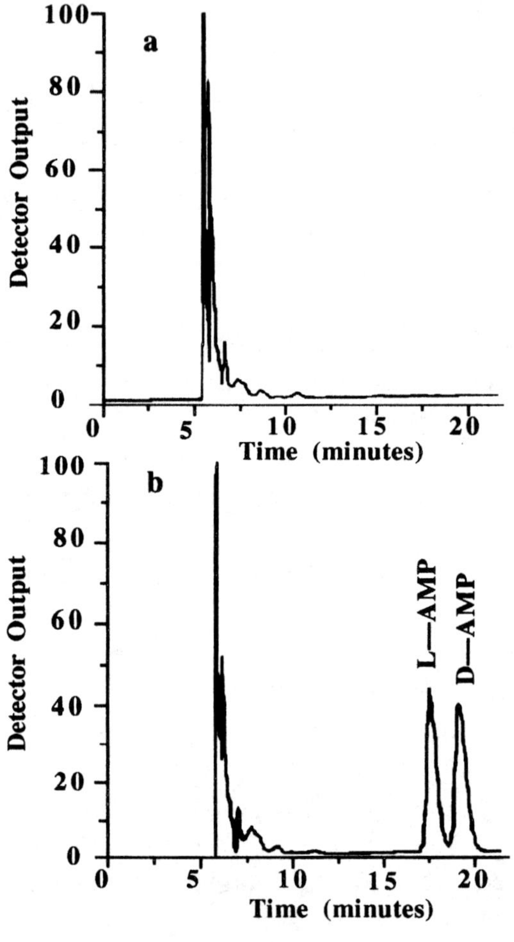

Courtesy of the Royal Society of Chemistry [ref. 10]

Figure 15 Chromatograms Obtained for Blank Urine (a) and Urine Spiked with 10 μg/ml of Racemic Amphetamine (b)

While the sample is being derivatized, the analytical column (LiChrospher 100 RP_{18}) 1.5 cm long, 4 mm I.D., packed with particles 5 μm in diameter, is being conditioned with an acetonitrile-methanol-acetate buffer mixture (0.05M sodium acetate adjusted to pH 7.5 with 2.5 M caustic soda solution) (46-6-48). After derivatization the two columns are connected and the mobile phase is passed through both columns and the enantiomers are separated and resolved in the second column. The elution profile of the derivatives is monitored with a fluorescence detector (1046 Series fluorescence detector by Hewlett Packard). An example of the separations obtained is shown in figure 15. It is seen that the two enantiomers are well resolved and are quite suitable for quantitative assessment. It is also seen that the majority of interfering substances have been removed The authors claimed good linearity and reproducibility in the 0.25 to 10 μg/ml concentration range. Recovery was about 88% ± 3% and the limits of detection was about 50 ng/ml for both isomers, a very pactical and useful level.

Determination of N-Methyl Carbamate Pesticides in Foods Using Solvent Extraction at Elevated Temperatures with Minicolumn Cleanup [11]

Various methods have been developed for measuring the levels of pesticides in food, a number of which have already been described. An interesting and novel method developed by Herráez-Herrández *et al.* [11] deserves special discussion as it is one of the few methods that exploit solvent extraction in a novel and very effective manner. These workers employed a heated solvent under pressure to extract the pesticides from the food matrix which was completed in less than 20 min. The extracts were cleaned up in a carboxylic acid mini-column and then separated by a conventional liquid chromatography system. The details of the analytical method are as follows.

A 500 g sample of the food was chopped in a conventional food processor for 5 min to obtain a well mixed homogenate. A 5 g sample of the homogenate was mixed with 6 g of Extrelut (particles of a diatomaceous earth size 160-800 μm) and ground in a mortar until the mixture was homogenous and, if appropriate, the sample was spiked with the chosen pesticide at this stage.

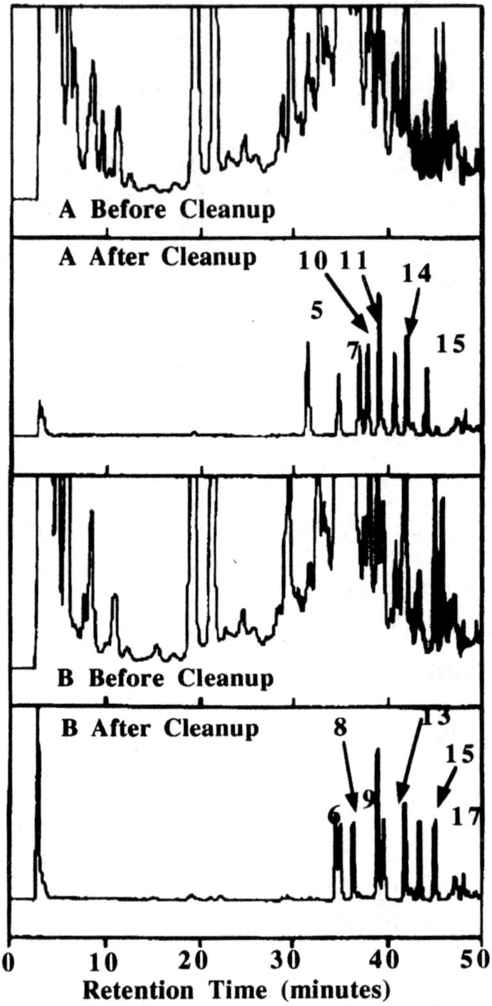

Courtesy of the Royal Society of Chemistry [ref. 11]

1. Butoxycarboxim*
2. Oxamyl*
3. Methomyl*
4. Dioxacarb*
5. Metolcarb*
6. Propoxur
7. Bendiocarb
8. Carbufuran
9. Carbaryl
10. Xylycarb
11. XMC
12. Pirimicarb
13. Isoprocarb
14. Trimethacarb
15. Fenobucarb
16. Methiocarb
17. Promecarb

* Not shown on the chromatogram

Figure 16 Before and After Cleanup Chromatograms of Grapefruit Spiked with 0.2 μg/g of Each Pesticide

The sample was then placed in a commercial extraction system (Dionex) and extracted at 2000 p.s.i. and 100°C with 60 ml of acetonitrile. The extract was transferred to a separating funnel and the container washed twice with 5 ml of acetonitrile and the washings added to the funnel. Two grams of sodium chloride were added, the mixture shaken vigorously and left for 1 hr to educe water from the acetonitrile.

The water portion was separated and discarded and anhydrous sodium sulfate added to dry the extract. The dry extract was evaporated to dryness at 40°C and the residue dissolved in 5 ml of acetone. The 500 mg carboxylic acid column was prerinsed with 10 ml of acetone and 10 ml of acetone-hexane mixture (3:7 v/v) and 1 ml of the extract was transferred to a centrifuge tub, 2 ml of acetone added and the mixture diluted to 9.8 ml with hexane. The extract was centrifuged and the supernatant liquid passed though the column and the column washed with 10 ml of an acetone-hexane mixture (3:7). The eluate and the washings were mixed and evaporated to dryness. The residue was taken up with methanol and diluted to 1 ml with water. The solution was used for liquid chromatographic analysis. For citrus fruits the cartridge was prerinsed with 10 ml acetone-hexane (1:9 v/v) instead of acetone-hexane (3:7 v/v).

The separation was carried out on an octadecylsilica column (STR-ODS II) 15 cm long, 4.6 mm I.D. and the separation developed for 10 min with a methanol-water mixture (18% methanol), from 10 to 40 min programmed from 18% to 70% methanol, from 40 to 43 min programmed from 70% to 90% methanol, from 43 to 60 min programmed maintained at 90% methanol and finally from 60 to 75 min changed to 18% methanol. The flow rate employed throughout the separation was 0.8 ml/min. Employing a post-column reactor and the fluorescent derivatizing agent *o*-phthaladhyde the eluted solutes were detected using a fluorescence detector. Two examples of the separation obtained that clearly illustrate the advantages of the cleanup procedure are shown in figure 16. The recoveries of the individual pesticides ranged from 70 to 100% with relative standard deviations of about 10%.

Ascorbic Acid Determination in Foodstuffs by Microdialysis Sampling and Liquid Chromatography with Electrochemical Detection [12]

Ascorbic acid (vitamin C) is a strong natural antioxidant present in many foods, particularly vegetables and fruit. It is thought to be important in disease prevention and is also often used as an additive to food to prevent deterioration in taste, color and odor. Due to its physiological importance fast, sensitive and routine methods for its assay are required. However, due to the ease in which the material oxidizes, special extraction-concentration procedures are necessary to avoid both oxidation and photodecomposition. This inherent instability is a serious problem in assaying milk, fruit juices and carbonated drinks. Mannino and Cosio [12] developed a microdialysis procedure for extracting ascorbic acid from a liquid medium without contact with air or oxygen.

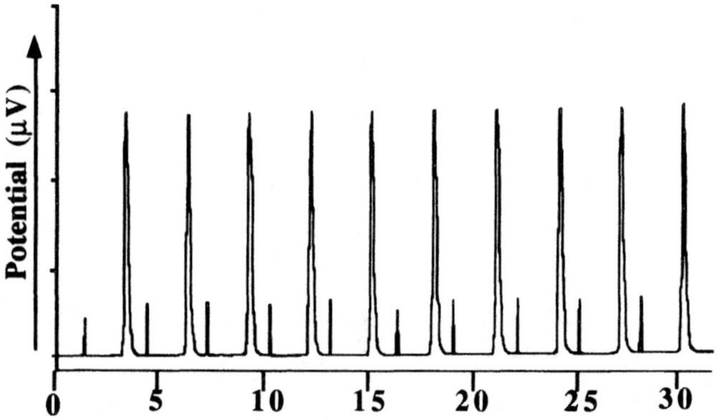

Courtesy of the Royal Society of Chemistry [ref. 12]

Figure 17 Repetitive Extraction by Dialysis of Ascorbic Acid at a Concentration of 100 mg/l

The microdialysis system consisted of a micro-injection pump connected through the dialysis system to an in-line column injector. The apparatus has been described elsewhere [13]. The microdialysis fiber was a Fitral 12 AN 69 HF (I.D. 200 μm and a molecular mass cut-off of 20,000 Da) and the perfusion liquid was de-ionized water pumped at appropriate flow rates for the sample being analyzed. The column used was 10 cm long, 7.8 mm I.D. (Fruit Quality, Bio-Raad Labs) and the mobile phase was 0.001 M sulfuric acid used at a flow rate of 0.7 ml/min. An electrochemical detector was employed to monitor the separation with a 750 mV potential relative to that of a Ag/AgCl cell operating at a current range of 0.5 μA. The volume of dialyzed solution used for analysis was 20 μl. A set of chromatograms demonstrating the repeatability of the system is shown in figure 17. It is seen that the repeatability of the system is excellent. Each sample contained 100 μg per liter of dialyzed solution. Chromatograms of milk and grapefruit juice containing progressively increasing concentrations of ascorbic acid are shown in figure 18.

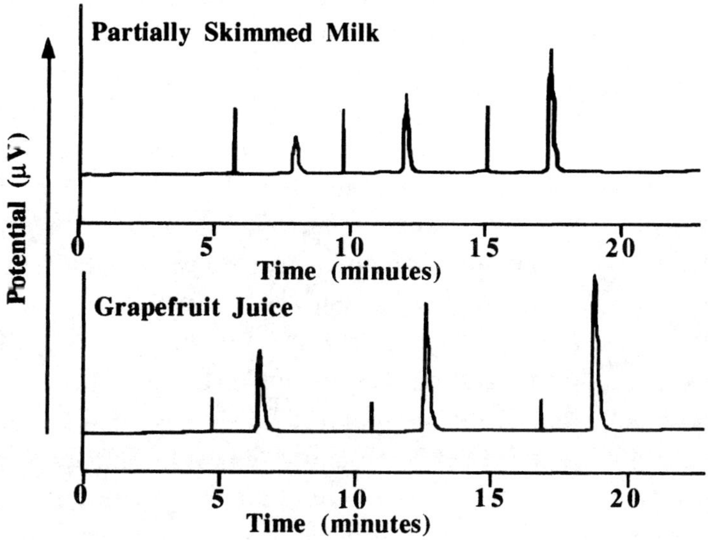

Courtesy of the Royal Society of Chemistry [ref. 12]

Figure 18 Chromatograms of Different Concentrations of Ascorbic Acid in Samples of Milk and Grapefruit Juice

The procedure was shown to be accurate, repeatable and relatively fast and was applied advantageously to the analysis of fruit juices, soft drinks and milk products. In addition, due to the high specificity of the electrochemical detector, each peak could be monitored with no interference from other eluting materials.

Synopsis

As already discussed, the technique of gas chromatography provides very high sensitivities and column efficiencies, and thus is normally the preferred technique. However, because liquid chromatography can handle involatile and labile substances, and there are far more assays involving involatile components than those that are volatile, liquid chromatographic methods are the more popular. Moreover, due to the lower sensitivities provided by liquid chromatography detectors, sample preparation usually involves solute concentration as well as isolation. In addition, derivatization procedures are often included to permit the use of more specific and thus more sensitive detectors. The literature contains many liquid chromatographic methods and analysts faced with novel applications will usually find supportive information in the journals *Analyst* or *Analytical Chemistry*. There are two separation problems that arise in liquid chromatographic analyses. First, the solutes of interest may be difficult to separate from one another, and, second, the solutes of interest are difficult to separate from the matrix compounds of no interest. The former is usually solved by the use of high efficiency columns, highly selective stationary phases or specific detectors, and the latter by matrix removal in sample preparation. For example, the proteins in human blood can block a column if not removed as well as interfere with the separation. In the analysis of bixin and norbixin in human blood the interfering proteins can be removed by preliminary precipitation with alcohol. The separation of enantiomers often require both high efficiencies and chirally selective stationary phases. Using a relatively simple sample preparation procedure, the enantiomers of methamphetamine in urine can easily be determined using a cyclodextrin stationary phase. The separation of the critical compounds from a complex matrix with accompanying concentration can often be achieved using membrane diffusion, where the membrane is designed to allow passage of the solutes of interest but to hold back the matrix materials and at the same time effect a concentration. This technique has been used for the determination of herbicides in ground water and extracted sediments. Determining trace materials almost always involves a concentration technique. For example, it is extremely important to be able to determine the inorganic ions in parenteral solutions and this can be achieved by pre-concentration from a large volume of sample on a chelating guard column. The chelated material is then displaced onto an ion exchange column and the individual ions separated. In most analyses standard columns and well-established stationary phases are used. As a result, fully characterized columns from manufacturers will be selected and good *reproducibility* and *repeatability* maintained. However, on occasion, certain special and unique stationary phases (usually adsorbed or attached to standard stationary phases or standard columns) are employed to meet special sample requirements. An example of this is the direct determination of theophylline and caffeine in blood using a standard reversed phase column coated with a bile acid which provides unique selectivity for the solutes of interest. Solid samples such as animal tissue can be homogenized and extracted with appropriate solvents, concentrated and injected directly onto the column. Solid phase exaction is an extremely efficient method of extracting and concentrating the solutes of interest from a complex matrix. This technique has been used for many applications, two examples being the determination of polynuclear hydrocarbons in tea infusions and the determination of alternariol in tomato products. Column switching, using two columns, one to extract and bulk the solutes of interest and the second (usually in line with the clean-up column) to separate the individual substances has also been used in the determination of the amphetamine isomers in urine. Solvent extraction carried out under pressure and at elevated temperatures has also been used to improve extraction efficiency and reduce preparation time. This technique has been used to assay the carbamate pesticides in foods. Finally, for particularly difficult extraction problems dialysis can be used to isolate the materials for analysis and, in conjunction with electrochemical detection, can provide extreme specificity and high sensitivity. This technique has been used to assay ascorbic acid in fruit juices and carbonated beverages.

References

1. L. W. Levy, E. Regalado, S. Navarrete and R. Hawkins, *Analyst*, **122**(1997)977.
2. O. Al-Dirbashi, N. Kuroda, F. Menchini, S. Noda, M. Minemoto and K. Nakashima, *Analyst*, **123**(1998)2333.
3. N. Mergersa and J. A. Jönsson, *Analyst*, **123**(1998)225.
4. G. Addunsson, *Anal. Chem.*, **58**(1986)1340.
5. A. A. Almeida, X. Jun and J. L. F. C. Lima, *Analyst*, **123**(1998)1283.
6. T. Umemura, R. Kitaguchi, K. Inagaki and H. Hara, *Analyst*, **123**(1998)1767.
7. A. A. Bergwerff. P. Scherpenisse and N. Haagsma, *Analyst*, **123**(1998)2139.
8. C. A. Fente, J. Jaimez, B. I. Vázquez, C. M. Franco and A. Cepeda, *Analyst*, **123**(1998)2145.
9. L. M. Seitz, in *Mycotoxins:Production, Isolation, Separation and Purification* (Ed., V. Betina) Elsevier, Amsterdam (1994)443.
10. M. D. Pastor-Navarro, R. Porras-Serrano, R. Herrández and P. Campins-Falcó, *Analyst*, **123**(1998)319.
11. R. Herráez-Herrández, P. Campins-Falcó and A. Sevillano-Cabeza, *J. Chromatogr.*, **679**(1996)125.
12. S. Mannino and M. S. Cosio, *Analyst*, **122**(1997)1153.
13. S. Mannino, M. S. Cosio and P. Zimei, *Electrodialysis*, **8**(1996)4.

Part 4

Thin Layer Chromatography

Chapter 10

Thin Layer Chromatography Apparatus

Thin layer chromatography is not the ideal chromatographic system for accurate quantitative analysis. The technique can be an excellent tool for qualitative work and for scouting different mobile phase systems for liquid chromatography [1]. It is also the only chromatographic technique which presents a complete profile of the separation and, as a consequence, reveals the extent to which clution is complete. The main advantage of the technique, however, resides in its low cost, but to ensure adequate accuracy and precision in quantitative assays, expensive scanning apparatus is usually necessary which undermines the intrinsic low cost advantages of the technique. Thin layer chromatography appears to have been first developed and utilized by Schraiber in 1939 [2], who, working with Izmailov at the Khar'kov Chemistry and Pharmacy Research Institute, employed the technique, largely qualitatively, for the analysis of pharmaceuticals. In addition to inventing the technique itself, Schraiber also was the first to employ fluorescence as the separation indicator or detection system. Unfortunately, Schraiber's work was not heeded and so no progress was made until the technique was rediscovered by Kirchner [3] in 1951.

Although, the equipment is far simpler and much less expensive, thin layer chromatography (TLC) phase systems are basically very similar to those used in liquid chromatography. Furthermore, as many separations can be carried out simultaneously by multiple spotting, the average analysis time can be much shorter. In some multiple separations, there may be 60 samples per plate which, in effect, means that each analysis may only take about 5 seconds to complete. Although the resolution obtained from TLC is far less than that obtainable by LC, the cost advantage has made the technique very popular. In fact, despite the many advances that have taken place in LC techniques over the past years, the use of TLC for routine analyses continues to grow.

There are, however, some severe disadvantages to TLC, one being a restricted resolving power due not merely to a relatively low plate efficiency, but more often to the restricted physical size of the plate. The size of the plate limits the number of sample spots that can be accommodated. Thus, samples containing multiple components cannot be separated by TLC due to restricted peak capacity. In TLC, all the solutes must be contained on the plate, whereas in LC, as the solutes are eluted from the column, the component capacity can be much greater. In addition, the minimum detectable mass of solute in TLC is much larger than that in liquid chromatography and thus the technique is not so suitable for trace analysis. Finally, the unique selectivity that is obtainable from reversed phases cannot be comprehensively realized in TLC because the reversed phases are not wetted by solvent mixtures having high water contents and consequently, for such, often essential solvent mixtures, plate development becomes impossible.

Elution Development in Thin Layer Chromatography

The development processes that take place on a thin layer plate can be complicated by the frontal analysis of the mobile phase itself, although the process is basically still an elution technique. The mobile phases used to elute the solutes in TLC are commonly multi-component, containing at least three different solvents, if not more. If the plate is not pre-conditioned with solvent, on elution there is an elaborate modification of the plate surface which is depicted for a ternary solvent mixture in figure 1. It is interesting to note that the separation is achieved (unless a single solvent is used as the mobile phase) by a *form* of gradient elution The edge of the plate is dipped into a tray of the solvent mixture which begins to migrate along the plate driven by surface tension forces. The different solvents array themselves on the surface in the manner shown in figure 1. The solvent that interacts most strongly with the stationary phase is extracted from the mixture and forms an adsorbed layer on the surface that corresponds to the area (X) in the diagram. The now binary mixture continues to migrate along the plate and the next solvent component that interacts most strongly with the stationary phase, solvent B, is adsorbed as a layer on the surface corresponding to the area (Y) in the diagram.

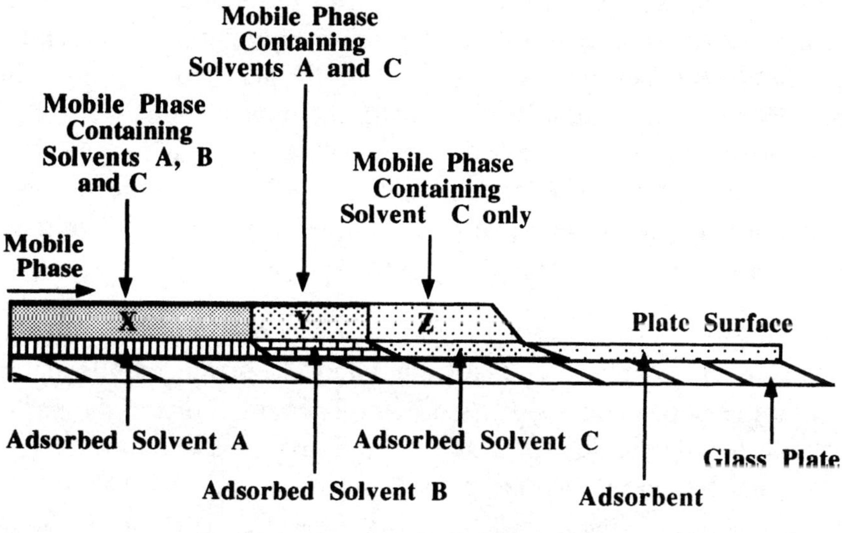

Figure 1 The Development of a Thin Layer Plate

Finally the remaining solvent, (C), with the weakest interactions with the stationary phase continues to migrate and cover the surface with a layer of solvent (C) in the area (Z). It is seen that the distribution system, which has resulted from the frontal analysis of the three components of the mobile phase, is now quite complex. Consider now the distribution systems with which the *solutes* will interact during the separation process. In the first section, (X), solutes will be distributed between the ternary solvent mixture (A), (B) and (C) and the surface covered with solvent (A). In the next section, (Y), the solutes will be distributed between a binary solvent mixture of (B) and (C) and a surface covered with solvent (B). Finally, distribution will take place in section (Z) between pure solvent (C) and a surface covered with solvent (C). Even this is an over-simplification, as the composition of the mobile phase in each section will not be constant but will decrease along the plate. Furthermore, as the separation progresses, the length of sections (X), (Y) and (Z) will continually increase. Such a system is extremely difficult to treat in a precise manner theoretically, particularly as the boundaries are not as sharp as those drawn in figure 1. In fact, the overall effect is as though the separation was carried out sequentially on three separate sections of a plate each section having a different stationary phase and mobile phase. In each section, the separation will then be achieved by elution development but the overall effect will be a form of gradient elution. The complexity of the system increases with the number of solvents used and, of course, their relative concentrations. This somewhat complicated elution procedure does not in any way affect the accuracy or precision of any quantitative assays but tends to make the optimum choice of phase system more difficult to predict.

The elution process can be simplified considerably by pre-conditioning the plate with solvent vapor from the mobile phase before the separation is started. Unfortunately, this only partly reduces the adsorption effect, as the equilibrium between the *solvent vapor* and the adsorbent surface will not necessarily be the same as that between the *liquid solvent* and the surface. It is clear that by forming a gradient by the frontal analysis of the mobile phase and carefully choosing the solvent mixture, very delicate *pseudo*-gradients can be formed which, in no small measure, accounts for the versatility, popularity, and success of TLC.

Thin Layer Chromatography Chambers

A diagram of two simple thin layer chromatography development chambers is shown in figure 2. It is important to prevent the solvent evaporating from the plate surface during development, as this would change the composition of the developing solvent. If the solvents were volatile and the evaporation rate became equal to the migration rate, then this would also result in a static "solvent front" and arrest the chromatographic development. In the simplest case the developing chamber can consist of a round or square glass jar fitted with a glass cap. Sufficient solvent mixture is placed in the chamber to raise the level about a centimeter from the base and the air space above the solvent is allowed to become saturated. The plate is then placed in the chamber with the end where the samples have been placed dipping into the solvent and the cap replaced. It is important to ensure that the sample spots themselves are not at, or below, the surface of the solvent mixture or they could be washed away from the plate. The solvent must rise up the plate and pass over the sample spots to develop the plate satisfactorily.

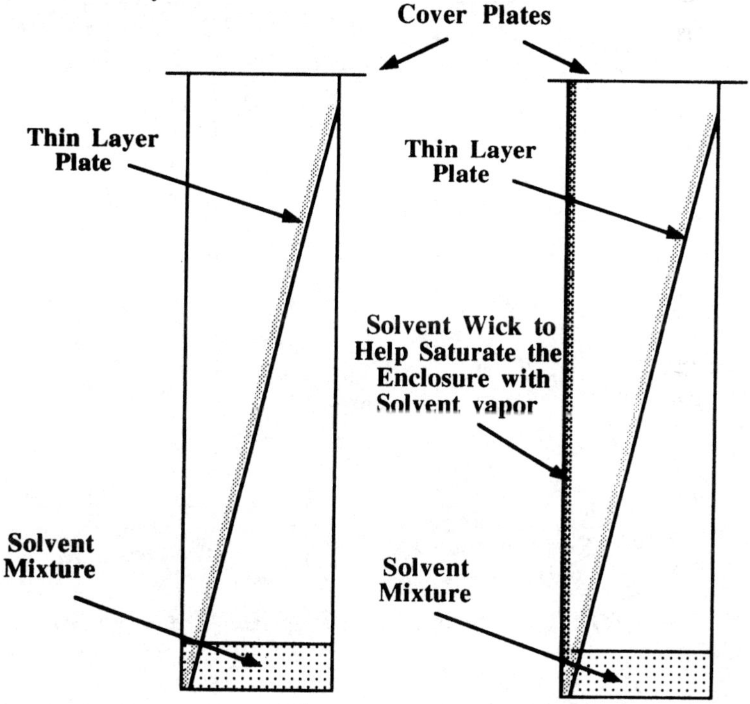

Figure 2 The Normal Method of Thin Layer Plate Development

To improve the air space saturation with solvent, the walls, or part of the walls of the chamber are sometimes covered with filter paper to act as a wick that soaks up the developing solvent and provides a greater surface area for evaporation. The use of a paper wick is depicted on the right-hand side of figure 2. The saturated solvent vapor in the chamber not only prevents solvent evaporation from the plate surface but also plays a part in controlling the retention mechanism by deactivating the surface of the stationary phase as already discussed. The components of the solvent are selectively adsorbed on the surface of the stationary phase causing the solutes to interact, not with the native silica surface, but with the silica surface covered with the most strongly interacting solvent. Again it should be emphasized that equilibrium between the *solvent vapor* and the plate will not be the same as equilibrium between the *solvent* and the plate.

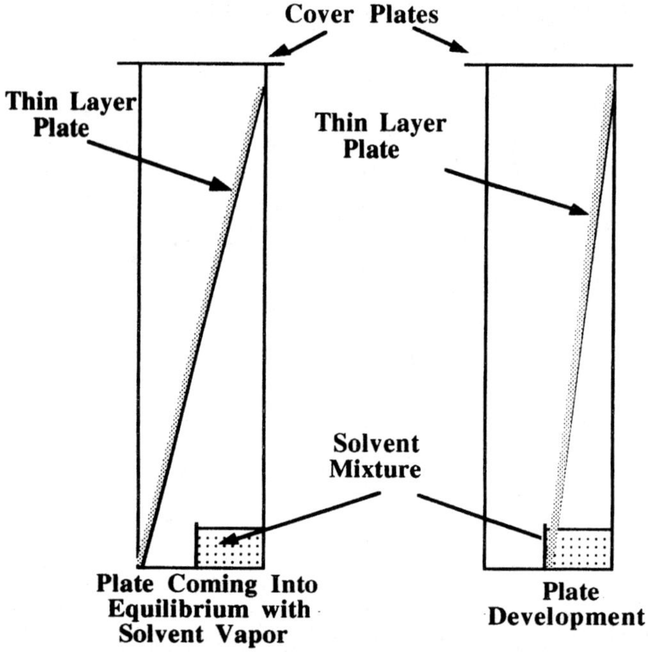

Figure 3 Apparatus for Establishing the Pre-Equilibrium of a TLC Plate

For example, if a binary mixture of solvents is considered for a non-equilibrium plate, and the concentrations of solvent in the gas phase are c_1 and c_2 respectively, then $c_1 \neq x_1$ and $c_2 \neq x_2$

where (x_1) is the molar fraction of solvent (1) and (x_2) is the molar fraction of solvent (2).

In fact, *Raoult's Law* applies and if (P_1) and (P_2) are vapor pressures of pure solvents (1) and (2) respectively,

$$\frac{c_1}{c_2} = \frac{x_1 P_1}{(1-x_1)P_2} \quad \text{where } x_2 = (1-x_1)$$

The apparatus necessary for the pre-equilibrium of a thin layer plate is shown in figure 3. The apparatus is very similar to that used for normal development but a separate reservoir is introduced which contains the developing solvent. The plate is placed in the enclosure and allowed to come into equilibrium with the solvent vapor for a few minutes (as already discussed this will not be the same equilibrium that would be established with the actual solvent mixture). The plate is then lifted and placed so that the end now dips into the developing solvent and the separation processed in the usual way. There are two main effects resulting from the pre-saturation of a TLC plate. These effects are depicted in figure 4.

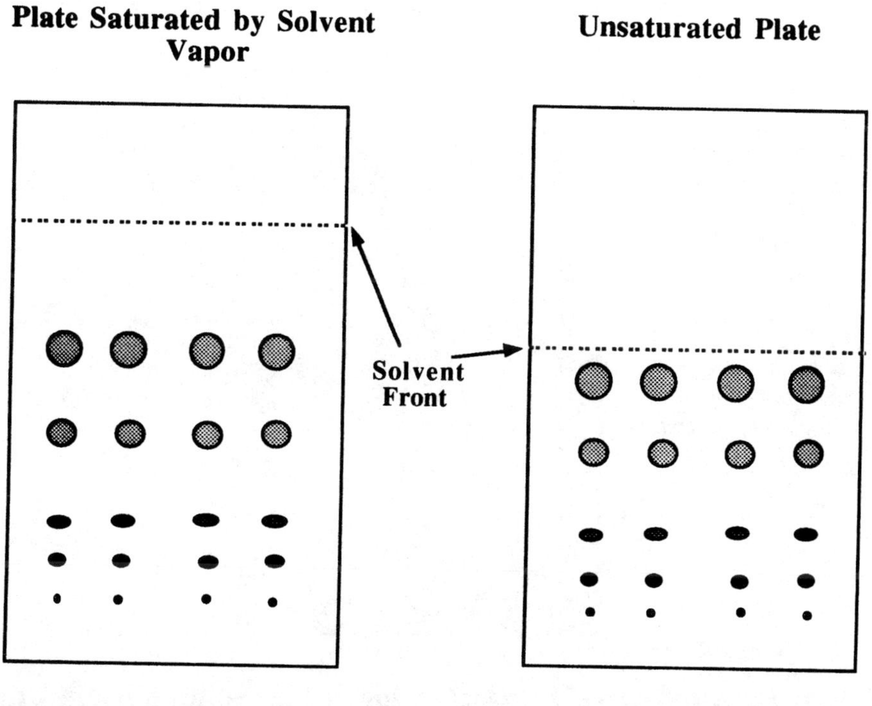

Figure 4 Effect of Plate Saturation on Plate Development

First, the velocity of the solvent front is increased relative to that of the unsaturated plate. Second, as a significant amount of the solvent at the front is adsorbed onto the silica surface of the unsaturated plate, the separated components are much closer to the solvent front. Consequently, the apparent advantage resulting from the faster moving front of the pre-saturated plate on the analysis time is offset by the closer proximity of the spots to the solvent front in the unsaturated plate. In most cases, the effect of pre-saturation on the actual separation is small but can be important for special mixtures where subtle changes in retention can make the necessary difference.

An alternative system for preconditioning a TLC plate that provides a different level of deactivation is to equilibrate the plate with a solvent other than that used for development, as shown in figure 5.

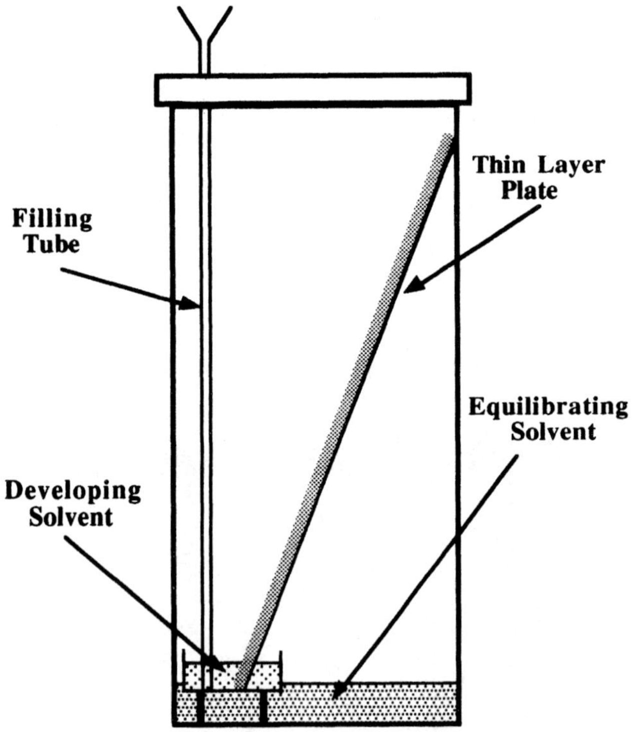

Figure 5 Alternative Apparatus for the Pre-Equilibrium of a TLC Plate

The plate chamber contains a separate small reservoir to contain the developing solvent with a tube through which it can be filled. The solvent with which the plate is to be equilibrated is placed in the bottom of the chamber and the thin layer plate placed in the small empty reservoir that will be used to contain the developing solvent. An appropriate period of time is allowed for the stationary phase on the plate to come into equilibrium with the solvent vapor. The developing solvent is then poured into the small reservoir through the filling tube and the plate developed in the normal way. The effect of equilibrating the plate with an alternative solvent to that used for development on the resulting resolution of the solute mixture is, again, generally rather subtle and the technique is not frequently used.

Continuous Plate Development

A number of other plate development techniques have been described, some of them, perhaps, more novel than useful and only marginally appropriate for quantitative assays. Nevertheless, for completion, some of the more established techniques will be described here. The normal development of a thin layer plate is limited by its physical dimensions but a continuous development procedure has been used employing special equipment.

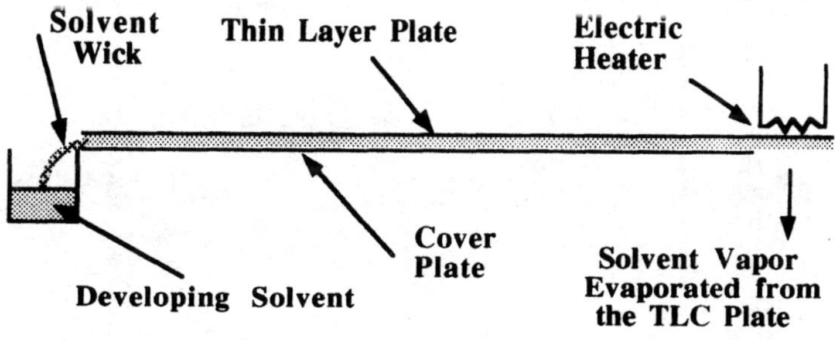

Figure 6 Apparatus for the Continuous Development of a Thin Layer Plate

One system that has been used successfully for the continuous development of a thin layer plate is shown in figure 6. The plate is held horizontal and inverted so that the stationary phase layer faces downwards and rests on a second glass cover plate. A wick transfers the solvent from the reservoir to the stationary phase coating which is sandwiched between the two glass plates as shown on the left-hand side of figure 6. The whole system is situated in a suitable chamber to prevent solvent evaporation from the reservoir. The solvent passes along the plate by surface tension forces in the usual way until it reaches the end of the plate. A small area at the end of the plate is exposed and heated electrically to evaporate the solvent as it arrives. In this way, development can be continued and the system now resembles an LC column of lamina shape flow driven by surface tension and evaporation. The value of this technique is a little questionable as its intent appears to be to simulate an LC column, in which case it would be far better to use an LC column in a normal liquid chromatograph in the first place.

Forced-Flow Development

A further technique of forced-flow plate development is even closer to normal LC techniques and virtually turns the thin layer plate into a lamina LC column. The apparatus used for forced flow thin layer chromatography is shown in figure 7.

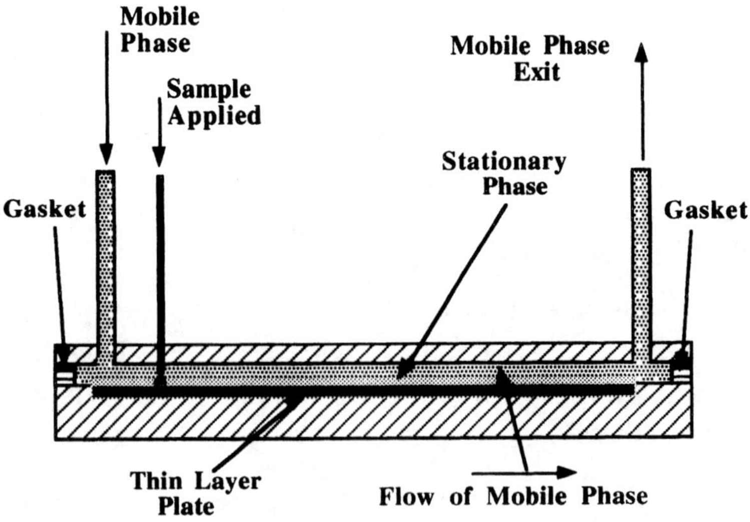

Figure 7 Forced-Flow Development

The TLC plate is sandwiched between two metal plates that are sealed by appropriate gaskets. The upper plate has three conduits. The first supplies mobile phase to one end of the plate, the second allows a sample to be placed onto the plate a little in advance of the solvent supply conduit and at the other end of the plate there is a conduit allowing the solvent to leave the chromatographic system. The forced-flow development arrangement in the single dimensional mode appears a little illogical as a normal LC column system would appear to provide a far more efficient and simpler alternative. If the forced flow system is employed in a two-dimensional manner, however, then the situation is quite different and a system is provided that cannot be simply emulated by normal LC systems. A two dimensional forced-flow TLC plate arrangement is shown in figure 8.

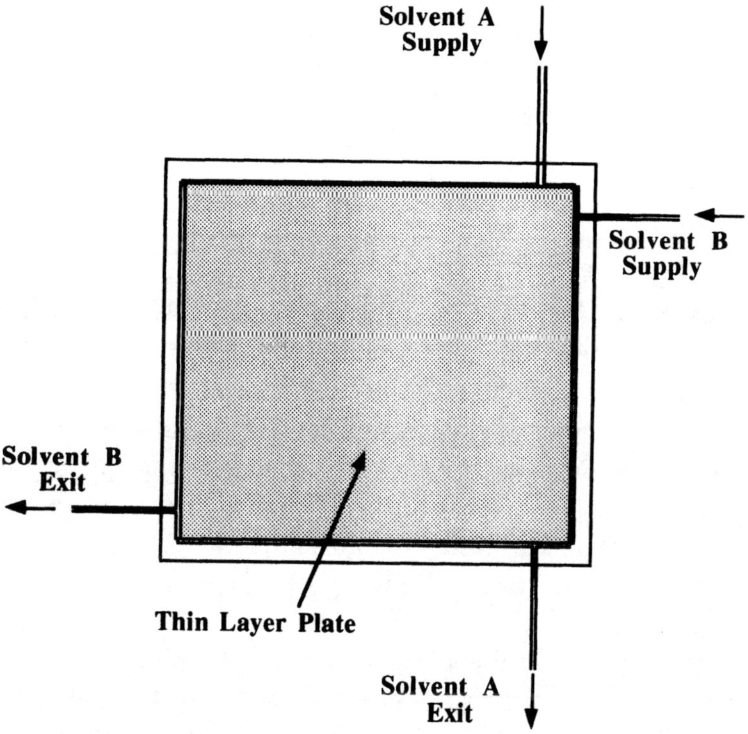

Figure 8 Two-Dimensional Forced-Flow Development

The apparatus is very similar to that used for the forced-flow method shown in figure 7 except that a square plate is used and there are two solvent inlets and two solvent outlets. The sample is initially spotted on the top right-hand corner of the plate and the plate developed by forced flow downwards. This produces a series of solute bands or spots down the right-hand side of the plate. After a predetermined time, the vertical downward flow is stopped and the plate is developed by a second solvent that is passed by forced flow from right to left across the plate. Two-dimensional separations can also be accomplished in LC but it requires *each* complex peak to be diverted to a second column and another separation carried out. Thus, although in two-dimensional TLC, the first separation on the first phase system is similar to that in LC, in the second development, all the separate composite peaks or spots are developed simultaneously (but all are developed by the same solvent mixture). In contrast, each composite solute band must be individually separated with two-dimensional LC, but a specific optimum solvent mixture or solvent program can be used for each composite peak. It is clear that although the resolution obtained using two-dimensional TLC can be very effective, the ultimate resolving power of two-dimensional LC will be far greater (but, of course, the equipment will be very much more expensive).

Sample Application

Samples must be applied to a TLC plate as a spot that is as small as possible. Normal TLC plates are usually dosed with 1 to 2 μl of sample but the so called 'high performance thin layer chromatography' (HPTLC) plates that are coated with particles 5-7 μm in diameter have a maximum loading of about 100-200 nl. Ideally, the sample should occupy a circular area no greater than 1 mm in diameter, on any HPTLC plate. Unfortunately, as this requires the use of samples a few nanoliters in volume, for many analyses, the sample will need to be concentrated. Manually, the sample is often applied with a micro-pipette which is filled by touching the end of the pipette with the sample solution and then discharging the contents of the pipette by touching the surface of the plate. If the solvent is then allowed to evaporate, a second sample can be placed on the top of the first and by a sequence of such operations a concentrated sample can be placed on the plate.

It is important to ensure that the solvent used to contain the sample does not contain components that will affect the elution of the solutes. For example, consider the case where a plate, coated with silica, is being developed using a mixture of n-hexane and methylene dichloride as the mobile phase and the sample is made up in a solvent containing significant amounts of methanol. The strong polar solvent would be carried along by the mobile phase and progressively deactivate the silica which would result in the rapid migration and poor resolution of the components of the mixture. This situation must be avoided and appropriate solvents used to dissolve the sample. However, the precision that is realized will depend both on the sample and the nature of the sample solvent.

Micro-syringes can also be employed to place a sample on to the plate. By careful adjustment, and a little local heating, sample concentration can also be accomplished. The contents of the syringe are slowly but continuously discharged onto the plate and at the same time the solvent is progressively evaporated. This procedure can be automated, using computer controlled syringes and in this way samples can be focused onto the plate and constrained to a very small area. It follows that such sampling procedures can provide the best resolution.

In general glass pipettes are inexpensive, need considerable skill in their use and, even then, are not particularly precise. Alternatively, micro-syringes, and mechanical sampling systems employing micro-syringes are far more precise, but are also more expensive. Neverthless, they are highly recommended for accurate quantitative analysis. It should be pointed out, however, they are also a little fragile in the sense that the needles of the syringes easily become blocked. For the maximum accuracy and precision the syringes need to be automated and if necessary controlled by a computer, in which case the sampling system can become very expensive indeed. There are various commercially available devices that will apply samples to a TLC plate, either individually or as multiple samples. An example of a device for automatically concentrating a sample and then placing it on a plate is shown in figure 9. The apparatus consists of a plate with a small indentation, at the center of which is an aperture that can be connected either to a vacuum or to a source of gas pressure. A polymer film is placed over the surface and a vacuum applied to suck the film into the indentation. Small quantities of sample solution are then placed in the indentation and the solvent allowed to evaporate.

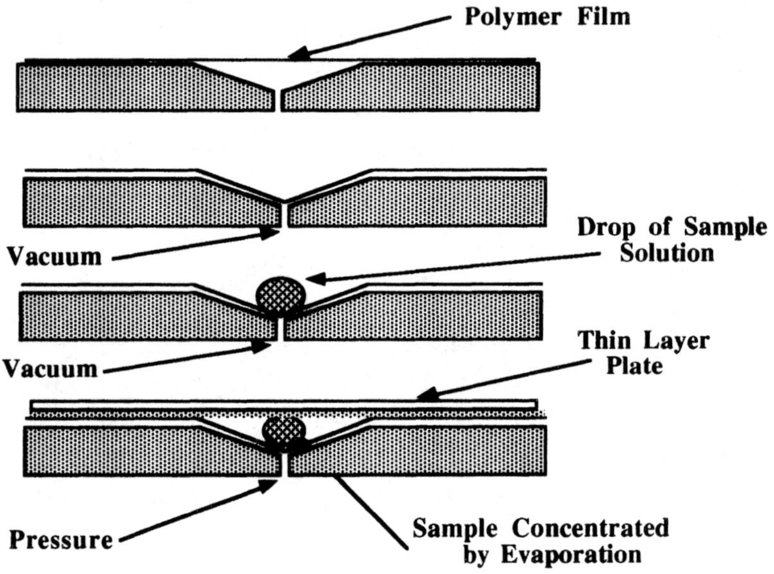

Figure 9 A TLC Sampling Device Incorporating Sample Concentration

This procedure is continued until it is estimated that sufficient sample is present on the film for a satisfactory TLC separation. It is important that the sample is *not* evaporated to *dryness* as the transfer of solid materials to the thin layer plate is *very inefficient*. When sufficient sample has been accumulated on the film and the droplet is still liquid, the thin layer plate is placed over the film and a positive pressure now applied to the aperture at the center of the indentation. The film extends to the surface of the TLC plate and the sample is transferred by contact as a result of surface tension forces and the adsorptivity of the TLC plate surface. The plate is then conveyed to the solvent chamber and the separation developed in the usual way. If larger samples are required for semi-preparative work, sample *bands* can be applied to the plate as opposed to sample *spots*. Sample bands can be applied by either using TLC plates with *concentrating zones* or alternatively by using *band applicators.* A diagram of a TLC plate that includes a concentrating zone is shown in figure 10.

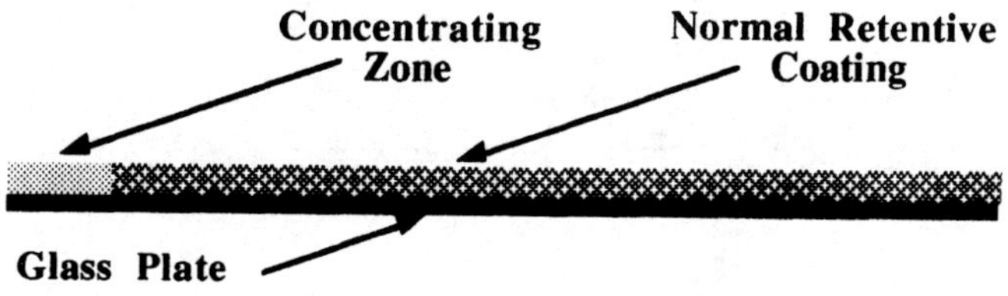

Figure 10 TLC Plate with Sample Concentrating Zone

The concentrating zone is about 2 cm wide and consists of a coating made from relatively large particles of silica with a relatively low surface area and consequently low retentive capacity. The concentrating band is coated closely adjacent to the normal retentive coating which, by definition, consists of the usual particles 5-7 mm in diameter, but with a much higher surface area and thus, much greater retentive capacity. Several samples of a few microliters or more can be placed sequentially onto the concentrating zone and the solvent allowed to evaporate until there is an adequate quantity of sample on the plate. The sample is now spread along the concentration zone in a fairly broad band. When the plate is developed the solutes move rapidly through the concentration zone due to its low retentive character to the interface between the layers. At the interface, the solutes are retarded by the more retentive layer and are thus focused as a sharp band at the front of the plate. As development proceeds, the solutes separate in the normal high retentive layer in the usual manner. This procedure has other advantages. If the sample is contaminated with salts or biological polymers, these will be trapped in the concentration zone and, thus, will not pass onto the separation region of the plate and affect the quality of the separation.

Band applicators operate in an entirely different way and are usually fully automated. The sample is atomized in a stream of air or nitrogen depending on the nature of the sample and its tendency to oxidation. A diagram of the type of atomizer used in band application is shown in figure 11.

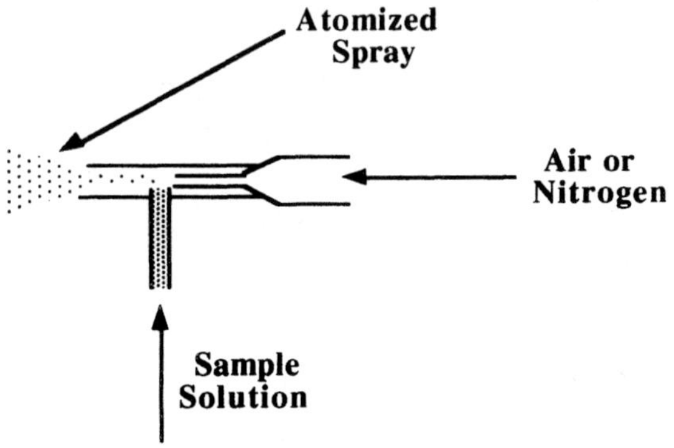

Figure 11 Atomizer for Band Applicator

The atomizer is mounted on a mechanical arm that can sweep from side to side while directing the atomized sample onto the surface of a TLC plate. The range of the sweep and the number of sweeps are usually controlled by a computer and the speed is adjusted such that the solvent evaporates from a given area of sample before it receives the subsequent dose. After dosing, the plate is developed in the normal way.

Some exceedingly novel and clever devices have been developed for TLC. Devices such as those for controlling mobile phase flow, the application of samples, sample removal, sample detection and for quantitative spot measurement. These devices indeed improve the performance of the TLC analysis but are also expensive and in many cases tend to make the TLC system more like a liquid chromatograph. *The great advantage of TLC is its low cost and its relatively high separating capability.* If the performance required to separate a sample is at the limit or beyond the capability of the technique, there is no point in trying to stretch the technique merely for the sake of the technique itself. The rational solution for the chemist or analyst would be to change to an alternative procedure such as liquid chromatography or to some other technique that is more appropriate.

Detection Techniques and the Quantitative Evaluation of TLC Spots

The first applications of thin layer chromatography involved the resolution of plant pigments which were colored Accordingly, the separation was easily observed with the naked eye. Unfortunately, the vast majority of compounds are not easily visible to the eye and consequently the spots must be developed to render them visible. However, the retinic response of the eye is not linear with light intensity, which, coupled with the variation of the iris with the light strength, makes the eye a very poor detecting system for quantitative estimation, even for clearly visible materials.

The eye, however, can be used very effectively as a null sensing device, where closely similar light intensities are being matched as in the use of a comparator. Consequently, if a set of standards covering a range of concentrations are separated on the same plate as the sample of interest and the spots simultaneously developed with an appropriate coloring reagent, an approximate estimation of the concentration of the unknown can be made by comparing its intensity with those of the standards. To improve the discrimination, a second set of standards can be used having closer intervals around the concentration identified as the most probable in the first assay, and the sample and standards again separated and processed and the intensities compared. In this way a more accurate assessment may be possible.

Nevertheless, for more accurate work, other methods of spot evaluation must be used and this involves scanning instrumentation. Firstly, however, the spot must be made to appear visible to the sensing method employed and there are a number of spot development methods that can be used. A common technique is to spray the developed plate with an appropriate reagent that will give color to the solute spots and allow them to be observed. The type of spray used for such a procedure is shown in figure 12. There are a number of different reagents that can be used and some of the more common are described as follows. However, only a limited number of these are suitable for quantitative assessment.

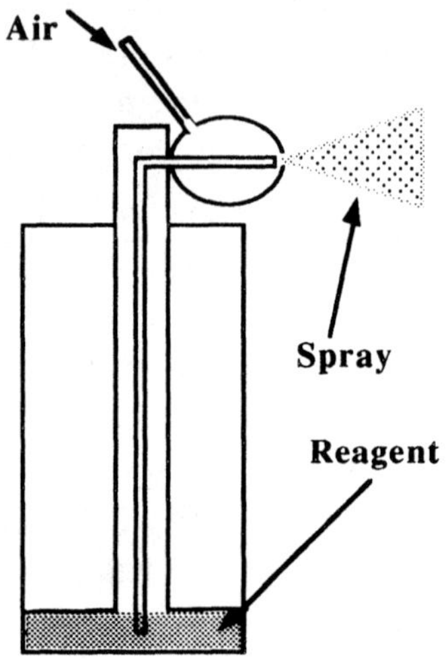

Figure 12 The TLC Reagent Spray

The Iodine Reagent

The majority of compounds become visible if exposed to iodine vapor. Although reaction can take place with some unsaturated compounds, in the vast majority of compounds, the iodine appears to be physically bound to the material and does not chemically react. The plate is placed in an enclosure containing iodine crystals and is preferably heated to about 50°C. Brown spots appear where the solutes are located. When the plate is removed from the enclosure the spots rapidly disappear and so the detection procedure appears to be largely non-destructive.

This detection method can be enhanced by exposing the plate to a very high concentration of vapor, removing the plate and allowing a few minutes for the excess iodine to evaporate. The plate is then sprayed with a 1% starch solution. The solutes then appear as blue spots on the plate. These methods will detect most organic substances and in particular unsaturated compounds.

The Sulfuric Acid Spray

This method is destructive and the solutes cannot be recovered after detection. The plate is sprayed with a 50% v/v aqueous solution of sulfuric acid in a fume hood. The hood will be exposed to extremely corrosive materials and thus all fittings should be made of glass or acid resistant plastic. After spraying, and while still in the fume hood, the plate is heated to about 239°C for about 10 minutes. The solutes are partially oxidized leaving behind a charred deposit of black carbon that is easy to discern. This method will detect most involatile organic compounds.

Chromic-Sulfuric Acid Spray

This method is a modification of the sulfuric acid method, but an oxidizing agent is added to facilitate the reaction and consequently, is also destructive. The plate is sprayed with a 5% w/v potassium dichromate solution in 40% v/v aqueous sulfuric acid solution. The plate is then heated for about 10 minutes at about 200°C to ensure charring is complete. This method will also detect most involatile organic compounds and it is thought to be more sensitive than the simple sulfuric acid spray.

Fluorescence

Many organic substances have a natural fluorescence or can be made to form fluorescent compounds by appropriate derivatization. The plate, after development, or after treatment with the derivatizing reagents, is exposed to UV light and the solutes can be clearly seen as fluorescent spots on a light blue background. This method will detect all materials that naturally fluoresce and those substance that can be made to form fluorescent derivatives.

Fluorescence quenching involves using a TLC plate that has been treated with a fluorescing reagent and thus the whole plate fluoresces. The plate is used to separate the solutes of interest in the usual way and then is exposed to UV light. The plate is seen as a bright fluorescent sheet and the solutes as dark non-fluorescent spots where the solutes have quenched the plate fluorescence. Most organic compounds and in particular aromatic and heterocyclic compounds are detectable by this procedure.

There are various special reagents that have been developed to detect certain classes of compounds. Acid/base indicators have been used to detect carboxylic acids. For example the alkaline form of bromocresol green (0.25% w/v in ethanol) when sprayed on the plate will give yellow spots on a green background for carboxylic acids. The reagent Ninhydrin (0.3% in butanol containing 3% of acetic acid) will provide purple spots on a white background for most amino acids and many amines. Aniline phthalate gives gray-black spots for reducing sugars and many natural products can be detected by spraying the plate with diphenylboricacid b-aninoethyl ester (1% in ethanol) which can give a variety of colors for different substances

Scanning Densitometry

In situ scanning of TLC plates employing optical instrumentation is relatively recent but, nevertheless, is now considered essential for both the accurate location of a spot and the precise quantitative estimation of its content. The surface of the plate can be examined using either reflected light, transmitted light or fluorescent light. In addition the incident light can be either adsorbed, diffusely scattered or transmitted. It is usual to measure the light scattered, reflected or generated by fluorescence from the spot and compare it electronically with light from a part of the plate where no sample has passed (the channel between the spots). Single beam and double beam instruments are available which are diagrammatically depicted in figure 13. The double beam is to be preferred; one sensor monitors the sample lane and the other monitors the blank region between the lanes. The difference signal is taken as that responding solely to the sample. By means of a monochrometer the incident light can be chosen to have wavelengths ranging between 200 and 700 nm. Halogen or tungsten lamps can be used to provide light at the higher wavelengths, whereas for light between 200 and 400 nm a deuterium lamp is to be preferred.

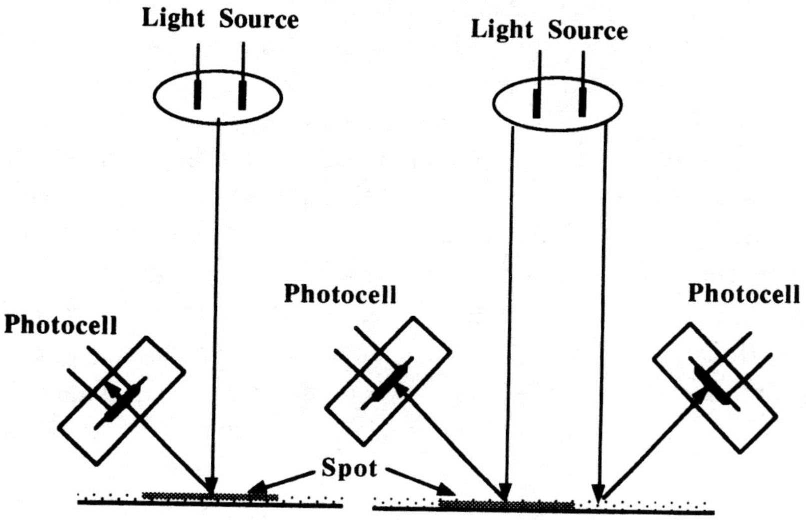

Figure 13 Single and Double Beam Densitometers

Sometimes, in order to induce fluorescence, lamps with higher energy outputs are necessary, such as the mercury lamp, which generates most of its emission at 254 nm and the xenon arc lamp, which generates light over a broad wavelength in a similar manner to the deuterium lamp, but can do so at much higher intensities. If a greater sensitivity can be obtained by using light having a narrow band of wavelengths, a monochromator is introduced between the light source and the plate to select the required wavelength and then either the tungsten or deuterium lamp can be used as the light source.

The sensitivity of a scanning densitometer depends on a number of factors, including the basic instrument design and, in particular, the quality of the optics. The plate surface is viewed by the scanner through a slit and the major factor affecting the overall sensitivity is the slit height to spot diameter ratio. Although the slit dimensions are usually selectable, as the spots along the plate will be of different size, it is not possible to adjust the slit to an optimum size for scanning the whole of the plate.

When measuring either the adsorbed light or the fluorescent light the sensitivity is inversely related to the scan rate. It follows that the slower the scan, the greater the signal. However, carried to the extreme, this approach can extend the analysis time considerably. The relationship between the adsorbed light and the concentration of solute in the spot is not linear, and so either calibration curves must be constructed or the signal must be electronically modified in an appropriate manner to render the output linearly related to solute concentration. In any event, standard solutions must be run for calibration purposes but with manual calibration many more calibration samples are necessary. In contrast when measuring fluorescence, the output is linearly related to solute concentration; in fact, the relationship can be linear over a concentration range of about three orders of magnitude (*e.g.* 0.1 to 100 ng). Calibration is still necessary but as linear curves are obtained, linear amplifiers can be used to electronically process the fluorescence signal.

The major sources of error that arise in scanning densitometry originate largely from sample manipulation. The sample must be carefully applied to the plate in a very reproducible manner and the diameter of the spot carefully controlled. The chromatographic conditions must also be kept constant, and although this is relatively easy in GC and LC, due to the nature of chromatographic development, it is more difficult in TLC. Finally in the measuring process extreme care must be taken to ensure the spot is located in the exact center of the measuring beam or, again, errors will result. In practice, it is sometimes difficult to obtain regular spots and irregular shapes lead to reduced precision and accuracy. Most scanners are suitable for scanning both TLC plates and also electrophoretic stains and can usually deal with objects 200 x 200 mm in size. Most instruments will allow the spot to be monitored by either reflected light, emitted light or transmitted light. The wavelengths available range from 190 nm to 800 nm and scanning speeds as fast as 100 mm/s are available. The spatial resolution can be selected from 25 to 200 mm. A diagram of the optical layout of the CAMAG TLC scanner is shown in figure 14 and would be typical of a TLC scanner.

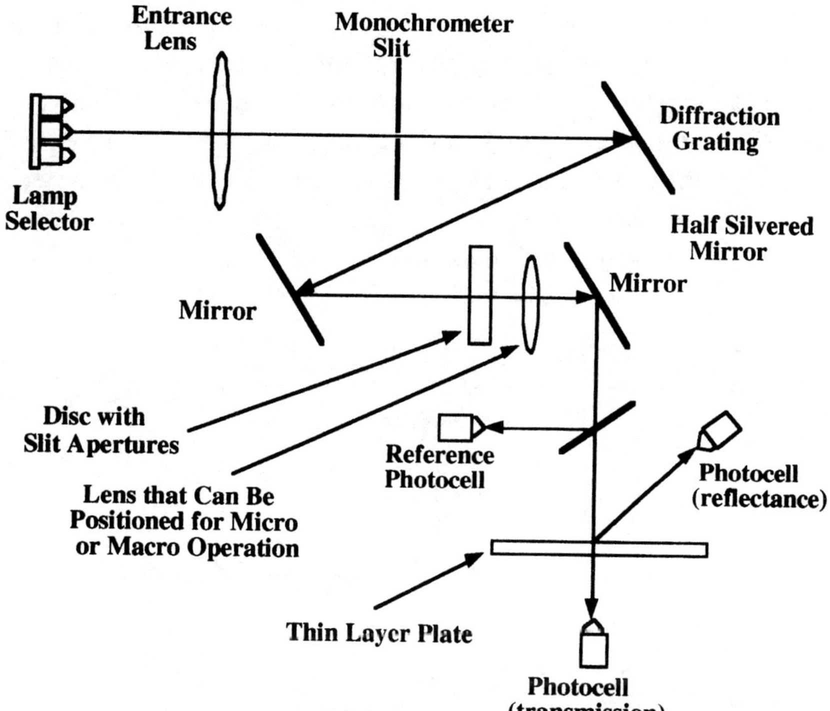

Courtesy of CAMAG Inc.

Figure 14 The Optical Layout of a Single Beam CAMAG TLC Scanner

There are three light sources from which to choose: a deuterium lamp providing light from 190 to 400 nm, a tungsten filament lamp providing light from 350 to 800 nm and a low pressure mercury vapor lamp, which provides high intensity line emissions at 254 nm and 578 nm. Light from the lamp passes through a lens that focuses it through a slit and onto a diffraction grating. The light from the diffraction grating is reflected by a plane mirror, through a selectable slit, and thence to another plane mirror and onto a half silvered mirror. Light is taken from the half silvered mirror to a reference photocell and the remaining light passes to the plate.

The light reflected from the plate is monitored by a photocell aligned 30° to the normal of the plate and the transmitted light is monitored by a photocell placed directly under the plate. The stage is driven by stepping motors in both the (x) and (y) directions and the reproducibility of positioning the monitor is ±50 μm in the (y) direction and ±100 μm in the (x) direction at a maximum scanning speed of 150 mm/sec. An example of the scan of a thin layer plate employing light of 220 nm is depicted in figure 15. The double beam scanner is basically two such optical systems that allows two parts of the plate to be scanned simultaneously.

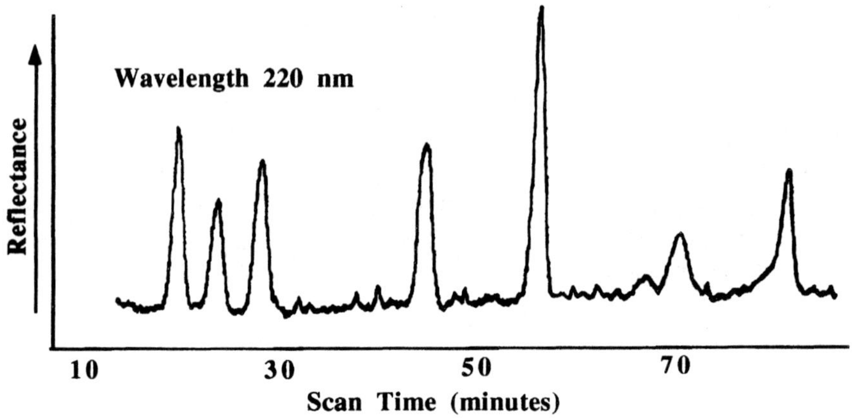

Courtesy of CAMAG Inc.

Figure 15 The Analog Calibration Curve for a Series of Sulfonamides Produced by Scanning a TLC Plate with the CAMAG Scanner

It is seen that the scanning time is over 80 minutes and the measuring points were averages from consecutive sample pairs. Nevertheless, a very respectable chromatogram is achieved from which quantitative data can readily be obtained. An example of the flexibility of multi-wavelength scanning is demonstrated in figure 16, where the separation of some sulfonamides and antibiotics is shown, having been scanned over a series of different wavelengths. The wavelengths have been previously identified as optimum for certain of the components, *i.e.* each component gave a maximum response to light of the chosen wavelength. The optimum wavelengths proved to be in the UV spectrum, *viz.*, 254 nm, 365 nm, 302 nm, 313 nm and 366 nm. This presentation is similar to that produced by the diode array UV–LC detector.

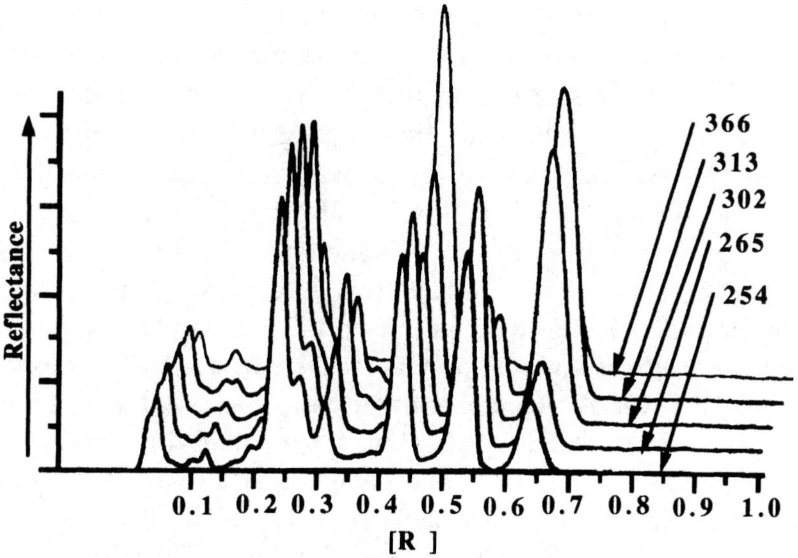

Courtesy of CAMAG Inc.

Figure 16 The Separation of Some Sulfonamides and Antibiotics from an Animal Feed Mix Monitored at Different Wavelengths

The three-dimensional presentation provides the maximum detailed information, with chromatographic characteristics displayed on one axis and spectroscopic characteristics displayed on the other. Nevertheless, the scanning procedure is time consuming and many of the advantages of TLC relative to LC are lost. Operating costs are still lower and solvent disposal problems will be significantly reduced, but the procedure will be slower and the monitoring equipment, although costing significantly less than the liquid chromatograph, is still relatively expensive.

Synopsis

Thin layer chromatography is not the ideal chromatographic technique for accurate and precise quantitative analysis. It is, in principle, an inexpensive procedure, but in order to achieve accurate and precise results a considerable investment in scanning equipment is usually necessary which detracts from its economic advantage. The development process is quite complex due to the selective adsorption of the more polar components of the mobile phase as the mobile phase front moves along the plate. This results in the progressive deactivation of the adsorbent and the parallel reduction in polarity of the mobile phase. This process becomes increasingly complex as the number of solvents in the mobile phase is increased. It also accounts for the subtle selectivity that can often be obtained from certain thin layer distribution systems. To reduce this effect, the plate is often deactivated prior to use by exposure to the mobile phase vapor and there are a range of different development chambers designed for this purpose. There are also a number of techniques and devices that allow continuous development of a thin layer plate including two-dimensional systems. This approach tends to simulate liquid chromatography which seems rather pointless, as it would be easier and better to use the latter technique in the first place. The precise orientation of the sample on the plate is important for accurate quantitative assays and a number of mechanical devices are available to serve this purpose. However, these devices are also expensive and also detract from the economic advantages of the technique. The first TLC separations involved colored substances and were easily observed with the naked eye, but for the majority of compounds the spots must be developed to render them visible. Unfortunately, the retinic response of the eye is not linear with light intensity, which makes the eye inadequate for quantitative estimation. The eye can be used as a null sensing device and if a set of standards are separated on the same plate as the sample and the spots simultaneously developed, an approximate quantitative assay can be made by spot comparison. For accurate work appropriate scanning instrumentation is necessary. The TLC plate can be developed for solely qualitative work by spraying with iodine, sulfuric acid, or chromic-sulfuric acid, but for quantitative work, reproducible derivatives must be used that fluoresce or adsorb light at an appropriate wavelength. There are two types of scanning methods, one uses a single light beam that simply measures the light reflected from, or is transmitted through, the spot; the other employs twin beams from the same source and compares the reflected or transmitted light from a spot with that from a solute-free portion of the plate. Fluorescence measurements can be taken in two ways, either by measuring the fluorescent light emitted by a fluorescent derivative of the spot, or by using a fluorescent TLC plate and measuring the quenching effect of the spot. The major sources of error arise from sample manipulation as the sample must be placed on the plate in such a manner to provide spots of carefully controlled diameter. The output of the scanner can be made to produce an analog curve of solute concentration against distance moved by the spot in the form of a clearly recognizable chromatogram. Such a chromatogram can be quantitatively evaluated in much the same way as a GC or LC chromatogram.

References

1. B. Buglio and V. S. Venturella, *J. Chromatogr. Sci.*, **20**(1982)165.
2. M. S.Schreiber, "75 Years of Chromatography-an Historical Dialogue", Elsevier Scientific Publishing Company, Amsterdam-Oxford-New York (1979)413.
3. J.G.Kirchner, "75 Years of Chromatography-an Historical Dialogue", Elsevier Scientific Publishing Company, Amsterdam-Oxford-New York (1979)201.

Chapter 11

Thin Layer Chromatography Applications

Quantitative analysis by TLC is most frequently carried out by the visual comparison of the intensity of the spot of interest to that of a standard. This, for example, would allow the sample concentration to be compared with that of a legal standard to ensure compliance with regulatory limits. For more accurate work, spot scanning techniques must be used. In general, it is the optical properties of the separated material that are used for the quantitative assessment of a thin layer spot. If the spot adsorbs in the visible range of wavelengths, then both reflected and transmitted light can be measured. If, however, the substance in the spot only adsorbs in the UV range of wavelengths, then only reflected light can be measured, as the silica base of the plate itself will adsorb UV light. If the substance fluoresces, then the fluorescent light can be directly measured as emitted light, or, alternatively, if a fluorescent plate is used, the quenching due to presence of the spot can be measured. Unfortunately, there is no simple linear relationship between spot transmittance or reflectance, and calibration curves must be obtained.

The relationship (with the exception of fluorescence) between the measured property and the mass of solute in the spot is obtained by arbitrarily fitting an appropriate function (usually a second or third order polynomial) to a set of calibration data. The calibration curve is then used for subsequent analysis. The calibration data are usually fitted to a polynomial function. However, care must be taken not to use too high an order of polynomial as the improved correlation will usually include fine structure in the curve fit which does not exist in practice and which will often lead to serious errors. To avoid this situation the order of the polynomial should never be greater than one-fifth of the number of data points. For example, to fit a second order polynomial there should be at least 10 calibration points that scan the total range of concentrations that are likely to be analyzed. If a third order polynomial is needed, then there should be at least 15 data points spanning the concentration range.

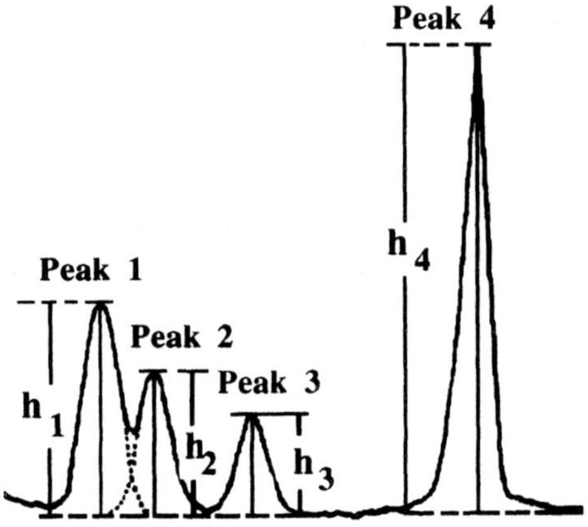

Figure 1 Peak Height Measurements on a TLC Plate

Fluorescence measurements have a number of advantages over transmittance and reflectance as the relationship between emitted light and solute mass is closely linear. In addition, the noise level is much smaller and thus the measurements more sensitive (the detection limit can be one to two orders of magnitude smaller than that obtained from transmittance or reflectance measurements) and as a result the measurements have a wider dynamic range. It is usually essential to insert a cut-off filter between the plate and the fluorescence sensor to remove all excitation light. The presence of excitation light provides a background signal which increases the noise and thus reduces both the sensitivity and the linear dynamic range of measurement.

Measurement Methods

There are two common methods of spot assessment: peak height and peak area measurements (including sliced area measurements), which have been discussed in earlier chapters. The measurement of peak heights is depicted in figure 1. The advantages of peak height measurement over peak area measurement have also been discussed in earlier chapters, but it is clear that the possibilities of error by peak overlap are much reduced if height measurements are employed. Examples of two types of peak integration are shown in figure 2.

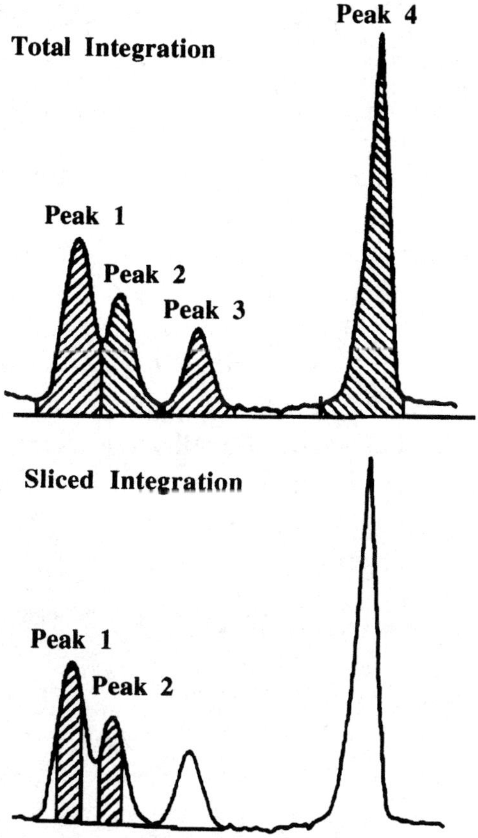

Figure 2 Different Forms of Peak Integration

It is clear that for accurate total area integration, adequate resolution must be achieved, or there will be significant errors introduced due to peak overlapping. However, complete resolution is far more difficult to achieve in thin layer chromatography than in other forms of chromatography and thus a modified form of peak integration is sometimes used, called sliced area integration. This procedure is depicted in the lower chromatogram in figure 2. Sliced area integration is a procedure that attempts to maintain the accuracy of area integration and maintain some of the advantages of peak height measurement. It is clear that slice integration does eliminate some of the errors caused by peak overlap. This procedure requires very careful selection of the limits used in the computer integration protocol, but if carefully carried out, can be a useful compromise between peak height and peak area measurements when using scanned data from a TLC plate.

There are a large number of applications employing TLC for quantitative analysis described in the literature, many of which, although applied to different materials, are from the point of view of the technique very similar in nature. The following examples are taken from reports mostly published either in the *Analyst*, *Analytical Chemistry* or the *Journal of Planar Chromatography* and these journals, together with the *Journal of Liquid Chromatography* are recommended as reliable sources of application information for those starting to use TLC for quantitative assays. The versatility of the technique is highlighted by the widely different application areas from which the examples are taken.

The Assay of Sulfamethazine in Pork Carcasses

Sulfonamides in general are widely used in veterinary practice as antibacterial agents, and sulfamethazine (4-amino-*N*-(4-6-dimethyl-pyrimdin-2-yl) benzene-sulfonamide) is one that is commonly used in pig husbandry.

Unfortunately, sulfamethazine can be potentially misused; for example, if there is an inadequate withdrawal period prior to slaughter, there may be a level of the drug in the meat that exceeds the minimum allowed by law. Consequently, a sensitive, rapid, on-site analytical procedure is required to ensure regulatory compliance. Shearan and O'Keeffe [1] developed a rapid technique with a TLC 'finish' that allowed the level of sulfamethazine to be determined as below, above or coincident with the statuary limit, at a rate of 20 samples per day.

The extraction procedure was similar to that developed by Shearan *et al.* [2]. 0.5 g of tissue was blended with 2 g C18 reversed phase adsorbent in a mortar and pestle for 30-45 s. The blend was then packed into the syringe barrel of a solid phase extractor and the contents washed with 8 ml of hexane. The blend was then extracted with 8 ml of dichloromethane and the extract evaporated to dryness in a stream of nitrogen and the residue taken up in 100 µl of dichloromethane.

The silica plates were 10 x 10 cm with a pre-absorbent strip for spot focusing. 40 µl aliquots of extracts and standards were applied to the plate using micro capillary pipettes. The plate was dried and placed in a chromatographic tank containing methanol. The solvent was allowed to advance to the top of the pre-absorbent strip and was then removed and dried. The plate was then placed in another tank filled with ethyl acetate and the solvent allowed to advance 2 cm beyond the pre-absorbent strip; the plate was then removed and sprayed with fluorescamine solution. After 15 min the plate was viewed under UV light (366 nm) to locate the fluorescent bands. Where it was necessary to resolve, sulfadiazine, sulfathiozole and sulfurmerazine from sulfamethazine, the ethyl acetate was allowed to advance 3.5 cm beyond the pre-absorbent strip.

The technique recovered about 80% of the antibacterial agents at levels between 50-100 ppb in porcine tissue, with a minimum level of detection of about 30 ppb. The standard used contained 0.2 µg per ml; less dense spots were reported as negative, containing the antibacterial agent at a level below that permissible; more dense spots were reported as positive and indicated that the antibacterial agent was present at a level above that permissible. This is a typical example of the more common type of quantitative analysis by TLC. The analysis determines whether a component is above or below the level of a standard and employs the eye as a comparator (for which it is most efficient) to give a pass or fail result.

The Determination of Aflatoxins in Palm Kernels

Aflatoxins are highly toxic metabolites produced by the molds *Aspergillus flavus* and *Aspergillus parasiticus*. These molds thrive on substrates rich in lipids and carbohydrates and particularly under the damp conditions resulting from the high relative humidity and elevated ambient temperatures common in tropical countries. The oil palm *Elaies guineensis Jacquin* occurs both wild and cultivated in the equatorial areas of Africa, South East Asia and South America. Palm oil is obtained from both the fruit and the nut, the latter containing as much as 50% of its mass as oil. The oil finds commercial use in the food and detergent industries and the residual pressed 'cake' provides a protein source for animal food. The aflatoxins are generated both before and after harvesting and their production is aggravated by cracking, insect activity and poor storage conditions. Due to the high toxicity of the aflatoxins, maximum levels of contamination have been set by most user-countries and consequently, efficient accurate and precise methods for their measurement had to be developed. Nawaz *et al.* [3] devised a practical analytical procedure that involved an efficient extraction procedure followed by a TLC separation. The plate was finally quantitatively assessed by automatically scanning the aflatoxin fluorescence.

Five hundred gram of finely ground palm kernel (appropriately sampled) were slurried with 750 ml of water in a blender at high speed for 3 min. A 100 g portion was then extracted by blending with 240 ml of acetone and filtered through a Whatman No. 1 filter. 5 ml of the crude extract was mixed with 5 ml of methanol, 1 ml of 20 %w/v lead acetate solution, 1 g of Celite and 63 ml of water. The lead salt was used to precipitate colloidal contaminants from the sample. The mixture was then passed through a solvated PH bonded phase cartridge at about 10 ml/min. (PH cartridge 608303 was solvated by treating with 10 ml of methanol followed by 10 ml of water). The cartridge was then washed with 10 ml of water and the aflatoxins eluted from the cartridge with 3 ml of chloroform.

The extract was dried by passing it through a tube packed with anhydrous sodium sulfate and collected in an 8 ml vial. The solid aflatoxins were obtained by evaporation at 45°C in a stream of nitrogen. The HPTLC plates (aluminum backed, 20 x 20 cm, Merk 5547) were immersed in methanol for 1 hr to remove any adsorbed material, dried for 5 min at 100°C and stored in a dessicator. The sample extract was dissolved in 300 µl of a benzene:acetonitrile mixture (98:2). 5 µl aliquots, occasionally interspersed with standards, were applied to the plate in the absence of light with an auto sampler. Sample spots were 5 mm apart along the top edge of the plate and one 10 x 20 cm plate could accommodate 30 sample spots and three standards along the 20 cm edge. Edge effects were eliminated by removing a strip of silica gel, 3 mm wide, along the plate edge parallel to the direction of development.

Before the analytical development was commenced, a further clean up was carried out on the plate by developing the plate in 20 ml of diethyl ether for 17 min in a continuous horizontal tank (the eluent reached the top of the plate in about 3 min). Development was continued for about 14 min., and, as a consequence, there was no solvent front. The plate was dried in a darkened chamber for 3 min in a stream of nitrogen. The interfering substances were contained at the plate edge and were removed by scraping 2 cm of silica from the edge. The plate was then rotated and developed in a normal tank and developed for 20 min with 20 ml of chloroform-xylene-acetone mixture (6:3:1). The plate was again dried for 5 min and the bottom portion (1 cm) of the silica was removed to decrease the development time and increase the resolution. The plate development was then continued with the same solvent mixture for 16 min. The plate was finally dried in a fan assisted oven at 100°C and scanned for fluorescence by the reflectance mode using the TLC II scanner (Camag 76610) and the SP 4270 integrator (Camag 76650). Scanning techniques are discussed in the previous chapter.

The procedure appears complicated and is certainly lengthy for this type of analysis, but recoveries in excess of 90% were readily obtained and detection limits for the aflatoxins AFB_1, AFB_2, AFG_1 and AFG_2 were found to be 3.7, 2.5, 3.0, and 1.3 µg/kg respectively. Compared with the British Standard Method, the procedure was found to be both more efficient and more precise.

Determination of the Platelet-Activating Factor and other Phospholipids in Human Tears

Phospholipids are important components of biological systems and the platelet-activating factor (PAF), 1-*O*-alkyl-2-acetyl-*sn*-glycero-3-phosphocholine, is particularly significant as it is an agent of immunization for a range of different diseases. PAF is difficult to determine in serum and other biological fluids as, following its secretion, it is immediately hydrolyzed by acetylhydrolase to form 1-hexade-canoyl-*sn*-glycero-3-phosphocholine (LysoPAF). In tears, however, acetylhydrolase is possibly absent or at least present at very low levels as in saliva, but there are also very small amounts of phospholipids present that offer the possibility of determining PAF in such samples. Nevertheless, due to the very low concentrations present, a concentrating technique is necessary. Ohyama *et al.* [4] developed a satisfactory analytical method using an extraction procedure followed by a TLC separation and a plate scanning 'finish'. The inorganic phosphate was liberated from the phospholipid molecules by an enzyme reaction which was followed by a high sensitivity color reaction that produced an ammonium molybdate-malachite green aggregate.

Tear samples were obtained by introducing 40 µl of physiological saline solution into the inferior conjunctival sac. After three gentle eye blinks, a 2 cm diameter circle of filter paper was placed near the inner *canthus* for tear collection. The filter paper was then soaked in 500 µl of chloroform-methanol-water mixture (6:69:50 v/v). The liquid was then passed through a diatomite column to remove proteins, sugar and polar inorganic substances and the lipids were then eluted with 6 ml of a chloroform-methanol mixture (95:5). The eluate was evaporated to dryness under vacuum and the residue dissolved in 50 μl of chloroform, 20 μl samples of the solution were used to spot the TLC plate.

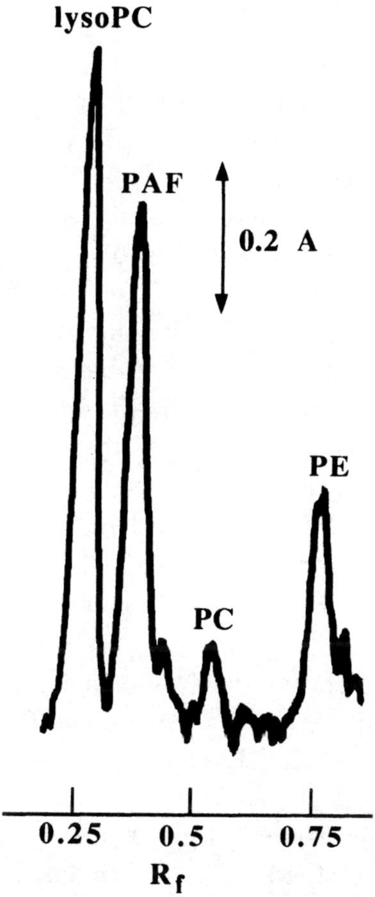

Figure 3 Typical Chromatogram of the Phospholipids in Human Tears

A high performance silica gel plate was employed (HP-K, Whatman) which was soaked overnight in a methanol water mixture (1:1 v/v) and then dried for 1 hr at 120° prior to use. The plate was developed in two dimensions; first with a hexane-diethyl ether mixture (4:1 v/v) for the removal of lipids other than phospholipids; second the plate was then developed at right angles with a chloroform-methanol-water mixture (65:35:7 v/v) to separate the phospholipids. After drying, the plate was sprayed with phospholipase C (PL-C from *Bacillus cereus*) and then alkaline phosphatase (Al-P from human placenta) to hydrolyze the phospholipids.

The volume of each enzyme solution used was equivalent to 2 ml/100 cm^2 of plate. The two enzymatic reactions were allowed to take place for 10 and 15 min respectively at 45°C. The plates were then sprayed with the Mo-MG reagent (200 ml of 38 mmol/l ammonium molybdate, 100 ml of 8.7 mol/l sulfuric acid and 200 ml of 3.2 mmol/l of malachite green and 500 ml of water) at a coverage of 2 ml/100 cm^2 of plate. The phosphates from the phospholipids reacted with the reagent to produce the intense blue-green spot of molybdophosphate malachite green. After 5 min. the plates were scanned at 620 nm with a CS-9000 two-dimensional densitometer (Schimadzu, Kyoto, Japan). A trace of the absorbance profiles (the chromatogram) is shown in figure 3. The quantitative assessment was carried out using peak area measurements in conjunction with calibration curves from standard phospholipid solutions. The peak area was found to be linearly related to concentrations over the range of 2-100 pmol per spot (RSD was 2%, n=7) About 86% of the PAF was recovered. The proportion of the individual phospholipids appeared to be sensibly constant for healthy humans and PAF, LysoPC, PC and PE were present at 26.2, 42.3, 10 and 19.7% respectively.

A Mass Spectrometric Measurement of Peptide-Like Materials Using a Hybrid Thin Layer Plate as the Transport Interface to a Matrix Assisted Laser Desorption Ionization System [5]

A hybrid TLC plate system has been employed as a transport medium to monitor a thin layer separation by means of a mass spectrometer. The combination has been used to determine substances of biological origin at extremely low levels of concentration. The essential transport medium was a thin layer plate, half of which was coated in the normal manner and on which the separation was developed. The other half consisted of a special coating that formed the base of the so called *matrix assisted laser desorption ionization source* for the mass spectrometer. A diagram of the hybrid plate is shown in figure 4. The plate was a modified form of the aluminum-backed silica gel plate manufactured by Merck (Cat. No. 1.0555). Silica was scraped from one side of the plate in such a way that resulted in a taper over a horizontal distance of 1 cm so that the thickness gradually decreased from the original 250 μm to zero over a distance of 1 cm, thus providing a gentle change from one coating to the other. About 5 g of the desorption medium containing 2,4,6 trihydroxyacetophenone or 2,5-dihydroxy benzoic acid was dispersed in a mixture of toluene and ethyl ether (60:40 for 2,4,6 trihydroxyacetophenone and 80:20 for 2,5-dihydroxy benzoic acid) and was sprayed onto the exposed plate backing.

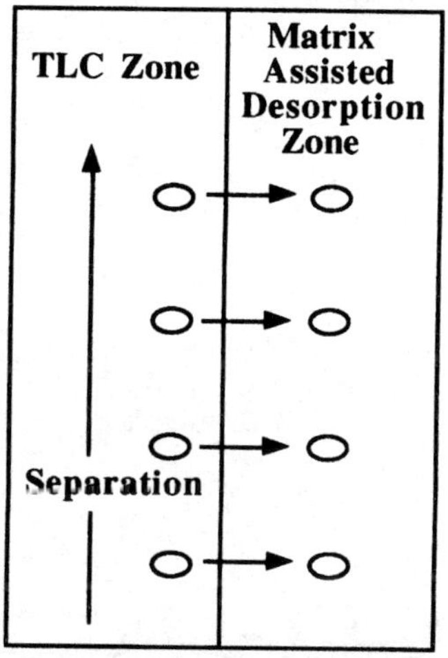

Figure 4 The Hybrid Thin Layer Plate that Acts as a Transport Interface to a Matrix Assisted Laser Desorption Ionization Source for a Mass Spectrometer

The silica gel layer was appropriately shielded from the spray by a microscope slide. The coating was carried out after the separation had been developed. The samples used to test the device were Valinomycin, cyclosporin, and the protected peptide BOC-Phe-Lue-Phe-Lue--Phe. The TLC separation was carried out in two stages, first using chloroform which only moved the BOC-Phe-Lue-Phe-Lue--Phe and then ethyl ether-toluene which only moved Valinomycin, which provided three separate spots suitably separated. The other side of the plate was then coated with the desorption material as described above and the plate rotated through 90°. The solute spots were eluted into the desorption zone by the solvent mixture chloroform-acetic acid-methanol containing 0.1% trifluoro-acetic acid, (1.5:1:1). After transfer of the spots by elution each spot area was treated with methanol-water mixture (1:1) to improve analyte-desorbing agent co-crystallization. After separation, the plate (now acting as a transport system) was transferred to the laser desorption mass spectrometer source and spectra taken in the usual manner. The spectrum obtained form the Valinomycin spot is shown in figure 5, where the results from the hybrid plate are compared with the more conventional contact pressing transfer procedure.

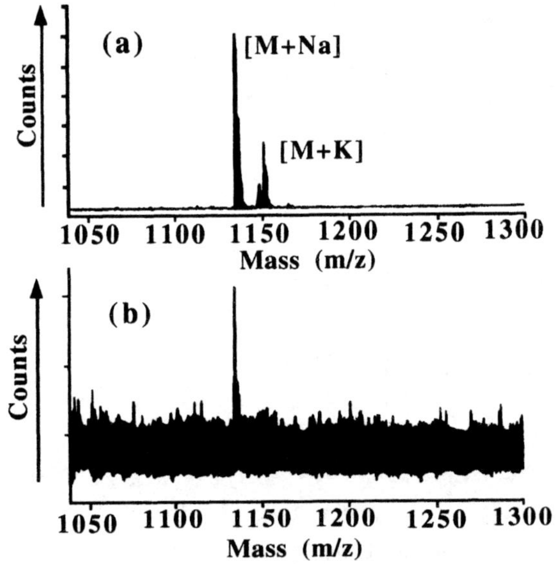

(a) Spectrum obtained from the Hybrid Plate
(b) Spectrum obtained from plate by contact plate pressing

Figure 5 Spectra Obtained for 4.5 pmol of Valinomycin from the Hybrid Plate and by Contact Pressing Transfer

It is seen from figure 5 that the use of the hybrid plate increases the signal to noise ratio significantly and consequently increases the sensitivity of the overall system by an order of magnitude or more. For quantitative work it is clear that an external standard will be necessary and the precision of quantitative assays is not reported. However, the technique offers the possibility of very high sensitivity, specific identification and a reasonably rapid assay.

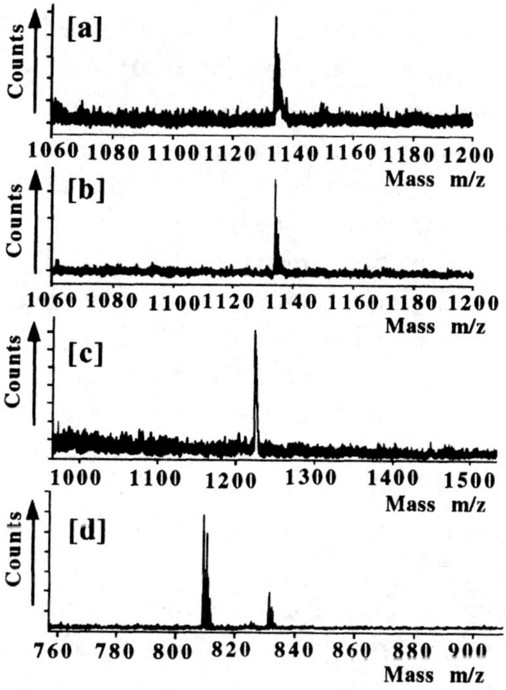

(a) 90 fmol of Valinomycin

(b) 0.83 pmol of Cyclosporin

(c) 6.3 pmol of BOC-Phe-Lue-Phe-Lue--Phe

(d) 2,4,6 trihydroxyacetophenone used as the matrix.

Figure 6 Spectra of Valinomycin, Cyclosporin, and the Protected Peptide BOC-Phe-Lue-Phe-Lue-Phe Obtained from the Hybrid Plates

The spectra obtained for the four test substances are shown in figure 6. It is seen that the sensitivity of the system ranges from about 90 fmol to about 6 pmol.

Determination of Caffeine Using Both Densitometry Measurements and an Image Analyzing System

A relatively recent method for scanning TLC plates that has been reported is by the use of an image analyzing system. This employs a digital camera to photograph the plate, the image is then stored in digital form in a computer and then processed by appropriate software. Vovk *et al.* [6] compared the relative precision obtainable from the image analyzing system to those obtained by normal scanning densitometry. The authors chose caffeine as the model substance and the separation was carried out on a 10 cm x 20 cm glass backed plate carrying a coated layer of silica gel 0.2 mm thick (Merck, Darmstadt, Germany). The samples were place on the plate with a Linomat IV applicator (Camag, Muttenz, Switzerland) which was equipped with a 100 µl syringe. The band was 10 mm wide, and 5 µl of sample was applied at a rate of 4 µl/s. Ten or twelve bands were applied per plate 15 mm apart and 15 mm from the bottom edge. The plates were developed with a solvent mixture consisting of dichloromethane:methanol (90:10 v/v) in an unsaturated glass twin trough chamber. The caffeine solute migrated 7.5 cm and the migration time was approximately 10 min. The plates were subsequently dried in a stream of warm air for about 5 min. A chromatogram of a separation monitored by densitometric scanning is shown in figure 7.

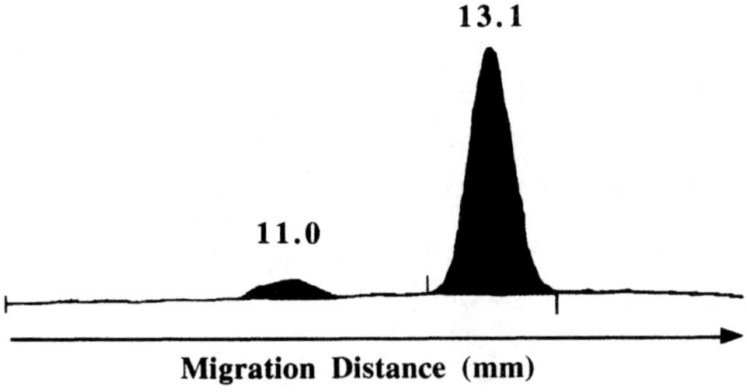

Courtesy of J. Planar Chromatogr. [ref. 6]

Figure 7 Chromatogram Obtained by Densitometer Measurements at 270 nm of a Caffeine Standard

The scanner was operated in the reflectance mode with slit widths of 0.8 mm and 6 mm respectively, the monochrometer band with was 30 nm (wavelength of maximum absorbance λ=270 nm). The densitometry processing was carried out with a Camag TLC scanner II, equipped with a built-in twelve bit analog-to-digital converter and controlled by an external computer with a RS32 interface. The software used was the QTLC-pack (KIBK-IFC, 1990).

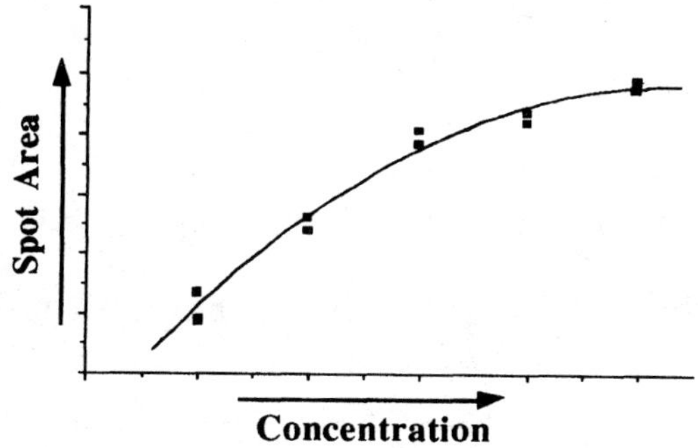

Calibration Curve from Densitometer Scanning

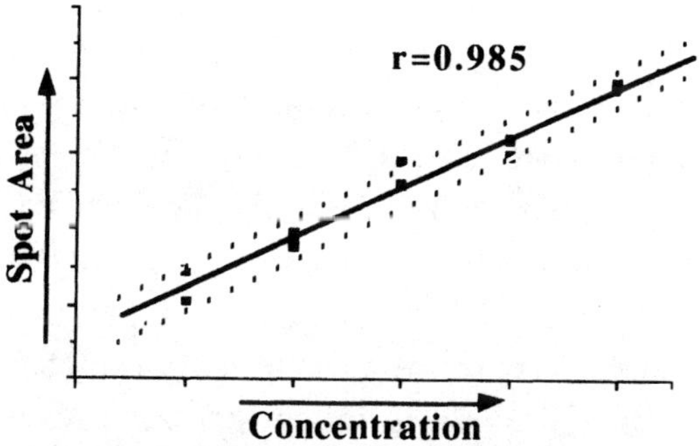

Calibration Curve from Digital Camera

Courtesy of J. Planar Chromatogr.[ref.6]

Figure 8 Calibration Curves Obtained by Densitometer Measurements at 270 nm of a Caffeine Standard and the Image Analyzing System at 259 nm

The image analyzing procedure commenced with the spots being captured on a high sensitivity CCD video camera (Hitachi Denshi Ltd.

model HV-C20). All images were obtained by exposure to direct UV light (λ=254 nm). The data were processed using the Camag Video Documentation system in conjunction with the Reprostar 3 and the Camag Video Scan 1.16 program for the evaluation of thin layer chromatograms. Calibration curves are shown for both data processing methods in figure 8. It is seen that the calibration curve for the image analyzing procedure is linear whereas the curve for densitometry measurements is not. However, despite this, the two procedures gave very similar precision and showed no significant difference in evaluation parameters.

The Use of Digital Auto-Radiography for the Analysis of Biological Samples and for Studying Drug Metabolism

Digital auto radiography is another technique that is used for quantitatively assessing thin layer plates when, for example, radioactive tracers are employed to follow the metabolic pathway of certain drugs. There are two basic methods of radioactive monitoring, both of which are extremely sensitive as, by the use of extended periods of measurement, they can identify very low levels of radioactivity. The first is contact radiography where, after separation and drying, a photosensitive film is placed in contact with the plate and exposed to radiation for an appropriate time. Obviously the longer the time of exposure, the lower the radioactivity that can be detected. The film is then optically scanned to obtain density measurements which will be proportional to the number of emission particles passing through the film. The second method is to use an automatic radioactive counter-scanner that will scan the plate, in much the same way as a optical scanner, but measure the frequency of particle emission across the plate directly. Again the sensitivity can be increased by slowing the rate of scanning, making a wide range of sensitivities available, while trading in time for sensitivity. Due to the first method requiring exposure time and scanning time, in general, the later is the faster method of evaluation.

Kjebovich *et al.* [7] demonstrated the successful use of auto-radiography to follow the metabolism of deramciclane (a new anxiolytic drug) in dogs, from urine samples analyzed by thin layer chromatography. An example of the radio chromatograms obtained are shown in figure 9.

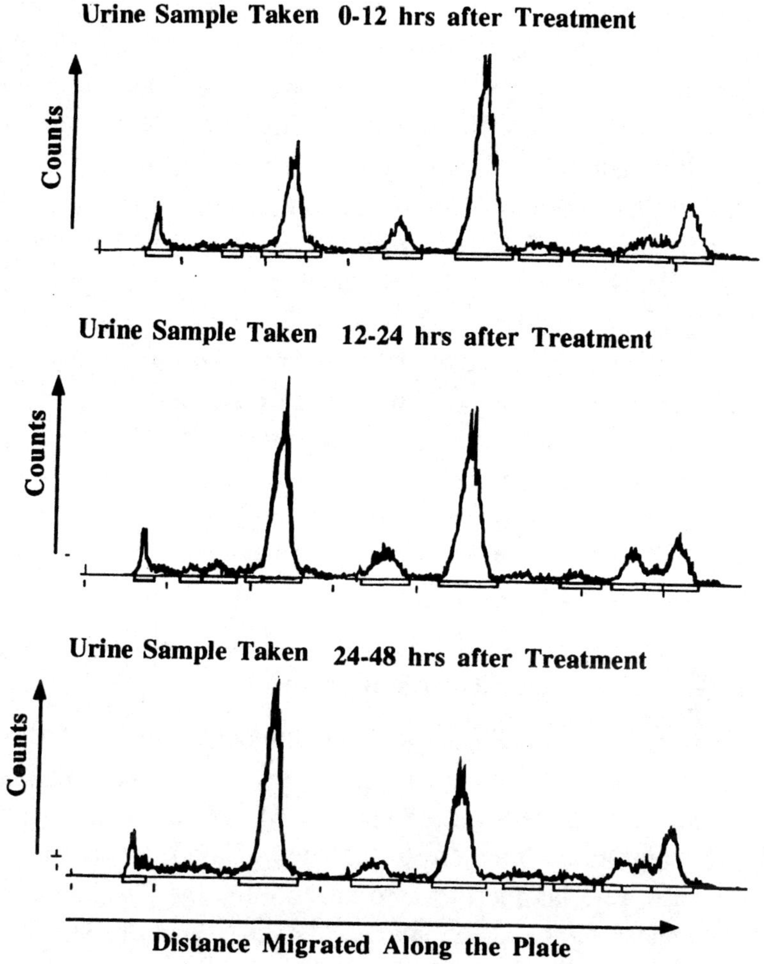

Courtesy of J. Planar Chromatogr [ref. 7]

Figure 9 Digital Auto-Radiograms of Dog Urine Samples Taken 6 hr, 18 hr and 36 hr After Treatment with [^{3}H]deramciclane

The dogs were treated both orally and intravenously with [^{3}H]deramciclane at doses of 3 and 10 mg per kilo respectively. Urine was collected between 10 and 12 hours, 12 and 24 hours and between

24 and 48 hours respectively. Samples were applied to the plate in 10 mm bands with a precision syringe and developed in a normal chamber to a distance of 16-18 cm. After development the plates were dried and scanned with a LB 287 type Berthold Digital Auto Radiograph in conjunction with Hewlett-Packard Vectra VL24/66 computer controlled with WinDAR software V.1.09. The instrument could scan an area of 20 x 20 cm which contained a 600 x 600 wire grid, two-dimensional, position-sensitive multiwire proportional chamber. A 9 to 1 mixture of argon and methane bubbled at 5 ml/min through methylal at 2.8°C was used as the counting gas and the potentials employed for ^{3}H- and ^{14}C-labeled compounds were 2040 and 1200 volts respectively. Signal analysis was performed by measuring 5 x 360000 detector cells per sec. The separation took about 6 hr and the detection took only 2.5 hr (compared with the two days required by contact auto radiography). Unfortunately, accuracy and precision were not reported but it was implied that both were or better or at least as good as that obtained from contact auto-radiography. The advantage that was emphasized was the significantly reduced analysis time.

Purity Measurements of Phthaloyl-Amiodipine Using Over-Pressurized Thin Layer Chromatography

Pressurized thin layer chromatography involves the application of a plastic membrane over the thin layer plate which is held there by hydraulic pressure applied to the opposite side of the membrane. By applying an appropriate pressure to the edge of the plate, the mobile phase is forced, in a lamina flow profile, through the adsorbent layer between the plate and the membrane. In practice, this procedure really constitutes normal liquid chromatography carried out in a column having a very thin rectangular cross section. The advantages of this procedure over the normal thin layer method appear to be improved resolution and shorter analysis time. The advantages of pressurized thin layer chromatography over a similar liquid chromatography system using conventional cylindrical columns, however, is not clear.

Phthaloyl-amiodipine (3-(ethoxycarbonyl)-2-[(2-phthalimidoethoxy)-methyl]-4-(2-chlorophenyl)-6-methyl-5-(methoxycarbonyl)-1, 4-dihydropyridine) is the final intermediate in the synthesis of amiodipine benzylate a highly important anti-ischemic and antihypertensive drug. The purity of the phthaloyl-amiodipine has a significant effect on the final product. Szikszay *et al.* [8] used both thin layer and pressurized thin layer chromatography to analyze phthaloylamidipine for impurities to assess the relative value of each technique for accurate and precise results. Separation by pressurized thin layer chromatography was carried out on fine particle silica gel plates (Merk 5548) and a mixture of *n*-hexane:butyl acetate:ethyl acetate:chloroform (60:15:15:20) was used as the mobile phase. An external pressure of 5 MPa was applied using 8.5 ml of mobile phase equivalent to a migration distance of 36 cm. The plate was scanned at 233 nm using a TLC Scanner 3 (Camag) in reflectance mode.

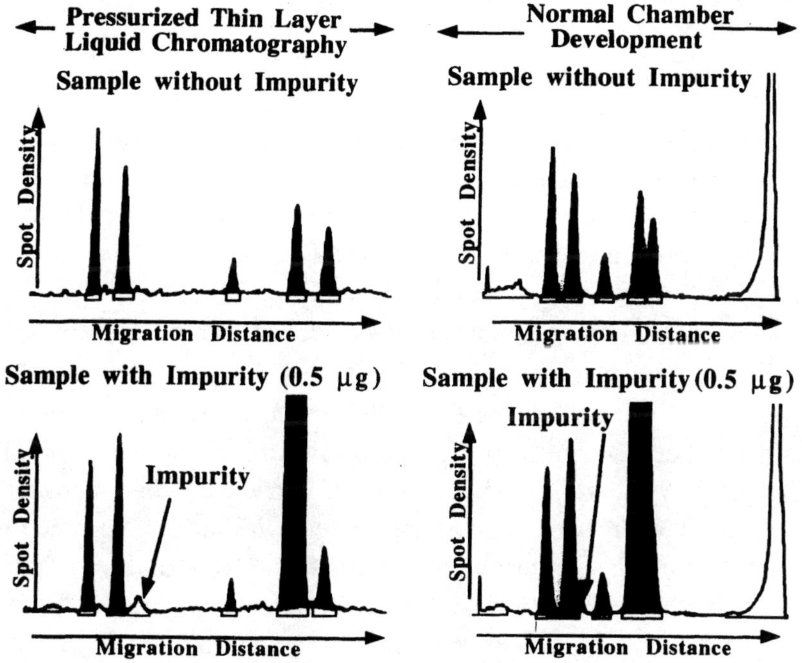

Courtesy of J. Planar Chromatogr [ref. 7]

Figure 9 Comparative Scanned Chromatograms Demonstrating the Improved Performance of Over-Pressurized Chromatography Relative to Normal Thin Layer Chromatography

The conventional thin layer separation was carried out on silica plates (Merck 5554) and a mixture of *n*-hexane:butyl acetate:ethyl acetate:chloroform (30:15:15:20) was used as the mobile phase. A normal chamber was employed with a saturation time of 30 min and a development distance of 15 cm. The plate was scanned in a similar manner to those obtained by normal thin layer chromatography. An example of some of the results obtained is shown in figure 9. The advantages of the pressurized system are clearly shown; the impurity is half masked by a major peak in the chromatogram obtained by normal thin layer chromatography. Out of seven replicates, the mean level of the impurity was determined as 0.847% and 1.111% by the pressurized and normal thin layer chromatographic procedures respectively. The precision, however, as measured by the coefficient of variation was 2.9% and 6.1% respectively. It is clear that the two methods differ both in accuracy and precision but the pressurized system provides the best precision.

Significant advances in thin layer chromatography have been few and far between over the last decade but progress continues, albeit at a reduced rate. To end this chapter on a positive note, a particularly interesting application was very recently described which involved a procedure that combines synthesis, separation, quantitative analysis and biological screening, consecutively carried out on a thin layer plate. This technique is particularly applicable to combinatorial chemistry and has been reported by Freemantle [9]. Details of the method have been described by Williams [10], an example of which is given as follows. Parallel syntheses were carried out in a series of spots on the baseline of a TLC plate coated with silica; the reaction was accelerated by microwave irradiation. Different synthetic conditions were used on each spot to identify the optimum conditions of the synthesis. The products were then separated by developing the plate with an appropriate solvent and a quantitative assessment carried out on each pertinent spot by suitable scanning techniques or visual comparison of the spots with a standard. Each spot was then screened *in situ* for microbial growth inhibition or other biological activity by standard agar-overlay methods directly on the plate. The value of this approach in combinatorial synthesis is clearly obvious as many different synthetic conditions can be investigated simultaneously and the biological activity of the products concurrently ascertained. This procedure is quite unique and is an excellent illustration of the extension of an ongoing and established technique to a challenging modern chemical problem.

Synopsis

The most common and inexpensive method of evaluating a thin layer plate quantitatively is by simple spot matching with the naked eye. The sample is separated with a standard (or set of standards) separated alongside. After appropriate processing, the intensity of the spot of interest is compared with that of the standard spot(s). A more accurate method is to scan the spot(s) by reflectance, employing light of appropriate wavelength or by fluorescence if naturally fluorescent or if fluorescent derivatives can be formed. Scanned chromatograms can be evaluated by using peak heights, peak areas or sliced areas. Sulfamethazine has been determined in animal tissue using the visual comparative technique at concentration levels in the original sample as low as 50 ppb. Thin layer chromatography has also been employed to determine aflatoxins in palm oils and kernels at levels of 1-2 mg/kg. Although a somewhat lengthy procedure, an extraction efficiency in excess of 90% was easily achieved. In a similar analysis which involved the determination of phospholipids in human tears. The extracted phospholipids were hydrolyzed by acetylhydrolase to form 1-hexade-canoyl-*sn*-glycero-3-phosphocholine and then sprayed with the Mo-MG reagent to produce the intense blue-green spot of molybdo-phosphate malachite green. Scanning at an appropriate wavelength, concentrations of 2-100 pmol were measurable with an RSD of about 2%. A hybrid thin layer plate (half covered with silica gel and half with a laser desorption matrix) has been used as a transport medium for a laser desorption sampling system of a mass spectrometer. After separation in one direction, the spots on the plate are transferred to the section coated with the desorption matrix by transverse elution with another solvent The plate is then dried and then inserted into the mass spectrometer sampling system. The sensitivity that was realized ranged from 90 fmol to 6 pmol. Thin layer plates can also be scanned by a digital imaging camera and the resulting image passed to a computer and the analysis processed by suitable software. Calibration curves obtained by this procedure are close to linear but there appears to be little difference between the precision and repeatability obtained by this method and that obtained by conventional scanning.

The use of radioactive tracers also opens up alternative methods of scanning a thin layer plate. One method is by contact radiography where, after development and drying, the plate is placed in contact with a photographic plate or film for a day or two. The film is developed and then scanned in the usual manner. Alternatively, the plate can be scanned with a radioactive counter-scanner and the radio chromatogram produced directly. This technique can be highly sensitive and at the same time provide good precision and accuracy. Pressurized thin layer chromatography, where the flow is energy-driven instead of by surface tension forces can provide better resolution and shorter analysis times. However, the accuracy and precision obtained form such techniques appear little better than those obtained from conventional thin layer chromatography. An interesting development of the thin layer technique is to first use the plate as a synthesis medium to produce the products of interest. The plate is then used to separate the products in the conventional manner. Finally the individual spots on the plate are used to test the biological activity of each separated component.

References

1. P. Shearan and M. O'Keeffe, *Analyst,* **119**(1994)2761.
2. P. Shearan, M. O'keefe and M. R. Smyth, *Food Addit. Contam.*, **11**(1994)7.
3. S. Nawaz, R. D. Coker and S. J. Haswell, *Analyst,* **11**(1992)67.
4. T. Ohyama, C. Matsurbara and K. Takamura, *Analyst,* **121**(1996)1943.
5. J. T. Mehl and D. H. Hercules, *Anal. Chem.*, **72**(2000)68.
6. I. Vovk, A. Gale-Wondra and M. Prosek, *J. Planar. Chromatogr.*, **10**(1997)416.
7. I. Kjebovich, J. Szünyog and I. Hazal, *J. Planar. Chromatogr.*,**10**(1997)399.

8. Z. Szikszay, Z. Végh and K Ferenezi-Fadar, *J. Planar. Chromatogr.*, **11**(1995)428.
9. M. Freemantle, *C&EN*, March 13(2000)9.
10. L. Williams, *Chem. Commun.*, (2000)4351.

Index